# Romanticism and the sciences

While the Romantic period has attracted a great deal of interest, there has been no general work since 1941 on the relationship between Romanticism and the sciences. This book presents a series of essays, each specially written by an expert in the area, which focus on the role of Romantic philosophy and ideology in the sciences, and on the role of the sciences in Romantic literature. The contributions are designed to give a systematic coverage of the whole field. They are written at a popular level, are well-illustrated and are accompanied by suggestions for further reading suitable for undergraduates and others.

Divided into four sections under the titles 'Romanticism', 'Sciences of the organic', 'Sciences of the inorganic' and 'Literature and the sciences', the book discusses various themes, movements and theories, as well as individual natural philosophers and writers (including Schelling, the brothers von Humboldt, Goethe, Ritter, Davy, Oersted, Kleist, Coleridge and Büchner). There is an editorial introduction prefiguring some of the concerns of the book. This original collection is designed to provide a balance of literary and scientific interests for students in both humanistic and scientific disciplines and for educated general readers.

# ROMANTICISM
# AND THE SCIENCES

*edited by*

ANDREW CUNNINGHAM *and* NICHOLAS JARDINE

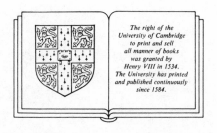

The right of the
University of Cambridge
to print and sell
all manner of books
was granted by
Henry VIII in 1534.
The University has printed
and published continuously
since 1584.

CAMBRIDGE UNIVERSITY PRESS

*Cambridge*

*New York   Port Chester   Melbourne   Sydney*

Published by the Press Syndicate of the University of Cambridge
The Pitt Building, Trumpington Street, Cambridge CB2 1RP
40 West 20th Street, New York, NY 10011, USA
10 Stamford Road, Oakleigh, Melbourne 3166, Australia

© Cambridge University Press 1990

First published 1990

Printed in Great Britain at the University Press, Cambridge

*British Library cataloguing in publication data*
Romanticism and the sciences.
1. Science. Theories. Scientific theories. Influence,
history of romanticism
I. Cunningham, Andrew II. Jardine, N. (Nicholas)
501

*Library of Congress cataloguing in publication data*
Romanticism and the sciences / edited by Andrew Cunningham and
Nicholas Jardine
p.    cm.
Includes bibliographical references.
ISBN 0 521 35602 4 – ISBN 0 521 35685 7 (pbk.)
1. Science – History. 2. Science – Philosophy. 3. Romanticism.
I. Cunningham, Andrew, Dr. II. Jardine, Nicholas.
Q125.R735 1990
509 – dc20    89-17310 CIP

ISBN 0 521 35602 4 hard covers
ISBN 0 521 35685 7 paperback

To Gerd Buchdahl

# Contents

List of illustrations                                          *page*     x
Notes on contributors                                                   xiii
Preface                                                                  xix
Short list of introductory reading in English                           xxi

*Introduction:* The age of reflexion
THE EDITORS                                                                1


## I ROMANTICISM

1   Romanticism and the sciences
    DAVID KNIGHT                                                          13

2   Schelling and the origins of his *Naturphilosophie*
    S. R. MORGAN                                                          25

3   Romantic philosophy and the organization of the disciplines:
    the founding of the Humboldt University of Berlin
    ELINOR S. SHAFFER                                                     38

4   Historical consciousness in the German Romantic
    *Naturforschung*
    DIETRICH VON ENGELHARDT (translated by Christine
    Salazar)                                                              55

5   Theology and the sciences in the German Romantic period
    FREDERICK GREGORY                                                     69

6   Genius in Romantic natural philosophy
    SIMON SCHAFFER                                                        82

## II SCIENCES OF THE ORGANIC

7  Doctors *contra* clysters and feudalism: the consequences of a Romantic revolution
NELLY TSOUYOPOULOS                                                      101

8  Morphotypes and the historical-genetic method in Romantic biology
TIMOTHY LENOIR                                                         119

9  'Metaphorical mystifications': the Romantic gestation of nature in British biology
EVELLEEN RICHARDS                                                     130

10  Transcendental anatomy
PHILIP F. REHBOCK                                                     144

11  Romantic thought and the origins of cell theory
L. S. JACYNA                                                          161

12  Alexander von Humboldt and the geography of vegetation
MALCOLM NICOLSON                                                      169

## III SCIENCES OF THE INORGANIC

13  Goethe, colour and the science of seeing
DENNIS L. SEPPER                                                      189

14  Johann Wilhelm Ritter: Romantic physics in Germany
WALTER D. WETZELS                                                     199

15  The power and the glory: Humphry Davy and Romanticism
CHRISTOPHER LAWRENCE                                                  213

16  Oersted's discovery of electromagnetism
H. A. M. SNELDERS                                                     228

17  Caves, fossils and the history of the earth
NICHOLAS A. RUPKE                                                     241

## IV LITERATURE AND THE SCIENCES

18  Goethe's use of chemical theory in his *Elective Affinities*
JEREMY ADLER                                                          263

19 Kleist's Bedlam: abnormal psychology and psychiatry in the
   works of Heinrich von Kleist
   NIGEL REEVES                                               280

20 Coleridge and the sciences
   TREVOR H. LEVERE                                           295

21 Nature's book: the language of science in the American
   Renaissance
   DAVID VAN LEER                                             307

22 The shattered whole: Georg Büchner and *Naturphilosophie*
   JOHN REDDICK                                               322

   *Index of names*                                          341

# Illustrations

1 Title-page of Schelling's *Ideas for a Philosophy of Nature* (1797).                                                       *page*  33
2 Statue of Wilhelm von Humboldt.                                               49
3 The 'steam-clyster', invented by Johann Kämpf.                       103
4 Table from Thomann's *Annales Instituti Medico-Clinici Wirceburgensis* (1799).                                                  109
5 Christoph Wilhelm Hufeland.                                                114
6 Andreas Röschlaub.                                                               115
7 Romanticized image of nature's 'gestation' revealed by geology, from Hugh Miller's *Footprints of the Creator* (1849).    136
8 Lorenz Oken.                                                                        148
9 Etienne Geoffroy Saint-Hilaire.                                             150
10 Amphitheatre of comparative anatomy, Jardin des Plantes.    151
11 The ideal typical vertebra, according to Richard Owen.         154
12 Archetype of the vertebrate skeleton, according to Richard Owen.                                                                         155
13 Profile of the Andes, from Alexander von Humboldt's *Essai sur la géographie des plantes* (1807).                            174–5
14 'Heights of the Old and New World, graphically compared', by Goethe (1813).                                            176–7
15 Frontispiece, *Alexander von Humboldt und Aimé Bonplands Reise* (1807).                                                      179
16 Map of isothermal lines by Alexander von Humboldt.            182
17 'Romantic' representation of Oersted's discovery of electromagnetism.                                                            229
18 Oersted's apparatus.                                                               230
19 Page from Oersted's laboratory notebook, July 1820.             237
20 Cave at Gaylenreuth, contemporary illustration.                    243
21 Caspar David Friedrich, *Kreidefelsen auf Rügen* (1820).        255

22  E. F. Geoffroy's table of affinities, 1718.                              266
23  Title-page of P. J. Macquer's *Elémens* (1753).                          267
24  Illustration by H. A. Dähling for 1811 edition of *Elective
    Affinities*.                                                            273
25  Anatomical drawings from Georg Büchner's doctoral
    dissertation.                                                          329
26  Büchner, drawn from life by Alexis Muston.                             337

## ACKNOWLEDGEMENTS

Permission to reproduce illustrations in their possession has been kindly
granted as follows: the Deutsche Akademie der Wissenschaften, Berlin, GDR,
for the cover picture; the British Library for figs. 1, 4, 7, 23 and 24; Ullstein
Bilderdienst, Berlin, FRG, for fig. 2; Münchner Stadtmuseum for fig. 6; the
National Library of Medicine, Bethesda, for figs. 9 and 12; Yale University
Library for fig. 15; the Library of University College London for fig. 16; the
Danmarks Tekniske Museum, Elsinore, for fig. 18; the Royal Danish Society
of Sciences for fig. 19; Museum Stiftung Oskar Reinhart, Winterthur, for fig.
21; the Syndics of Cambridge University Library for figs. 13, 14, 19, 25 and 26;
the Whipple Library, Cambridge for figs. 20 and 22.

# Notes on contributors.

JEREMY ADLER is a Reader in German at Queen Mary and Westfield College (University of London). His London Ph.D thesis of 1977 was on Goethe's *Elective Affinities*. He has been awarded studentships by the Herzog August Bibliothek (Wolfenbüttel), and has been a Fellow at the Centre for Advanced Studies in Berlin. He has published articles on seventeenth-century poetry, Goethe, Hölderlin, Rilke and twentieth-century writing, and translations of Hölderlin, Stramm and Stadler. Apart from science and literature, his research interests include visual poetry, and with Ulrich Ernst he published *Text als Figur* in 1987.

ANDREW CUNNINGHAM is Wellcome Lecturer in the History of Medicine at Cambridge University. He publishes in history of medicine and is currently working on the origin, career and demise of natural philosophy.

DIETRICH VON ENGELHARDT is Director of the Institute for History of Medicine and Natural Sciences at the Medical University of Lübeck. Amongst other books he has published *Hegel und die Chemie* (Wiesbaden, 1976); *Historisches Bewusstsein in der Naturwissenschaft von der Aufklärung bis zum Positivismus* (Freiburg, 1979); and *Die inneren Verbindungen zwischen Philosophie und Medizin im 20. Jahrhundert* (with H. Schipperges) (Darmstadt, 1980).

FREDERICK GREGORY is Associate Professor of History of Science in the Department of History at the University of Florida. His research interests lie in German science of the eighteenth and nineteenth centuries, especially as it relates to the intellectual currents of society. In addition to his continuing work on a biography of Jakob Fries, he is currently completing a study of the reception of Darwin among German theologians.

L. S. JACYNA is Acting Director of the Wellcome Unit for the History of Medicine at the University of Glasgow. His research interests lie in medical science in the nineteenth century. He is co-author, with Edwin Clarke, of *Nineteenth-Century Origins of Neuroscientific Concepts* (Berkeley, Calif., 1987)', and has a book on the Edinburgh medical school in the nineteenth century forthcoming.

NICHOLAS JARDINE is Reader in the History and Philosophy of Science at Cambridge University. His books include *The Birth of History and Philosophy of Science* (Cambridge, 1984), *Fortunes of Inquiry* (Oxford, 1986) and *Scenes of Inquiry* (forthcoming). He is Editor of *Studies in History and Philosophy of Science*.

DAVID KNIGHT is Reader in the History of Science in the University of Durham, where he has taught since 1964. From 1981 to 1988 he has been Editor of *The British Journal for the History of Science*. His most recent books are *The Age of Science* (Oxford, 1986) and an alphabetical *Companion to the Physical Sciences* (London, 1989). He is working on a history of ideas in chemistry and on a life of Humphry Davy and he is General Editor of a new series of *Scientific Biographies*, which will be published by Basil Blackwell.

CHRISTOPHER LAWRENCE is Senior Lecturer in the History of Medicine at the Wellcome Institute for the History of Medicine in London. He has worked extensively on the history of the Scottish Enlightenment. He is co-author, with Daniel Fox, of *Photographing Medicine: Images and Power in Britain and America since 1840* (New York, 1988).

TIMOTHY LENOIR is the author of *The Strategy of Life: Teleology and Mechanics in Nineteenth Century German Biology* (Dordrecht, 1982). He is presently completing a study entitled *The Eye as Measuring Device: The Nativism–Empiricism Controversy and the Politics of Practical Interest in Germany, 1847–1870*, which focuses on the role of Helmholtz's work in physiological optics in reshaping the foundations of knowledge and in creating a scientific ideology supportive of the pursuit of material interests. Other projects include an investigation of extra-university, profit-oriented research institutes in Germany at the turn of the twentieth century. He has been at Stanford University since 1987, where he is a member of the Department of History and Program in the History of Science.

TREVOR H. LEVERE is Professor of the History of Science at the University of Toronto. He is a Fellow of the Royal Society of Canada, a Foreign Member of the Dutch Society of Sciences (Haarlem) and a Corresponding Member of the International Academy of the History of Science (Paris). His research interests are primarily in nineteenth-century science, including the history of chemistry, science and Romanticism, scientific instruments and science in Canada. He is currently writing a book on science in the exploration of the Canadian Arctic prior to 1920. He is the author of *Affinity and Matter: Elements of Chemical Philosophy 1800–1865* (Oxford, 1971) and *Poetry Realized in Nature: Samuel Taylor Coleridge and Early Nineteenth-Century Science* (Cambridge, 1981) and co-author of *Van Marum's Scientific Instruments in Teyler's Museum* (Leyden, 1973) and *A Curious Field-Book: Science and Society in Canadian History* (Oxford, 1974).

SUE MORGAN is a graduate of Cambridge University, where she studied History and Philosophy of Science. She is currently working on Schelling and his *Naturphilosophie*.

MALCOLM NICOLSON is a Research Fellow at the Wellcome Institute for the History of Medicine, London. He works on a variety of topics in the history of medicine and biology. His recent publications include 'The Metastatic Theory of Pathogenesis and the Professional Interests of the Eighteenth-Century Physician', *Medical History*, 32 (1988), and (with C. MacLaughlin) 'Social Constructionism and Medical Sociology: A Study of the Vascular Theory of Multiple Sclerosis', *Sociology of Health and Illness*, 10 (1988).

JOHN REDDICK is Professor of German and Head of the Department of German at the University of Liverpool; his current focus of research interest is Georg Büchner, and he has a monograph on Büchner due to appear with Oxford University Press in 1990.

NIGEL REEVES is Professor of German and Head of the Department of Linguistic and International Studies and Dean of the Faculty of Human Studies at the University of Surrey. He has published a monograph on Schiller and his debts to the medicine and psychology of his time (with Kenneth Dewhurst) and translated (with David Luke) Kleist's short stories for the Penguin Classics series. He was Alexander von Humboldt Fellow at the University of Tübingen 1974–5 and at the University of Hamburg 1986. Professor Reeves is also President of the National Association of Language Advisors and of the Association of Teachers of German, Chairman of the National Congress on Language in Education and Vice-Chairman of the Conference of University Teachers of German. In addition to his interests in German literature and the history of ideas he has been an active advocate of the practical application in the UK of foreign language skills to overseas trade and was awarded the OBE in 1987 for services to export. In October 1988 he published, with Professor David Liston, *The Invisible Economy: A Profile of Britain's Invisible Exports* (London). He was awarded the Goethe Medaille by the Goethe Institute for services to the German language overseas and for the promotion of international cultural cooperation in 1989.

PHILIP F. REHBOCK was educated at Stanford University and received his doctorate in history of science from The Johns Hopkins University. He is the author of *The Philosophical Naturalists: Themes in Early Nineteenth-*

*Century British Biology* (Madison, Wisc., 1983) and co-editor (with Roy MacLeod) of *Nature in Its Greatest Extent: Western Science in the Pacific* (Honolulu, 1988). He is currently editing the *Challenger* letters of Joseph Matkin, under the auspices of the Scripps Institution of Oceanography, La Jolla, California. He also edits the *Pacific Circle Newsletter*. Dr Rehbock teaches in the history and general science departments of the University of Hawaii in Honolulu.

EVELLEEN RICHARDS is a Senior Lecturer in the Department of Science and Technology Studies at the University of Wollongong, Australia. She teaches and publishes in the areas of the social history of evolutionary biology, the sociology of medical knowledge and gender and science. She is currently at work on a book, *Vitamin C and Cancer: Medicine or Politics?*, to be published by Macmillan.

NICHOLAS A. RUPKE, FGSA, FRHistS, is Senior Fellow at the Research School of Social Sciences, Australian National University, Canberra. He holds a Ph.D from Princeton University and has held research fellowships at the Smithsonian Institution, the University of Oxford, the University of Tübingen, the Netherlands Institute for Advanced Study, the Wellcome Institute for the History of Medicine, London, and the National Humanities Center. His many publications deal with topics in both marine geology and the history of Romantic and Victorian science. He is the author of *Distinctive Properties of Turbiditic and Hemi-pelagic Mud Layers* (1974) and *The Great Chain of History* (1983). He is also the editor of *Vivisection in Historical Perspective* (1987) and *Science, Politics and the Public Good* (1988). He is currently writing a biography of Richard Owen.

SIMON SCHAFFER lectures in history and philosophy of science at the University of Cambridge. He is co-author, with Steven Shapin, of *Leviathan and the Air-Pump: Hobbes, Boyle and the Experimental life* (Princeton, 1985), and co-editor of *The Uses of Experiment: Studies in the Natural Sciences* (Cambridge, 1988). He is currently preparing a book on the public culture of experimental natural philosophy in late-eighteenth-century Britain and France.

DENNIS L. SEPPER is Associate Professor of Philosophy at the University of Dallas, and recently held a fellowship at the Institute for Research in the Humanities of the University of Wisconsin (Madison). His major research interests are in the philosophy of science, the history of modern and early modern physical and perceptual sciences and hermeneutic and phenomenological philosophy. He is the author of *Goethe contra Newton: Polemics and the Project for a New Science of Color* (Cambridge, 1988) and is presently writing a book on the role of imagination in Descartes's science and philosophy.

ELINOR S. SHAFFER, Reader in the School of Modern Languages and European History at the University of East Anglia, has published and lectured extensively on English and German Romanticism and on topics in literature and science. Her books include '*Kubla Khan' and the Fall of Jerusalem: The Mythological School in Biblical Criticism and Secular Literature, 1770–1880* (Cambridge, 1975). Relevant papers include 'Coleridge's Natural Philosophy' (*History of Science*, 1974) and 'Archeology and the New Contextualism' in *Hartford Studies in Literature and Science*, ed. G. S. Rousseau (1986). She has held an American Council of Learned Societies Study Fellowship for research into the idea of 'organism' in literature and science around 1800; she is British representative of the Society of Literature and Science. A forthcoming volume (number 13) of *Comparative Criticism: An Annual Journal*, which she edits, will be devoted to 'Literature and Science'. She is working on a study of Coleridge and *Naturphilosophie*.

H. A. M. SNELDERS is Ordinary Professor of the History of Science at the University of Utrecht and Extraordinary Professor at the Vrije Universiteit in Amsterdam. His graduation thesis was on 'The Influence of Kant, Romanticism and *Naturphilosophie* on the Inorganic Natural Sciences in Germany' (1973). He has written extensively on the history of physics and chemistry in the Netherlands and on Romantic *Naturphilosophie*. His latest book is an edition of the letters from Gerrit Jan Mudler to Justus Liebig, 1838–1846 (Amsterdam, 1986). He is now working on a history of physical chemistry in the Netherlands.

NELLY TSOUYOPOULOS is Professor of History and Philosophy of Medicine at the Westfälische Wilhelms-Universität, Münster. She also teaches philosophy in the Department of Philosophy. She works on the interrelations between philosophy, medicine and politics in nineteenth-century Germany. Her publications include *Andreas Röschlaub und die Romantische Medizin* (Stuttgart, 1982), which is a study of the philosophical background of a medical movement and of the influence of John Brown on German thought.

DAVID VAN LEER is Associate Professor of English and American Literature at the University of California, Davis. He is the author of *Emerson's Epistemology: The Argument of the Essays* (Cambridge, 1986) and articles on American literature, gender criticism and popular culture. He is currently at work on a book-length study of Edgar Allan Poe and science.

WALTER D. WETZELS is Professor of Germanic Languages at the University of Texas at Austin. His publications include *Johann Wilhelm Ritter: Physik im Wirkungsfeld der deutschen Romantik* (Berlin, 1973), *Myth and Reason* (edited with an introduction, Austin, 1973) and *Literature and*

*History* (co-edited with Leonard Schulze, Lanham, Md., 1983). He has published a number of articles on the relation between literature and science, including the popularization of science; other essays are about German literature of the eighteenth century, concentrating on Goethe, Herder and the Romantics.

# *Preface*

Goethe's Faust spurns knowledge 'extorted with levers and screws', longing instead for a grasp of Nature's secret elements, her hidden active forces, the harmony of her whole and parts. In the course of Goethe's long life there were many, especially in the German lands, who like his Faust turned against what they perceived as the soul-less mechanical natural philosophy of the Enlightenment, seeking rather a spiritual and dynamic insight into the natural world. This book is concerned with the Romantic modes of understanding of Nature, both as they figured in the various natural sciences and as they were manifested in the literature of the period.

In the course of the nineteenth century strong opposition arose to the new Romantic approaches in the sciences as exemplified by Goethe's colour theory and the Nature Philosophy (*Naturphilosophie*) of Schelling and his followers. 'Black Death of our century' and 'Walpurgis-nightmare' are typical of the jeers of such empirically minded luminaries of the natural sciences as Justus von Liebig and Emil du Bois-Reymond. Until quite recently historians of science have generally followed this lead, dismissing Romanticism in the natural sciences as an aberration from the path of healthy scientific progress. The past fifteen years have, however, seen a revival of scholarly interest and a series of reassessments. There is now a widespread recognition of the importance of particular Romantic contributions to the natural sciences. Moreover, there is increased recognition that through the creation of new disciplines and the reorientation of old ones, Romantic approaches to the study of Nature played a major role in the so-called 'Second Scientific Revolution' through which natural science in the modern sense became established as a disciplinary category and as a force in society. Further, there is a now a significant body of scholarship that shows appreciation of the importance of Romantic attitudes in the sciences for our understanding of Romantic ideology and of the literature of Romanticism. But despite the expansion of scholarly interest, there has been no English-language book on the topic since Gode-von Aesch's *Natural Science in German Romanticism* (New York, 1941). Further, most of the recent English-language articles in this area have been addressed to specialists in the history of science or literary history. The essays in the present volume are intended to be fully accessible to students and amateurs both of the arts and of the sciences.

'Romanticism and the Sciences' is an immense topic. In planning the volume our aim has been to convey both its extent and the liveliness and international character of the recent scholarship it has provoked. Inevitably certain issues have been emphasized at the expense of others equally important. In part this is deliberate, for we have tried to avoid too much overlap with the excellent recent collections of essays *Hegel and the Sciences* (edited by R. S. Cohen and H. W. Wartofsky, Dordrecht, 1984) and *Goethe and the Sciences: A Reappraisal* (edited by F. Amrine, F. J. Zucker and H. Wheeler, Dordrecht, 1987). As editors we urged all our contributors to make their essays as accessible as possible, to keep their footnotes and references to a minimum (sometimes against their scholarly instincts) and to provide indications of further readings in English on their particular topic. Since we believe in the value of a multiplicity of historical approaches (and did not wish to set ourselves up as cultural commissars), we offered only general editorial suggestions to our contributors. But we did urge them to concentrate on the contents and the historical and literary contexts of Romantic doctrines and approaches and not to devote undue space to discussion of their validity by present-day scientific criteria; we also urged them to be cautious in the use of terms such as 'science' and 'scientist' which acquired their modern connotations only in the period covered by this volume. We are most grateful to our contributors for the speed and efficiency with which they all delivered their splendid contributions.

In the spring of 1987 at the Department of the History and Philosophy of Science at Cambridge, we convened an informal seminar, 'Romanticism and Idealism in the Sciences'. We both contributed papers, as did Nelly Tsouyopoulos, Simon Schaffer, Andy Bowie and Gerd Buchdahl. It was the liveliness of this seminar that inspired our plans for the present volume. We thank all those who took part.

Others whom we wish to thank for valuable advice include Martin Bernal, Mike Dettelbach, Emory Elliott, Johanna Geyer-Kordesch, Sue Morgan, Barry Nisbet and Mikuláš Teich. We offer special thanks to Simon Schaffer (of whom, as of William Whewell, it may be said that omniscience is his foible) for generously sharing his knowledge with us. We would particularly like to thank Maureen Street for her outstandingly sensible and sensitive sub-editorial suggestions.

Finally we wish to pay a tribute to a former colleague, Gerd Buchdahl. For many years, in a sometimes hostile environment, he has urged the importance of the topics addressed in this work. As editors, we dedicate it to him.

ARC
NJ

*Cambridge 1989*

# Short list of
# introductory reading in English

## GENERAL WORKS

Hugh Honour, *Romanticism* (London, 1979)
F. Clandon, *The Concise Encyclopaedia of Romanticism* (Toronto, 1986)
L. Eitner, *Neoclassicism and Romanticism, 1750–1850: Sources and Documents* (Englewood Cliffs, N.J., 1971)
H. Eichner (ed.), *'Romantic' and Its Cognates: The European History of a Word* (Toronto, 1972)
H. G. Schenk, *The Mind of the European Romantics* (London, 1966)
J. J. McGann, *The Romantic Ideology: A Critical Investigation* (Chicago, 1983)

## TO CAPTURE THE MOOD OF ROMANTICISM
### Music

Beethoven, Symphony no. 6, 'The Pastoral'
Schubert, 'Winterreise' (song-cycle)
Weber, 'Der Freischutz' (opera)
Berlioz, 'Symphonie fantastique'

### Painting

Michel Le Bris, *Romantics and Romanticism* (London, 1981)
W. Vaughan, *German Romantic Painting* (New Haven, 1980)

### Poetry and drama

Goethe, *Faust, Part I* (many editions)
Georg Büchner, *Woyzeck* and *Danton's Death* (many editions)
D. Wright, *The Penguin Book of English Romantic Verse* (Harmondsworth, 1968)
Graham Hough, *The Romantic Poets* (London, 1953, reprinted many times)

### Prose

Goethe, *Sorrows of Young Werther* (many editions)
Hölderlin, *Hyperion* (English translation by W. R. Trask, New York, 1965)
Novalis, *The Novices at Sais* (English translation, New York, 1949)
E. T. A. Hoffmann, *Tales of Hoffmann* (many editions)
Mary Wollstonecraft Shelley, *Frankenstein, or, the Modern Prometheus* (many editions)

## POLITICAL AND SOCIAL CONTEXTS OF ROMANTICISM

Henry Braunschwig, *Enlightenment and Romanticism in Eighteenth-Century Prussia* (trans. F. Jellinek, Chicago, 1974)

Marilyn Butler, *Romantics, Rebels and Reactionaries: English Literature and its Background, 1760–1830* (Oxford, 1981)

Raymond Williams, *Culture and Society, 1780–1950* (Harmondsworth, 1958), chs. 1–4

Roy Porter and Mikuláš Teich, *Romanticism in National Context* (Cambridge, 1988)

H. Reiss, *The Political Thought of the German Romeutics, 1793–1815* (Oxford, 1955)

## ROMANTIC PHILOSOPHY, AESTHETICS AND THEOLOGY

F. W. J. Schelling, *Ideas for a Philosophy of Nature* (trans. E. E. Harris and P. Heath, Cambridge, 1988)

S. T. Coleridge, *Biographia Literaria* (many editions)

F. Schleiermacher, *Soliloquies* (trans. H. L. Friess, Chicago, 1926)

Bernard Reardon, *Religion in the Age of Romanticism* (Cambridge, 1985)

Peter le Huray and James Day, *Music and Aesthetics in the Eighteenth and Early Nineteenth Centuries* (Cambridge, 1981)

Kathleen Wheeler (ed.), *German Aesthetic and Literary Criticism: The Romantic Ironists and Goethe* (Cambridge, 1984)

Edward Craig, *The Mind of God and the Works of Man* (Oxford, 1987), ch. 3

M. H. Abrams, *The Mirror and the Lamp: Romantic Theory and the Critical Tradition* (Oxford, 1953)

## ROMANTICISM AND THE SCIENCES

A. Gode-von Aesch, *Natural Science in German Romanticism* (New York, 1941; repr. New York, 1966)

R. S. Cohen and M. W. Wartofsky (eds.), *Hegel and the Sciences* (Dordrecht, 1984)

F. Amrine, F. J. Zucker and H. Wheeler, *Goethe and the Sciences: A Reappraisal* (Dordrecht, 1987)

David Knight, *The Age of Science: The Scientific World-View of the Nineteenth Century* (Oxford, 1984), ch. 4.

Robert Darnton, *Mesmerism and the End of the Enlightenment in France* (Cambridge, Mass., 1968)

An important bibliography of secondary literature is D. von Engelhardt, 'Bibliographie der Sekundarliteratur zur romantischen Naturforschung und Medizin 1950–1975', in R. Brinkmann (ed.), *Romantik in Deutschland* (Stuttgart, 1978); useful bibliographies are to be found in J. Neubauer, *Bifocal Vision: Novalis' Philosophy of Nature and Disease* (Chapel Hill, 1971) and in Amrine, Zucker and Wheeler, *Goethe and the Sciences*, cited above.

# INTRODUCTION

## *The age of reflexion*

### THE EDITORS

One succeeded – he lifted the veil of the Goddess at Sais. But what did he see? He saw – wonder of wonders – himself.[1]

Around 1800 the self stood in unprecedentedly high esteem. What is God? What is man? What is nature? What is a work of art? What are the sciences? On what principles do they rest? What are their limits? How are they to be taught, learned, practised? In this 'Age of Reflexion' all these came to be perceived as questions of self-understanding.

Historians of the sciences increasingly acknowledge the radical differences of orientation, agenda and practice between, on the one hand, eighteenth-century natural philosophy, natural history and mathematics and, on the other hand, natural science, *science naturelle* and *Naturwissenschaft* of the nineteenth and twentieth centuries. Indeed two 'Scientific Revolutions' are now commonly recognized – a first revolution around the turn of the sixteenth century, in which new mathematically and experimentally oriented branches of natural philosophy were created, and a second revolution around the turn of the eighteenth century, in which was formed the federation of disciplines that we call 'science'. Science, in our sense, once held to be more than two thousand years old, is now credited with less than two hundred years of history.

Self-understanding certainly played an important role in the methodology and historical reflexion of the first scientific revolution. Late-Renaissance treatises on method, discovery and classification of the arts and sciences were premised on the elaborate faculty psychologies of the period; and in the writings of Mersenne, Bacon, Descartes and others an examination of the powers and limits of the faculties was presented as prerequisite for the establishment of the methods and scope of the sciences. Further, new kinds of cultural self-understanding became evident in sixteenth- and seventeenth-century concerns with the history of learning, especially in the context of the 'battle of the Ancients and Moderns' over the superiority of ancient learning.

The second revolution, however, is marked by enterprises in individual and cultural self-understanding that are both more radical and more explicit. To

1

start with we may note that the major critical movements of the period which affect the arts and sciences are all enterprises in self-understanding. This is manifestly so in the case of the critical philosophies which explore the scope and limits of human knowledge on the basis of an examination of the nature and powers of the human cognitive faculties – the philosophies of Hume, Kant and Johann Heinrich Lambert, for example. The critique of educational methods and institutions that reaches its high point in the great controversy over the constitution and curriculum of the new University of Berlin (1810) can be seen as an exercise in social and mental reflexion. Self-understanding is central to the period's new subject-oriented aesthetics, in which the character-istic terms of approbation – spontaneity, originality, integrity, sincerity, intensity – relate not, as in earlier canons of taste, to the imitation in the work of art of an extrinsic reality, but to the passions, faculties and powers of the artist. Self-understanding is the name of the game also in the historical critique of the period. For Romantic historians in the wake of Herder, as for cultural historians of the Göttingen school, what we – the human race, the nation, the speakers of a given language, the exponents of a discipline – are, is inextrica-bly bound up with our history. And the distinctive new types of history-writing of the period – history of language, institutional history, local and folk history – are precisely those conducive to the self-understanding of nations, societies and disciplines.

However, to appreciate the centrality of self-understanding in the Roman-tic practice and interpretation of the sciences we need to consider, in addition to the critical roles of self-understanding, certain more specifically Romantic views concerning the relation between self and nature. Once, in a Golden Age, in the Garden before the Fall, man was at one with himself and nature. Now he is divided in himself, his once harmoniously united faculties at war with each other and at odds with nature. He is 'the sour grape in Nature's sweet clusters'.[2] How is the lost unity of self and nature to be restored? At least in this life, we cannot hope to recover the original innocent state – as Kleist declares, 'Paradise is bolted and barred, and the cherub behind us'; if we are to find our way back to Paradise, our route must be indirect, 'We must make a journey round the world and see whether it may not be open from the back.'[3] The new state of grace will not, however, be simple and instinctive but rather, Hölderlin says, complex and self-conscious:

There are two ideals of our being: one a state of the highest simplicity in which through the mere organization of nature, without our help, our desires are in tune with each other and with our powers and with everything to which we are related; the other a state of the highest civilisation [*Bildung*] in which the same would come about with endlessly multiplied and strengthened desires and powers through the organization which we ourselves establish. The eccentric way by which mankind, in general and individually, passes from one point (that of more or less pure simplicity) to the other (that of more or less complete cultivation) shows itself to be the same in its essential direction.[4]

What is this 'journey round the world', this 'eccentric way' to a new harmony of self and nature? There are almost as many answers as there are Romantic authors. The dominant prescription is poetic and aesthetic. In such landmarks of Romantic aesthetics as Schiller's *Letters on the Aesthetic Education of Mankind* (1795), 'Jean Paul' Friedrich Richter's *School for Aesthetics* (1804) and Coleridge's *Biographia Literaria* (1817) it is the creation and appreciation of works of art that can restore fragmented mankind to unity, reintegrate the spontaneity of sense and feeling with the discipline of reason and restore the isolated individual to an organic relationship with his social and natural surroundings.

Though works of art mark the high road to human redemption, other roads are recognized. In particular, there are aesthetic modes of contemplation of nature and poetic modes of research into nature: these too may lead to reconciliation. For Henrik Steffens, study of the developmental history of the earth and its creatures ennobles us because it allows us to grasp the unity of spiritual process in nature and ourselves: 'Do you want to know nature? Turn your glance inwards and you will be granted the privilege of beholding nature's stages of development in the stages of your spiritual education. Do you want to know yourself? Seek in nature: her works are those of the self-same spirit.'[5] For Lorenz Oken, a philosophical study of natural history can 'unify the German people with themselves and the world', as well as 'giving them an understanding of their own nature and their relations to plants and animals'.[6] For Schelling, the philosopher of nature can attain by artifice what the artist reveals spontaneously: a view of 'the holy of holies, where burns in eternal and original unity, as if in a single flame, that which in nature and history is rent asunder, and in life and action, no less than in thought, must for ever fly apart'.[7]

The Romantics were certainly hostile to the mechanical natural philosophy and descriptive natural history that they inherited from the Enlightenment. This antipathy is symbolized by Goethe's attack on Newton's colour theory, by the famous toast 'Newton's health, and confusion to mathematics', proposed by Charles Lamb, and by Keats's lines in *Lamia*:

> Do not all charms fly
> At the mere touch of cold philosophy?
> There was an awful rainbow once in heaven:
> We know her woof, her texture; she is given
> In the dull catalogue of common things.
> Philosophy will clip an Angel's wings,
> Conquer all mysteries by rule and line,
> Empty the haunted air, and gnoméd mine –
> Unweave a rainbow . . .

This hostility runs deep: for the Romantics mechanistic natural philosophy is the culmination of the analytic and judgemental approach responsible for our fall from grace with nature; but their attitude was rarely one of outright rejection. Goethe's Faust spurns bookish learning and the cold-hearted attempt to extort knowledge from nature 'with levers and screws'. Yet Goethe's own ventures into the sciences are both methodical and solidly founded in observation and experiment. Oken and Steffens follow Schelling in attempting to 'overcome' the static and analytical natural history of their predecessors.[8] But the dynamic and synthetic histories of the development of nature which they themselves set out, are proposed as complements to and completions of – not as replacements for – descriptive natural histories. In a letter to Wordsworth, Coleridge writes of 'the philosophy of mechanism, which, in everything that is most worthy of the human intellect, strikes *Death*'.[9] But on reading the *Optiks* of the arch-mechanist Newton he declares his delight 'with the beauty and neatness of his experiments, and with the accuracy of his *immediate* deductions from them'.[10] Nor does he envisage an outright refutation of the mechanical philosophy at the hands of his new 'philosophy of life and intelligence'; instead the mechanical view will be shown to embody only a superficial and partial insight.[11] Though the natural history and natural philosophy of the Enlightenment typify for the Romantics the attitudes responsible for our fall from grace with nature, they are also seen as a necessary stage on man's 'eccentric' way to redemption. What is to be the next, more reflexive, stage in the contemplation of nature? What disciplines for the study of nature are to supersede descriptive natural history and mechanistic natural philosophy?

We may look for indications of the Romantic answer to this question in Novalis's *The Novices at Sais*, a paradigm of High Romanticism. 'The ways of considering nature are innumerable', he writes (p. 208);[12] and the central theme of the work is precisely the multiplicity of the roads to the novices' goal – reunion with nature. Novalis (otherwise Friedrich von Hardenberg), 'the Prophet of Romanticism', was an official in the directorate of the salt-mines of Saxony. He had studied under Abraham Gottlob Werner at the Freiberg *Bergakademie* from 1797 to 1799, and was widely read in medicine and the sciences. The opening pages of *The Novices at Sais* portray an academy-cum-museum in which a Master encourages each novice to find his own way to redemption. The body of the work, entitled *Die Natur*, starts by rehearsing the fundamental Romantic tenets: the original unity of man and nature in a Golden Age ('everything seemed human, familiar and companionable, there was freshness in all their perceptions, each one of their utterances was a true product of nature, their ideas could not help but accord with the world around them and express it faithfully', p. 205); the subsequent separation of man from nature and the fragmentation of the human faculties ('which may well be compared to the splitting of a ray of light', p. 205); the interpretability of the history of the universe in human, spiritual, terms ('to treat the history of the universe as a history of mankind, to find only human happenings and relations everywhere', p. 206); the possibility of salvation through the contemplation of nature. There follows a discussion of the variety of human moral attitudes and aesthetic responses to nature. At the heart of the work is the allegorical tale of Hyacinth and Rosenblütchen, in which the lover, separated from his sweetheart by bookish learning, finds her again at the end of his pilgrimage. Next is an interlude in which the disgruntled museum specimens complain about mankind, who has misplaced and misranked them: 'If only he would learn to feel!' (p. 219). Then appear travellers returned from a quest. Again the discourse is of nature, but now at the level of specific disciplines for her contemplation and study. The work concludes with an account by the travellers of their search for traces of the original language of nature and mankind, the *Ursprache*, and an address by the Master on his role as a 'prophet of nature'.

It is the conversation of the travellers that is of special interest to us. After light refreshments served by friendly children, four of the travellers take part in two rounds of 'lively discourse'. In his opening speech the first traveller sets the strongly reflexive tone, talking of the necessity for an understanding of the powers and mechanisms of our faculties if we are to decipher nature and reach the essences of things. He talks of the possibilities of the deliberate refinement, extension and multiplication of our sensory capacities so as to control phenomena at will and create a world with which we will be more in harmony. This echoes Novalis's own programme for a 'Magic Idealism'. The traveller touches also on the creative powers of the human imagination, and in his second speech he declares:

In order to understand nature we must allow it to develop inwardly in its full sequence. . . The thinking man returns to the original function of his existence, to creative contemplation, to the point where knowledge and creation were united in a most wonderful mutual tie, to that creative moment of true enjoyment, of inward self-conception. If he immerses himself entirely in the *contemplation* of this original appearance, the history of the creation of nature unfolds itself before him in newly emerging times and spaces like a tale that never ends. (p.225)

The language here is redolent of the idealist philosophers Fichte and Schelling. In particular, the unfolding of a generative history of nature through an 'original intuition' is precisely the aim of Schelling's *Naturphilosophie*.

The second traveller doubts the value of attempts to construct a unified system of nature, speaking of the harmonious interaction of innumerable worlds, in a manner which may well reflect the cosmological speculations of J. H. Lambert and Kant.

The third traveller starts by defending systems of nature on the grounds that they enable us to produce 'beautiful and useful phenomena' and to 'improvise on nature as a great instrument' (p. 222). However, far from yielding knowledge, such systems have 'debased nature to the level of a uniform machine, without past and future'; true knowledge of nature is 'the province of the historian of nature'. In his second speech he takes up again the theme of an active artistic improvization on nature, involving creative and imaginative experiment. These remarks are complemented by a later passage which talks of a poetic approach to chemistry 'in which laboratories would become temples' (p. 229) and by passages in Novalis's other writings in which the experimental disciplines (*Experimentallehren*) of physics and chemistry are characterized as forms of artistic production. The active approach is contrasted with a passive study of nature which enjoys 'experiencing rather than making, receiving rather than giving' (p. 227). This passive approach is evidently that of amateur naturalists, whose work is of inestimable value to the researcher (*Forscher*) who 'follows their steps and gathers every treasure they let fall'.

The fourth traveller speaks first of the poet's intimacy with nature and then, in exalted vein, of the mystic reunion of man and nature in the primordial fluid of the world-soul. After drinks of 'cooling flame from crystal goblets', the travellers tell the tale of their quest for traces of the primaeval language. This is the language of the oldest and noblest race of men. It is also, as indicated at the beginning of the work, nature's own language – 'that great cipher which we discern written everywhere, in wings, egg-shells, clouds and snow, in crystals and in stone formations, on ice-covered waters, on the inside and outside of mountains, plants, beasts and men, in the lights of heaven, on touched or stroked discs of pitch and glass or in iron filings around a magnet'.[13] Finally the Master tells of his role as a 'prophet of nature', one who seeks to systematize and impart the innumerable roads to understanding of nature suited to the manifold types of men.

Novalis's 'many roads' are indicative of the entire range of the disciplines for the study of nature fostered by the Romantics: *Naturphilosophie*, cosmology and cosmogony; developmental history of the earth and its creatures; the new science of biology; investigations of mental states, conscious and unconscious, normal and abnormal; experimental disciplines concerned to unveil the hidden forces of nature – electricity, magnetism, galvanism and other life-forces; disciplines concerned to read essences, inner natures and ideal types from outward and visible appearances, signs and symptoms – physiognomy, phrenology, meteorology, mineralogy, 'philosophical' anatomy, etc. Such are the disciplines explored in the essays of the present volume.

What was the fate of these new approaches to the study of nature? It has long been acknowledged that Romantic attitudes were of major importance in the development of genuinely historical sciences of the cosmos, the earth and the earth's inhabitants.[14] However, the approaches most directly linked with Romanticism – those that presupposed spiritual development and ideal plans in nature – are generally thought to have perished at the hands of such luminaries of the empirical natural sciences as Justus Liebig, Hermann von Helmholz, Emil Du Bois-Reymond and Thomas Henry Huxley.

It must, of course, be conceded that many of the central doctrines of *Naturphilosophie* and of idealistic natural history have perished almost without trace. There may also be internal reasons for the failure of certain of the enterprises promoted by the Romantics: Novalis's 'Magic Idealism' seems so solipsistic as to rule out communal enterprise; the quests for a universal history of the heavens, earth and mankind of Gotthilf Heinrich von Schubert and Karl Hieronymus Windischmann seem too speculative to allow attainment of consensus; the overtly mystical concerns of the later works of Schelling and Franz Xavier von Baader seem too esoteric to form disciplines. There is surely an element of truth in Steffens's judgement (in a letter to Ludwig Tieck) about the first heady days of the Romantic movement:

Certain as it is that the time in which Goethe, Fichte, Schelling, the Schlegels, you, Novalis, Richter, and I dreamt in unison, was rich in *seeds* of all kinds, yet there was something pointless about it all. A spiritual tower of Babel was to be erected to which all spirits should come from afar. But the confusion of speech buried this ambitious work in its own debris.[15]

However, the standard account of the extinction of *Naturphilosophie* is questionable. To start with, it may be noted that the issue of continuity in scientific disciplines hinges much more on the preservation of practices and agendas of inquiry than on maintenance of doctrines; so the general abandonment of idealist assumptions in the latter part of the nineteenth century by no means proves that the disciplines associated with *Naturphilosophie* were themselves displaced. In this connection experimental physiology is of particular interest. Its spokesmen in the 1850s, 1860s and 1870s held it up as a model member of the natural sciences and were vociferous in their denunciations of the speculative excesses of *Naturphilosophie*. However, a major role in the establishment of the new German physiology institutes was played by sympathizers with *Naturphilosophie*;[16] and the topics pursued in those institutes were remarkably close to the physiological interests of the *Naturphilosophen*.[17]

Secondly, it may be observed that just as the Romantics wrote histories which portrayed the natural history and natural philosophy of their predecessors as limited and outmoded, so in turn the new breed of professional natural scientists of the 1840s, 1850s and 1860s used history to promote their own ideology and to discredit Romanticism in the sciences.[18] We may well suspect that both the stereotype of the Romantic sciences as speculative, fantastic, mystical and ill-disciplined, and their alleged defeat by the empirical natural sciences, are polemical constructs rather than the fruits of unbiased historical research.

Indeed, it is arguable that the ideology and institutions of the new natural science owed much to Romanticism and *Naturphilosophie*. For all the metaphysical differences between *Naturphilosophie* and the new natural science there is a striking coincidence in the range of disciplines they sought to unite. Thus it is that new institutional arrangements for the sciences associated with *Naturphilosophie* – notably the *Gesellschaft deutscher Naturforscher und Aertzte* (German Association of Naturalists and Physicians) set up by Lorenz Oken and the establishment of *Naturforschung* in the curriculum of the Humboldt University of Berlin – were able to serve as models for the institutionalization of the new natural science. Moreover, the self-image of the new 'men of science' was to be largely constituted by Romantic themes – scientific discovery as the work of genius, the pursuit of knowledge as a disinterested and heroic quest, the scientist as actor in a dramatic history, the autonomy of a scientific elite.

The political creeds which dominate our world, our views of human agency and morality, our arts and our conceptions of artistic creativity, all are rooted in the Romantic movement. What of our science? Is it, too, in substantial measure a product of Romantic reflexion?

## NOTES

1  Novalis, *Werke*, ed. H.-J. Mähl and R. Samuel (Munich, 1978–87), II, p. 234.
2  Hölderlin, *Hyperion*, ed. J. Schmidt (Baden-Baden, 1979), p. 58.
3  Kleist, *Über das Marionettentheater*, in his *Werke*, ed. H. Sembdner (Munich, 1977), II, p. 342; cited in E. Craig, *The Mind of God and the Works of Man* (Oxford, 1987), p. 143.
4  Hölderlin, *Fragment von Hyperion*, in *Hyperion*, ed. Schmidt, p. 202 (our translation).
5  Steffens, 'Ueber die Vegetation', in his *Alt und Neu* (Breslau, 1821), II, p. 102.
6  Oken, *Gesammelte Schriften*, ed. J. Schuster (Berlin, 1939), pp. 258–9.
7  Schelling, *System of Transcendental Idealism* (1800; trans. P. Heath, Charlottesville, 1978), p. 231.
8  Schelling, *Sämtliche Werke*, ed. K. F. A. von Schelling (Stuttgart and Augsburg, 1856–61), III, pp. 62–8; Oken, *Lehrbuch der Naturphilosophie*, 3 vols. (Jena, 1809–11); Steffens, *Grundzüge der philosophischen Naturwissenschaft* (Berlin, 1806).
9  Coleridge, *Letters*, ed. E. H. Coleridge (Cambridge, Mass., 1895), II, p. 649; cited in M. H. Abrams, *The Mirror and the Lamp: Romantic Theory and the Critical Tradition* (Oxford, 1953), p. 170.
10  Coleridge, *Collected Letters*, ed. E. L. Griggs (Oxford, 1956–  ), II, p. 709.

11 *Biographia Literaria*, ed. J. Shawcross (Oxford, 1907), I, pp. 169–70.
12 References here, and in the following quotations, are to Novalis, *Werke*, ed. H.-J. Mähl and R. Samuel (Munich, 1978–87), vol. II. We have based our translations on *The Novices at Sais* (New York, 1949). In interpreting Novalis we have been helped by H. J. Balmes's commentary and notes in Novalis, *Werke*, vol. III, and by the reading of the work in H. D. Schmid, 'Friedrich von Hardenberg (Novalis) und Abraham Gottlob Werner' (unpublished doctoral dissertation, Tübingen, 1951).
13 According to H. J. Balmes (Novalis, *Werke*, III, p. 120), 'Figures seen on touched or stroked discs of pitch and glass' is a reference to E. F. F. Chladni's study of the figures produced by violin-bowing of sand-covered plates: *Entdeckungen über die Theorie des Klanges* (1787).
14 See, for example, O. Temkin, 'German Concepts of Ontogeny and History around 1800', *Bulletin for the History of Medicine and Allied Sciences*, 24 (1950), 227–46.
15 Cited in U. Birch's 'Introduction' to Novalis's *The Disciples at Sais* (London, 1903), p. 9.
16 See, for example, R. Kremer, 'Building Institutes for Physiology in Prussia: Contexts, Interests and Rhetoric', in A. Cunningham and P. Williams (eds.) *Medicine and the Laboratory*, forthcoming.
17 See C. A. Culotta, 'German Biophysics, Objective Knowledge, and Romanticism', *Historical Studies in the Physical Sciences*, 4 (1975–6), 3–38.
18 See D. von Engelhardt, *Historisches Bewusstsein in der Naturwissenschaft von der Aufklärung bis zum Positivismus* (Freiburg, 1979), parts III and IV.

# PART I

# ROMANTICISM

# I

# Romanticism and the sciences

## DAVID KNIGHT

'Romanticism' refers to a period and not just to a state of mind. While therefore one can classify the men of science of any period into the Romantic and the classical, we shall be concerned only with those active in approximately the life-span of Humphry Davy, 1778–1829. We all have an idea of what a Romantic author, painter or musician was like, but the various groups which they formed or which were identified by friends or critics were shifting. Consistency was not seen as a virtue, and personal relations were important; so one cannot define members of the Romantic Movement as one might members of a political party.[1] There was no membership card; and as the movement affected different arts in different countries at different times we cannot expect to find it easy to generalize, or to say who exactly was or was not influenced by Romanticism: almost everybody was, in some degree.

The sciences also lacked sharp and natural frontiers. Polymaths were not uncommon: John Dalton not only published an atomic theory but also worked on meteorology and colour-blindness, and Thomas Young proposed a wave theory of light, identified astigmatism in the human eye, published tables of chemical affinities, did fundamental work in mechanics and began the decipherment of the Rosetta stone. Neither of these were what one would think of as a Romantic man of science; taking a wide range was an acceptable and expected thing to do. Specialization was indeed beginning, but not yet necessary for success in the sciences.

Again, some of the most popular sciences of the day have not survived, such as mesmerism and phrenology. Then as now what looks most showy may collapse soonest, like William Beckford's gothick tower at Fonthill Abbey; but it is important that we do not convict of credulity or absurdity those who adhered to what seemed to their contemporaries exciting disciplines on the frontiers of knowledge. They did not have our benefit of hindsight; and to know in one's own time what is going to turn out well is a great gift. Excessive caution at any period may lead to one's missing out on important discoveries; playing safe is not necessarily the best strategy.

Around 1800 'science' was not opposed to 'arts'; there was nothing like the 'Two Cultures' of C. P. Snow's famous essay. Indeed the then current classification of subjects would have put engineering among the arts, a useful rather than a fine art, while almost all other subjects now taught in universities, such as chemistry, history and theology, would have been sciences. The real division was between the realm of science, governed by reason, and that of practice, or rule of thumb; and apostles of science hoped to replace habit by reason in the affairs of life. Some aspects of natural science, such as pharmacy, were seen as professional: to discuss them in public would have been impolite, 'talking shop'; but discoveries in the sciences were generally very appropriate for civilized conversation even when it included women.[2]

We can take as our text a poem by Davy, not published by him but in a notebook:[3]

> Oh, most magnificent and noble Nature!
> Have I not worshipped thee with such a love
> As never mortal man before displayed?
> Adored thee in thy majesty of visible creation,
> And searched into thy hidden and mysterious ways
> As Poet, as Philosopher, as Sage?

Here nature is not 'it' but by implication is 'she'; personified and active, 'natura naturans' rather than 'natura naturata', in process rather than complete: God is working his purpose out.[4] Davy did not see the universe as an enormous clock; and insofar as there is any Romantic natural science, it seems to involve rejecting mechanical metaphors in favour of organic ones. Under S. T. Coleridge's tutelage, the young Davy had had 'an experience of nature' outside in the sunshine; a mystical feeling of oneness in which it would have pained him to pull a leaf from a tree.

Davy shared with William Paley, the natural theologian, a love of fishing, unaffected by this experience;[5] and indeed records an anecdote about Paley being bullied by the Bishop of Durham to finish his great book which would put atheists to silence, and replying that he would get down to it again as soon as the fly-fishing season was over. Paley's *Natural Theology* of 1802 begins with the finding of a watch: the reader sees how all the components fit together, and cannot accept that the particles which compose the watch could have come together by chance. The rest of the book is taken up with the argument that the world is like a great watch, the work of a Designer. Paley's rhetoric is very skilful, but his argument is very old, going back to the great clocks of the Renaissance:[6] and to Davy and anybody associated with the Romantic movement, Paley's mechanical world-view (like his Utilitarian moral philosophy) was unwelcome.

Nature if she is magnificent and noble does not respond to the mere watch-repairer or anatomist, accustomed to taking to pieces things which do not work, and perhaps moved to praise God because he could not have done a better job himself. Instead, an attitude of admiration, love and worship is what Davy calls for: that is, a personal response. Coleridge[7] called his time 'the age of personality', referring to the large egos he found amongst his contemporaries; but, in other senses too, that is what it was. Davy was impressed by the alchemists' humble search for wisdom, with the belief that knowledge is only given to those who deserve it: natural science is a personal interaction with nature, not an autopsy.

This is expressed in Coleridge's 'Dejection Ode', in the lines:

> O Lady, we receive but what we give
> And in our life alone doth nature live,
> Ours is her wedding garment, ours her shroud.

Coleridge had got the idea from Plotinus that our minds are active, and that we project onto nature what we expect to find in her. The Kantian doctrine of categories imposed by us upon experience went well with such ideas; and indeed Kant's work was interpreted by early reviewers, including Davy's mentor Thomas Beddoes, as a revival of Neoplatonism. In the sciences, Alexander von Humboldt called his book about his travels in Central and South America a *Personal Narrative*; this classic work formed a model for Charles Darwin when writing up his diary kept on board HMS Beagle. Humboldt sought to balance the subjective and the objective; and his later *Aspects of Nature* was an attempt at word-pictures of exotic scenery, informed by botany, geology and zoology. Descriptive natural science in Humboldt was passionate, in line with Davy's poem; and particularly united the aesthetic and the scientific.

Davy's searching into the hidden and mysterious ways of nature may be prosaically rephrased as seeking a satisfactory explanation of phenomena. In 1830, John Herschel in his famous *Preliminary Discourse*[8] urged that explanation must be based upon a 'vera causa', that is analogy from known causes which produce similar effects elsewhere. Those rejecting a mechanical world-picture would find some analogies unpersuasive; in the natural sciences there is always the danger that one may seek to explain the known and familiar in terms of the unfamiliar or the incomprehensible, and here personal and cultural attitudes determine which is which. For Hegel, 'when philosophy paints its grey on grey, a form of life is over: the owl of Minerva flies only at the coming of the dusk', but this view of natural science as a post-mortem examination was not shared by all who admired organic analogies and sought what they called a dynamical science, based upon forces and the understanding of process.

Davy saw himself in his poem first as poet; and indeed both Coleridge and Robert Southey had been impressed with Davy in this role, and believed that

he might have been a great poet if he had not become a great chemist.⁹ The
poetry he has left behind does not give much support to this claim; but there
was one poet of the first rank who spent a good deal of time on investigations
in natural science, and that was Goethe. While he never thought of himself as a
Romantic (even in his period of 'Sturm und Drang'), loving Italy and the
classical tradition, his natural philosophy is resonant with Romanticism.

In natural history, in his studies of plants and skeletons, Goethe sought
unity of plan.¹⁰ Behind the variety of nature, he believed he could perceive the
Ur-plant: the simplest plant, upon which nature had, as it were, played a series
of variations. Flowers were thus to be understood as modified leaves. The
development of flowers was a kind of transcendental evolution; but this idea is
not like Darwin's, because Goethe was not seeking an explanation of how
plants came to be the way they are, but an understanding. To see the Ur-plant
in the rose or the daisy is akin to seeing the 'form' of something in Plato's or in
Bacon's philosophy: one is not in the business of genealogy, constructing
family trees, but searching for natural kinds. In the same vein, Goethe
discovered the inter-maxillary bone in the human skull. This was not like
finding a little bone which nobody had noticed before: rather it was a matter
of identifying a part of our skull with something much more prominent in
other animals. Again, it was a perception of underlying unity, based upon
careful looking.

This work created much less stir than Goethe's experiments in optics. In
Italy he had been impressed by the colours he saw in shadows; and he took up
some textbooks to see how light and colours were explained. He found there
ambiguous descriptions of Newton's 'crucial experiment' with prisms.¹¹
Newton had caused a narrow beam of light to fall upon a prism, and to be
refracted onto a screen some considerable distance away; he found that the
rays were not merely bent but that the image was elongated into a spectrum.
When he put a second prism near the screen in such a way that rays of only one
colour went through it, he found that they were refracted but not further
decomposed into colours. The second prism put near the first but the other
way up would recombine the rays so that a circular white image was formed
on the screen. Newton concluded that white light was therefore a mixture of
rays of the other colours.

Goethe misunderstood the descriptions he found, and looked through the
prism instead, when he found that he saw colours only when there was a sharp
boundary between black and white in his visual field. He believed that this
was a fundamental or primordial phenomenon, an *Urphänomen*; and
although Newtonians assured him that it could be explained in terms of
current physics, he believed that he had shown that Newton's work was not
based upon open-minded experiment but upon dogma: an axiomatic theory, a
mathematical abstraction based on the notion of rays, which had been

perfunctorily tested. The innocent eye recording phenomena was the proper foundation for natural science; in optics Goethe believed that Robert Boyle had worked this way, but that Newton had propelled the science in quite the wrong direction. This objection to overweening confidence and premature abstraction is quite different from Keats's feeling that Newton had destroyed all the associations of the rainbow, reducing the world to a prosy factuality. Goethe wanted a science of optics, but its aim would be to cast light upon what we see. Its basis was that colours were generated at boundaries of black and white, and he had experiments to show it.

Goethe went on to do various experiments in perception; but as Dennis Sepper argues in his *Goethe contra Newton: Polemics and the Project for a New Science of Color*,[12] the most interesting feature of his attack upon Newton was that he sought to undermine the whole notion of a 'crucial experiment'. For Bacon, the crux was the signpost at a crossroads, and a crucial experiment was one which indicated the most promising line of inquiry to follow. Newton, with the conviction that certainty could be found in geometrical demonstration, believed that a crucial experiment could be used to prove a theory: and this has become the modern use of the term. Given a theoretical proposition, some consequences could be deduced from it, preferably unexpected, surprising or almost paradoxical ones; and if these could be demonstrated experimentally, then the theory must be right. This process was illustrated in our century when Einstein's astonishing prediction that because space was curved, some stars actually behind the sun would be visible beside it in a solar eclipse, was verified in 1919, thereby falsifying Newton's own system of physics.

For Goethe, there were more or less plausible interpretations of experiments, but no experiment could entail a theoretical interpretation. Goethe hoped in optics, and in the sciences generally, for competing positions and critical debate, dreading dogma: this was something which in his own day he saw in chemistry, where all was in flux and where crowds came to hear public lectures. Goethe's ideal of natural science was thus anarchic; it was personal knowledge, not gained at second hand but based upon flashes of insight or disclosures, and tested by its comprehensiveness, unity and truth to experience. Those engaged in natural science should never take the groundwork upon authority: natural philosophy would then be a real education, an essential part of their individual *Bildung* (self-cultivation), ethics and politics. This is in some respects a curiously modern vision, and an attractive one in many ways, though hardly doing justice to the masses of information which constitute elementary natural science; but Goethe failed to persuade his contemporaries. This was largely because his tone was so polemical; feeling ignored, he assailed the Newtonians in the tones of a heresy-hunt or an election, and thus made sure that outside the narrow ranks of a few disciples he made no converts. His insights into natural philosophy were lost; and though some of his psychological optics and his advice to artists on the use of colour were taken up in the nineteenth century, his work was generally seen as a dreadful warning of what can happen when able but untrained people venture into unfamiliar territory.

Goethe also used concepts from the natural sciences, especially chemistry, in writing literature. His *Faust* is full of alchemical ideas; but one could hardly say that these were widely current in the chemical community of 1800. They were from an occult tradition, in which texts have meanings on different levels, and are therefore open to both the ignorant and the initiated. Alchemy had earlier links with poetry, for instance in George Herbert, so there was nothing Romantic about using it. In his novel *Die Wahlverwandtschaften* (*Elective Affinities*), however, Goethe did something very different. The idea that metals and other substances had their particular and quantifiable elective affinities was a strong feature of the chemistry of the eighteenth century, and lasted into the nineteenth with Young and in elementary works, perhaps especially in Britain.

When a solution of what we call barium chloride is added to a solution of a sulphate, say sodium sulphate, then a 'double decomposition' happens: barium sulphate comes down as a white precipitate, and sodium chloride is left in solution. This is one of the standard tests for sulphates. Goethe picked up this idea in writing about a marriage and what happens when another man and woman are brought into the household. Goethe exploits the general interest in chemistry of the epoch, the time of A. L. Lavoisier, Joseph Priestley and C. W. Scheele, to set the scene with great skill, and uses chemical metaphors right through. But whereas in the test-tube there is no pain and the reaction goes speedily and quietly to completion, in life everything is very different.[13] The book is the history of a disaster, which makes the reader reflect how different people are from particles of matter. A book begun with metaphors from the inorganic world shows how inappropriate and reductive they are. This is indeed a Romantic message, conveyed with subtlety and intensity in a masterpiece of fiction. The cobbler should perhaps have stuck to his last: Goethe's novel is much more successful than his *Zur Farbenlehre* (*On the Doctrine of Colours*), but of course it raises questions of natural philosophy less directly.

Goethe may stand then for our poet; but Davy's next category was 'philosopher'. Around 1800, this term included the 'natural philosopher', and on board HMS Beagle Darwin was 'Philos' to his shipmates. Indeed the word by itself might almost mean what we call 'scientist', except that it lacked any professional connotations. But there were philosophers in our sense of the word who were involved in Romanticism and in natural science.[14] Friedrich Schelling in his *Naturphilosophie* presented a world of opposed polar forces, in which apparent rest was in truth only dynamic equilibrium, and solid bodies only in reality endured in the way waterfalls or columns of smoke do: his was a Heraclitean world where all is flux. His most eminent disciple among men of science was Hans Christian Oersted, who believed that electricity and magnetism, both polar powers, must be effects of one underlying polar force and be interconvertible: and proved it so with his discovery of electromagnetism. Oersted also wrote philosophical essays, later translated into English as *The Soul in Nature*, setting out his metaphysical beliefs; Davy described him on the strength of some of his earlier papers as a German metaphysician, an epithet which he did not mean to be polite. Another disciple of Schelling was J. W. Ritter, who discovered the ultra-violet rays in the spectrum. It is possible to talk of Romantic men of science in Germany as a genuine group.

Hegel was not strictly a Romantic, but one of his great interests was in classifying the sciences: as a part of his project for an encyclopaedia, and generally in his scheme of circles, levels and transitions between them.[15] He was strongly opposed, like Romantics, to reduction: he did not believe that the sciences of inorganic matter were at the top of the hierarchy, but at the bottom. One moved upwards through chemistry and electricity to the sciences of life and then on to psychology; and confusion must result when ideas appropriate to one level are unthinkingly applied on another. In Hegel's scheme, getting appropriate general ideas was as important as getting authenticated facts; but both were important. In both Schelling and Hegel, we find dialectic emphasized; truth is realized through resolution of the clash of apparent opposites. Similarly, chemical synthesis results from the fusion of opposites, as water is generated when oxygen and hydrogen are sparked together – this being then a rather new piece of chemistry. Men of science in Germany were aware of their use of and need for metaphysics,[16] while in Britain empiricism prevailed, at least in public.

The Romantic period was one of rising interest in physiology, the science of life: and in Britain the great man here had been John Hunter, who had raised surgery from a craft to a science. One of his well-known investigations concerned self-digestion. Our stomachs are made of flesh, and yet contain juices which are capable of digesting raw meat if we care to gulp it down. After death, it does sometimes happen that the fluids in the stomach do begin to attack it; and for Hunter and his disciples this showed that dead and living matter are subject to different laws. Only after death do the ordinary laws of inorganic matter supervene, and decomposition is the result.

Hunter was also impressed with the way we endure despite the flux of the material particles which compose our bodies; and in the Animal Chemistry Club in the first decade of the nineteenth century, which included Davy, William Allen the Baconian and Quaker pharmacist and Benjamin Brodie the physician, these questions were discussed. One of Brodie's experiments concerned Lavoisier's idea that in respiration oxygen is absorbed and converted into carbonic acid, with production of heat. Brodie found that when oxygen is blown into the lungs of a corpse then carbonic acid is produced, but the body cools down quicker than if left alone: which was evidence for him that analogies between the living and the dead are to be suspected. Philippe Ariès notes how medical men at this period sought to make a sharper distinction between life and death.[17] Concern with the processes of life was very appropriate for Romantics; John Keats was a surgeon, and Mary Shelley's *Frankenstein*, with its portrait of the man of science as a sorcerer's apprentice, starts from the assumptions of a materialistic physiology.

Davy was probably the most important man of science in Britain who can be described as a Romantic. His abilities were first recognized by Gregory Watt, son of James Watt, who got him invited to Bristol as assistant to Thomas Beddoes at the Pneumatic Institute where 'factitious airs' (synthetic gases), recently discovered by Priestley, were to be administered to invalids, especially sufferers from tuberculosis. Here he met Coleridge, Southey and William Wordsworth, and might therefore be said to have become a Romantic, though this could never be more than one characteristic among others: it is more important in understanding Davy that he had always been a storyteller, and it was through rhetoric, in public lectures, that this small dark man from the Celtic fringe made himself the best-known natural philosopher of his day, and eventually President of the Royal Society.

Appointed first Lecturer and then Professor at the Royal Institution in London, Davy became the apostle of applied science, seeking to understand, and then perhaps to improve, the processes of tanning and of fertilizing crops in the years of Continental Blockade. From these rather unromantic and smelly bits of natural science, Davy returned in 1806 to work he had begun in 1800 in Bristol: the study of electrochemistry. Alessandro Volta's view had been that the mere contact of dissimilar metals would generate electricity, but this Davy, like most of his compatriots, could not believe. He saw electricity as the power responsible for chemical affinity, and therefore believed that a chemical reaction was necessary to generate electricity in the 'pile' of Volta. William Nicholson and Anthony Carlisle had in 1800 applied the terminals of a pile to water, and found that they got oxygen and hydrogen at the positive and negative poles respectively. Later investigations showed that the water around the positive pole became acidic, and that round the negative pole alkaline, so it seemed that some complex reactions must be going on. Davy refused to believe that electricity could be more than an agent of analysis, and in 1806 went on refining his experiments until they fitted his expectations – not the sort of thing Goethe would have approved of. For Davy, this meant using gold, silver and agate apparatus and then putting it in an atmosphere of hydrogen, so that dissolved nitrogen could not react with the nascent oxygen and hydrogen round the poles, forming nitrous acid and ammonia.

Having proved to his satisfaction that chemical affinity and electricity were manifestations of one power, Davy investigated the chemical effects of charging metals with electricity. He found that negatively charged metals are inert, and positively charged ones reactive, something he was later able to use in proposing cathodic protection for the copper bottoms of warships. He was particularly delighted with this finding, because it chimed in with his belief that chemical properties are not determined by the nature of components. For Lavoisier, oxygen was the generator of acids; acidity was a property one could expect if and only if oxygen were a component. Davy could not accept this; like his hero Newton, and in a reaction to the materialism of Priestley and the atheism of the French *philosophes* characteristic of his generation, he believed that matter was inert rather than active. God had chosen to impose powers upon brute matter, which otherwise could not think, live or even form crystals. Davy demonstrated at Bristol how different are laughing gas, air and nitric oxide, though all composed of oxygen and nitrogen; in London that the acid from sea salt contains no oxygen, but the substance which he called the element chlorine; and in Florence, after his marriage to a wealthy widow, that charcoal and diamond, which look so different, are chemically identical, both burning to give the same quantity of carbon dioxide. It was not the components, but the powers associated with them, which gave character to substances.

At the end of his life, Davy put together earlier thoughts into a series of dialogues about natural science and life, which was published posthumously as *Consolations in Travel*:[18] here Davy emerged as the sage, passing on experience to the younger generation. But although the book sold well, later editions having attractive engravings after drawings by Lady Murchison, it does not seem to have had much influence. It is mentioned as something someone was reading in a novel by Anne Brontë, and some of Davy's remarks on geology were taken up by Charles Lyell in his *Principles of Geology*, perhaps because Davy was safely dead and could not challenge Lyell's extreme uniformitarianism. Davy's research programme was developed by Faraday,[19] and his insight that chemical affinity was electrical is still a fundamental tenet of chemistry. This was a special case of the Romantic belief that all force was one, which led some men of science in the next generation towards the conception of conservation of energy, and the creation of classical physics.

And yet Davy's electrochemistry was very similar to the ideas developed independently by J. J. Berzelius in Sweden, who was not in any obvious way a

Romantic; and in the later development of conservation of energy some of those involved were not Romantic in any way, an example being James Joule.[20]

Romantic ideas about the unity of organisms went with a transcendental evolutionary scheme, which like the idea of the conservation of energy brought unity into the various branches of natural science which were otherwise separating into specialized disciplines; but Darwin's great synthesis in the *Origin of Species* was not rooted in Romanticism but in the very different tradition of Paley and Thomas Malthus.[21] The theory of the conservation of energy and evolutionary theory in the mid-nineteenth century developed in Germany and in Britain, where Romantic natural science had been strongest, and not in France; and this was a factor in the relative decline of French science in the course of the nineteenth century. We can no longer simply assent to Justus von Liebig's view that *Naturphilosophie* was the Black Death of the nineteenth century.

In scientific illustration, we also see the transition from the portrayal of stiffly mounted corpses on studio stumps to lively pictures of birds and animals in their habitats; but this development of visual language in natural science depended partly upon new methods of reproduction, wood-engraving and lithography, and partly upon the growth of zoos and the simplification of travel which made it almost as easy to get the artist to the creature as to get the corpse to the artist.[22] A pioneer like Thomas Bewick does not fit into our idea of the Romantic movement, though perhaps William Swainson might and Edward Lear[23] and J. J. Audubon in their different ways do so.

We are back where we started: Romanticism was not something to which one formally subscribed, and while it was not a polar opposite of natural science as some have supposed, one cannot outside Germany draw up a table of Romantic men of science and expect them to form a coherent group. Collective biographies, prosopographies, which can illuminate disciplines, societies and other institutions, do not seem very promising for so indefinite a group; but in this period of political and intellectual ferment, we shall find that some men of science and some writers and painters found that they had things to learn from one another. The biography of Davy or of Alexander von Humboldt must include references to Romanticism if it is to be adequate; and similarly anybody seeking to understand Goethe, Coleridge, J. M. W. Turner or William Martin will have to refer to some natural science. It is only because we are still dominated by the idea of the Two Cultures that we find this surprising; in the early nineteenth century natural science was a component of the one culture of western Europe. Romantics preferred some parts of it, the organic rather than the mechanical; but understanding nature was not something quite separate from other concerns, and was still open to those without formal training. In the Romantic period, natural science could still be fun.

## NOTES

1 J. R. Watson (ed.), *An Infinite Complexity: Essays in Romanticism* (Edinburgh, 1983).

2 See my 'Accomplishment or Dogma: Chemistry in the Introductory Works of Jane Marcet and Samuel Parkes', *Ambix*, 33 (1986), 94–8.

3 J. Davy (ed.), *Fragmentary Remains, Literary and Scientific, of Sir Humphry Davy* (London, 1858), p. 14.

4 H. W. Piper, *The Active Universe* (London, 1962).

5 [H. Davy], *Salmonia*, in J. Davy (ed.), *The Collected Works of Sir Humphry Davy*, 9 vols. (London, 1839–40), IX, p. 11; D. Knight, 'Davy's *Salmonia*', in S. Forgan (ed.), *Science and the Sons of Genius: Studies on Humphry Davy* (London, 1980), pp. 201–30.

6 T. Cosslett, *Science and Religion in the Nineteenth Century* (Cambridge, 1984), pp. 25–45; C. M. Cipolla, *Clocks and Culture, 1300–1700* (new edn, New York, 1977); O. Meier, *Authority, Liberty and Automatic Machinery in Early Modern Europe* (Baltimore, 1986).

7 T. H. Levere, *Poetry Realized in Nature: Samuel Taylor Coleridge and Early Nineteenth-Century Science* (Cambridge, 1981).

8 J. F. W. Herschel, *A Preliminary Discourse on the Study of Natural Philosophy* (originally published 1830; with Introduction by A. Fine, Chicago, 1987), pp. 144ff.

9 T. H. Levere, 'Humphry Davy, "the Sons of Genius", and the Idea of Glory', in S. Forgan (ed.), *Science and the Sons of Genius*, pp. 33–58; J. Heath-Stubbs and P. Salman (eds.), *Poems of Science* (Harmondsworth, 1984).

10 See W. D. Wetzels, 'Art and Science: Organicism and Goethe's Classical Aesthetics', in F. Burwick (ed.), *Approaches to Organic Form* (Dordrecht, 1987). On forms in Bacon, see his *Advancement of Learning*, II, 5, and *Novum Organum*, esp. II, 11, on the form of heat: it is interesting that S. T. Coleridge, in his celebrated comparison of Bacon and Plato in *The Friend* (ed. B. E. Rooke, London, 1969), I, pp. 482ff., does not refer to this aspect of Bacon's philosophy.

11 See G. Cantor, *Optics After Newton: Theories of Light in Britain and Ireland, 1704–1840* (Manchester, 1983), esp. chs. 1 and 2.

12 Cambridge, 1988.

13 See my 'Chemistry and Poetic Imagery', *Chemistry in Britain*, 19 (1983), 578–82.

14 D. von Engelhardt, 'Romanticism in Germany', in R. Porter and M. Teich (eds.), *Romanticism in National Context* (Cambridge, 1988), pp. 109–33.

15 R.-P. Horstmann and M. Petry, *Hegels Philosophie der Natur* (Stuttgart, 1986).

16 D. von Engelhardt, 'Bibliographie der Sekundarliteratur zur romantischen Naturforschung und Medizin 1950–1975', in R. Brinkmann (ed.), *Romantik in Deutschland* (Stuttgart, 1978), pp. 307–30.

17 G. Averly, 'The "Social Chemists": English Chemical Societies', *Ambix*, 33 (1986), 99–128, esp. pp. 101f. P. Ariès, *The Hour of our Death* (trans. H. Weaver, Harmondsworth, 1981), ch. 9.

18 J. Z. Fullmer (ed.), *Sir Humphry Davy's Published Works* (Cambridge, Mass., 1969); *Consolations* is reprinted in Davy's *Collected Works*, IX, pp. 213–83.

19  D. Knight, 'Davy and Faraday: Fathers and Sons', in D. Gooding and F. A. J. L. James, *Faraday Rediscovered* (London, 1986), pp. 33–49.

20  See T. S. Kuhn's paper 'Energy Conservation as an Example of Simultaneous Discovery', reprinted in his *The Essential Tension* (Chicago, 1977).

21  J. C. Greene, *Science, Ideology, and World View* (Berkeley, Calif., 1981), esp. ch. 1.

22  W. Blunt, *The Ark in the Park: The Zoo in the Nineteenth Century* (London, 1976); D. Knight, *Zoological Illustration: An Essay towards a History of Printed Zoological Pictures* (Folkestone, 1977); A. Moyal, *A Bright and Savage Land: Scientists in Colonial Australia* (Sydney, 1986).

23  S. Hyman, *Edward Lear's Birds* (London, 1980); M. J. S. Rudwick, 'A Visual Language for Geology', *History of Science*, 14 (1976), 149–95; D. Knight, 'William Swainson: Naturalist, Author and Illustrator', *Archives of Natural History*, 13 (1986), 275–90.

# 2

## Schelling and the origins of his Naturphilosophie

### S. R. MORGAN

### TWO REVOLUTIONS

Whilst in a neighbouring great state the most remarkable Revolution goes on which has perhaps ever taken place on the scene of the Earth, (a change effected in part by *philosophers*!), we are wandering around in the labyrinth of Metaphysics, squabbling about things which are a trouble to understand . . . Without now looking into the causes of this characteristic by which we distinguish ourselves from other nations, and without deciding whether it redounds to our advantage or disadvantage, I will only remark, that that was also the case for the Greeks.[1]

On 20 April 1792, the fourth year of Liberty, the Legislative Assembly of revolutionary France voted to declare war on Francis of Austria, shortly to be crowned Holy Roman Emperor.[2] The other powerful state of the Empire, the protestant state of Prussia, was bound and ready, due to its recent *rapprochement* with Austria, to act with it against the revolutionary nation. The ministers of the two great powers were not expecting a long war. A Prussian minister told a group of officers not to 'buy too many horses' – the 'army of lawyers' would be 'annihilated in Belgium'. Austria and Prussia were pursuing a war in the interests of territorial expansion; the other states of the Empire (of which there were many) worried greatly about the compromising of imperial interests. And in fact the war against the nation of free men, which became a Republic in September, was to last almost one quarter of a century, and one of its first victims was the Holy Roman Empire.

The first step came in 1795, when Prussia, totally bankrupt, signed a separate peace with the Republic. The Treaty of Basel split the Empire in two, by means of a demarcation line going through the middle of it. The Prussians also agreed in principle, and in secret, to the secularization of ecclesiastical states, in the game of territorial compensation. Austria, forced to make peace two years later, behaved no better. A secret article in the treaty signed at Campo Formio in October 1797 revealed that even catholic Austria would concede to secularization. And in Rastatt the lack of interest of both powers in the fate of the Empire became obvious. Gathered together to conclude a peace treaty between the Republic and the Empire, the delegates of the smaller states were informed that the Austrians had been persuaded (by a young man called Bonaparte) to withdraw from Mainz, the seat of their archbishop. The representatives of the Empire were scandalized.

On 30 December 1797, the day of the handing over of Mainz, at 3 o'clock in the afternoon, there died peacefully and blessedly, at the ripe old age of 955 years, five months and 28 days, as a result of total enervation and a final stroke, completely conscious and comforted by all the holy sacraments, the Holy Roman Empire of ponderous memory.

So wrote Joseph Görres in 1798.[3] The Empire finally came to an end nine years later, in 1806.

It was within this world, behind the demarcation line, and settled in the Lutheran University of Jena situated in the Duchy of Saxe-Weimar-Eisenach, that Schiller produced his response to the revolution of the French, and created the 'Jena manifesto', in the form of *The Horae* (*Die Horen*), which he ran between 1795 and 1797. If man were 'ever to solve that problem of politics in practice', he said in the second of his *Letters on the Aesthetic Education of Man* (the opening work of his journal), then he would 'have to approach it through the problem of the aesthetic, because it is only through Beauty that man makes his way to Freedom'. In a similar spirit the Jena historian Karl Ludwig Woltmann, writing in the first volume of *The Horae*, said that France must first become a Garden of Beauty, before the fruits of freedom could ripen there.[4] Schiller, in the construction of his idea of beauty and of freedom, availed himself not only of the ideas of the Greeks, but of the principles of the philosophy of a Prussian, Immanuel Kant.

Within the same University of Jena, but also elsewhere, were others who were trying to complete, in the face of the activities of the French, the revolution Kant had started in 1781 with a *Critique of Pure Reason*. Those in Germany who would have called themselves Kantian or critical were very numerous in the last years of that Empire. Above all they were together in its many universities – and most particularly in the protestant ones – in that 'Fourth Faculty', that of Philosophy. Jena was a nest of Kantianism and, according to Crabb Robinson, the 'most fashionable Seat of the New Philosophy'. A journal, a *Philosophical Journal of a Society of German Scholars*, had been set up there in 1795, to help this Kantian revolution along. Friedrich Immanuel Niethammer was in charge of it, a pupil of Karl Leonhard Reinhold: a pupil, that is, of the man who had preached Kant's message in Jena for seven years, between 1787 and 1794, and who had claimed, according to Schiller, that within a hundred years Kant would have the reputation of Jesus Christ. Niethammer wrote, in his journal, of the point of the enterprise: there being no question that Kant's 'critical philosophy' had opened up the way to have philosophy as a science, there remained the task of finally completing it as one. The Science of all sciences, the Science of the Necessary and the General in all concepts and judgements: out of this Philosophy should then spring, in the manner of deduction, the fundamental principles of all particular sciences. Which was to say that out of the original laws of the human spirit, the structure of which Kant had begun to unfold, should come the whole system of human knowledge. As Clement Brendaṇo (*sic*) and a 2d Rate Genius (*sic*) explained to Crabb Robinson: 'There is but *one* philosophy Sir – "And *where* is that?" In your Head Sir if you know how to get at it, our Theory is not taught by Books it is generated within – '

Johann Gottlieb Fichte, the most radical of the interpreters of Kant, was precisely of this opinion. In working out his own philosophy, which was to set out what Kant (in his genius) had presupposed but not said, Fichte was not just going beyond the letters of the Kantian philosophy and grasping its spirit. The principle, which he believed the fundamental one, out of which the whole business of the human spirit should develop, was the 'I', which posited itself: and this was a truth which could only be intuited, and never found in words. Fichte came to Jena in 1794, to replace Reinhold. He still had hopes for the French Republic in 1795, and was considering, in the spring of that year, taking himself there to work out in peace the rest of his system of philosophy, his *Wissenschaftslehre*, the Lore of Ultimate Science (as Coleridge called it) – were that nation to give him a pension to do so. His system, he explained,[5] really belonged to the Republic, because as he wrote about their Revolution, the first hints of his own system had come to him. And more than that: his system was 'the first system of freedom', and 'as that nation tore man loose from his outer chains' so his system 'tears man from the fetters of the things-in-themselves, from the outside influences which are more or less bound around him in all other previous systems, including the Kantian, and describes him in his first principles as an independent being'.

Here were the beginnings, in the 1790s, of the kind of reaction to the political revolution of the French which later Germans (such as Marx) were to look back on in despair: a revolution of the people accompanied by only the abstract machinations of the spirit. Critical philosophers of the time doubtless saw it as Reinhold did, when thinking of the possible fruits of the Kantian philosophy: of securing the 'sublime vocation' of Germany as 'the *future School of Europe*'. Schelling, himself engaged in the *ideal* reaction to the French, maintained in an obituary for Kant that it was the same spirit which had freed itself in the two revolutions, the Kantian and the French. And it was, he said, a consequence of the Kantian philosophy that in Germany a judge-ment formed so quickly about the French Revolution, just as it was a consequence of the new beliefs in Rights and Constitutions that the Kantian philosophy became a necessity even for statesmen and men of the world. So there was, in the 1790s, a German theory of freedom, which had its sources in the philosophy of Kant, and out of this came the idea that there were different ways of being free. The distinction between *Freiheit* and *Liberté*, which on the one hand was manifested, in the form of the French army, on the banks of the Rhine and then beyond, also manifested itself in the disciplines and theories which were created behind the demarcation line during the last years of the Empire. Schelling's philosophy of nature was one of them, which celebrated, like the theories of Fichte and Schiller, an *inner freedom*, by supposing that nature was the symbol of the path to it.

## A THEORY OF FREEDOM

A Philosophy then, which sets up as its first principle the claim, that the essence of man exists only in absolute freedom, that man is no thing, no object of an arbitrary will, and in his proper being no object at all . . .

(Schelling, *Of the I as the Principle of Philosophy* (1795))

Schelling in 1795 was twenty, and in the last months of his education at the *Tübinger Stift*. This was the main theological seminary in Württemberg, a state with a strong pietist tradition, which lay in the southwest, surrounded by catholics. It was also the place which had produced, two years earlier, G. W. F. Hegel and Friedrich Hölderlin. Not one of them, however, entered the church of the state which produced them, and so the seminary is better known as a birth place of 'idealism'. Schelling was also a contributor to Niethammer's *Philosophical Journal*. Niethammer, also a product of the *Stift*, had invited him to join in sometime in the summer of 1795. And not without reason. Within the walls of the *Stift* Schelling had already written two books bearing the stamp of the new philosophy. The first, a small book he called *On the Possibility of a Form of Philosophy in General*, he had finished in September 1794. The second, a larger undertaking, he completed the following March, and called it *Of the I as the Principle of Philosophy, or On the Unconditional in Human Knowledge*. The origin of the two works owed much to Schelling's belief in the promise of the new philosophy of Fichte. 'We expect everything from philosophy', wrote Schelling to Hegel in January 1795. Fichte was to complete the work, and Schelling wanted to be one of the first to greet 'the new hero' in 'the land of truth'. His first book was an attempt to show, like Fichte, that the fundamental principle providing the form and content of all philosophy was the 'I', which posited itself. His second made explicit that it was the autonomy of the human spirit which was to be gained by accepting it. It also made clear his debt to Spinoza, whose ideas were then so fashionable. 'The orthodox concepts of God are also no longer for us', declared Schelling to Hegel in February. For Schelling God was the absolute 'I', a God which was unorthodox, much like Spinoza's, because it was neither personal nor outside the world. The absolute 'I', like Spinoza's *Deus sive Natura*, was the immanent cause of everything which is. Yet this God, the absolute 'I', Schelling had chosen because of its freedom. That which was the 'highest principle of all philosophy' was 'not at all conditioned by Objects, but was posited through Freedom'. Its essence was freedom, because it posited itself (and thus all reality in itself) through its own absolute power. With that was decided the nature of man. The absolute freedom of the immanent cause of the world, the impersonal 'I', guaranteed the freedom of the finite personal ones, because they participated in it; they only differed from it because their freedom came up against its limits, in a world of objects.

Faced with a world of objects, the little 'I' could not realize itself: but it was a moral law that one should attempt to do so, to break through the boundaries of being finite, and become absolutely free. This was the message which Schelling repeated in the first of his contributions to Niethammer's *Philosophical Journal*, in his *Philosophical Letters on Dogmatism and Criticism*. The point of all finite activity was a return to the divine – that 'most holy thought of antiquity'. And so a finite subject had to strive, eternally, toward an unconditional freedom, an unlimited activity, and thereby approach the divine within. But this was not a truth which was easily had. It required the activity of 'a secret and wonderful faculty', an *intellectual intuition*, which could only be brought about by freedom, a faculty 'alien and unknown' to anyone whose freedom, 'overwhelmed by the penetrating power of objects, is hardly enough to bring about consciousness'. His philosophy, Schelling admitted, was in itself esoteric, and not for those who were only interested in the letters of the truth: it was only for those who could make themselves free.

Schelling's *Letters* were also an expression of his frustration with the theologians of his *Stift*, and were written to attack them, because they used Kant's moral philosophy to support a belief in the orthodox God. But by the time he had finished his *Letters*, in January 1796, he was also finished with Tübingen, and had taken up a position as a tutor to two young barons, whom he was to accompany to the University of Leipzig in Saxony, in time for the summer term of 1796. He missed, in doing so, the occupation of his Fatherland by the French.


## FREEDOM IN NATURE

If, that is, the human spirit is originally autonomous, then it is a being which carries within itself not only the *ground* but also the *border* of its being and its reality, for which this border can be determined by nothing external to it, it is a totality, contained within itself and complete in itself (a monogram, as it were, of freedom constructed out of the infinite and the finite).

Thus wrote Schelling in 1797, in Niethammer's *Philosophical Journal*. Before he came to Leipzig he had considered the world of objects only negatively, which was natural for someone so concerned with a philosophy of freedom.

The object was that against which the subject strove to exert its freedom. In Leipzig, the city of the book fair, Schelling was exposed to a different world, and the houses of other professors were opened up to him. He attended lectures given by the Faculty of Medicine, and he must have heard Karl Friedrich Hindenburg lecturing on Experimental Physics, using as a text Johann Christian Polykarp Erxleben's *Anfangsgründe der Naturlehre* (*Elements of the Doctrine of Nature*), which Lichtenberg had edited and commentated. He therefore became acquainted with the new French chemistry, introduced by Lavoisier in 1789, which had caused such a stir. Towards the end of that year Schelling had started work on a new book, which was to appear at the Easter book fair in 1797, with the title *Ideas for a Philosophy of Nature, Books One and Two*. That same winter of 1796, he took on the task of reviewing, for the 'new' *Philosophical Journal* (which was, from January 1797, also to be edited by Fichte), the 'newest in philosophical literature'.

It was here in the *Philosophical Journal* that Schelling first said that the human spirit gave the object its *matter*, as well as its form. The object was not at all something which was given to us (formless) from the outside, as the orthodox Kantians thought: it was 'only a product of an original spiritual activity of the self, which, out of opposed activities, creates and brings about a third, common to both (κοινόν in Plato)'. This was indeed the consequence of having the god within us. Schelling had taken that thought to its limits. The individual (the little 'I') was a being in which the infinite and the finite were united. It was a being in which an infinite activity strove, eternally, to become conscious of itself, which it could only do by limiting itself, its own activity. In determining itself, *in creating its own boundaries*, the spirit within had to be at the same time active and passive. There was, then, an original strife within us, between an infinite freedom and a passivity (the negation of the spirit's own activity), and the product of this strife was something finite. The strife *gave rise* to the variety and wealth of a world of objects, because an infinite spirit, having made itself finite in its attempt to become conscious, nevertheless made this attempt again and again. Thus the free spirit *separated* itself from its product, and left it as an *object* behind. This was the birth of matter out of spirit, and also the beginning of time. It was nothing less, in fact, than creation *ex nihilo*: the product of the god within. In separating itself from its product, the spirit also created the means of becoming conscious of itself, because in this separation lay the opposition between *inner* and *outer*. Consciousness, the inner freedom, was only possible in opposition to an outer world of objects: because only an activity which had stepped out of itself, and set its own limits, could freely return into itself and open up an inner world.

So the outer world of objects was a product of the divine within, and was there for a reason: so that the infinite and absolutely free spirit (the absolute 'I') might become conscious of its freedom (of, that is, itself). And because

Nature consisted only of the products of the spirit, it meant that the spirit 'indicated, through its own products, not noticeable to the common eye, but clearly and determinately to the philosopher, the way by which it gradually reached self-consciousness'. And so Schelling concluded that 'The outer world lies opened out for us, in order to find within it the history of our spirit.' Nature was, that is, a history of the path to freedom.

The whole story runs like one of the creation myths Schelling had looked into when he was in Tübingen,[6] retold in the language of the new philosophy. And just like in Genesis and the other ancient myths of the golden age, the story carried with it a message about a Fall – the separation of the spirit from its own product, and thus the creation of subject and object. This itself contained its message for those who investigated nature, because those who taught that the object was a thing-in-itself, forever independent of the subject, were supposing that separation to be permanent. For them, that is, there could be no return to the divine, no chance of becoming absolutely free.

The belief that the spirit intuited, in its products, nothing but its 'own self-developing nature', meant that for Schelling the properties of the spirit – its freedom and limitation – were expressed in nature, and there for the *Naturforscher* (literally, 'researcher into nature') to see. So there was organization in nature, because the spirit, in producing its own representations, was cause and effect of itself: 'it will therefore intuit itself as an object *that is mutually cause and effect of itself*, or, what is the same thing, as a self-organizing nature'. Furthermore:

Because there is in our spirit an infinite striving to organize itself, so in the outer world must a general tendency to organization reveal itself. It is so. The world system is a kind of organization, which has formed itself from a common centre. The powers of chemical matter are already beyond the boundaries of the merely mechanical. Even raw materials which separate out of a common medium crystallize out as regular figures. The general formative drive [*Bildungstrieb*] in nature loses itself finally in an infinitude, which even the prepared eye is unable to measure. The perpetual and fixed passage of nature to organization betrays clearly enough a lively drive which, as it were, wrestling with the raw matter, now winning, now losing, breaks through it now in free forms, now in limited. It is the general *Spirit* of nature, which gradually [in]forms the raw matter into itself [*der allmälig die rohe Materie sich selbst anbildet*]. From moss, in which the trace of organization is hardly visible, to the noble *Form* [*Gestalt*], which seems to have shed the chains of matter, there rules one and the same drive within, which strives to work according to one and the same Ideal of Purposefulness, strives to express *ad infinitum* one and the same Original Image [*Urbild*], the Pure Form of our Spirit.[7]

Schelling had said what nature was, and what it was there for: it was a product of the spirit, and there so that the spirit should become conscious of its freedom. He would never have dared to do so had not the new chemistry, the new investigations into the nature of the living, and Kant's theory of both, provided him with the means of filling those claims out – which was what he was doing, in writing the *Ideas* and then *On the World Soul*. 'We will not rest in philosophy', wrote Schelling in the *Philosophical Journal*, 'until we have accompanied the spirit to the goal of its striving, to self-consciousness. We will follow it from representation to representation, product to product, to where it first tears itself free of all product, grasps itself in its pure act, and now intuits nothing but *itself* in its absolute activity.' This was the point of a philosophy of nature.

The *Ideas* was a book which started at the beginning: with matter and its different qualities. Matter was a product of the opposed activities of the spirit, and it appeared to the *Naturforscher* as the product of the opposed powers of repulsion and attraction. The different qualities of matter, which we could sense, were just differences in the relation between the two powers. Schelling was in good company here because Kant had said something very similar in 1786, in his *Metaphysische Anfangsgründe der Naturwissenschaft* (*Metaphysical Foundations of Natural Science*). Like Kant, Schelling was attacking the fallen men of the mechanical philosophy, who said that matter was composed of atoms or corpuscles, which were absolutely impenetrable and indivisible, and forever independent of the spirit. Kant's little book was the source of the terminology with which Schelling inaugurated his new *dynamical* philosophy of nature. (It was also a source of truth for Franz von Baader and C. A. Eschenmayer, both of whom wrote, at just this time, books using the principles of Kant's philosophy of nature.) It was with these principles that Schelling assessed and redescribed the new chemistry imported from the French, and expressed the hope that its elements would give rise to a new system of nature. And in the new system the *Bildersprache* (figurative language) of the chemist would be replaced by the language of the *higher* science of *Naturphilosophie*. For only the *Naturphilosoph*, with access to the truth about the absolute spirit, could rightly say what Nature was. The lower sciences could only ever deal with the symbols of that reality.

In his book *On the World Soul* Schelling made use of the new chemistry, its offspring, the 'chemical physiology', and the work of C. F. Kielmeyer and J. F. Blumenbach, to show that the whole system of Nature was a product of the dualism of the spirit (its activity and passivity, its freedom and limitation), and that it was 'one and the same principle' which was at work in both inorganic and organic nature and which bound all of it 'into a general organism'. This was also the being which the 'most ancient philosophy' had divined as 'the *common Soul of Nature*', and which some of them had held to be one with the 'shaping and forming Aether (the part of the most noble natures)'. It was a great bonus, as far as Schelling was concerned, that this was what the ancients had thought. It meant a beginning of the return to the oldest and most holy of beliefs about nature, when man was at one with it. And so here, as with so

1   Title-page of Schelling's *Ideas for a Philosophy of Nature* (1797). At the touch of Apollo, god of wisdom, the tree of life yields the cornucopia of plenty.

many of his generation, Schelling revealed the tyranny of Greece over Germany. He also revealed in *On the World Soul* his delight (to be so typical of the Romantic generation) in turning the world of the *Naturforscher* upside down: because even in the atmosphere one could see a sketch of life on earth, and chemical processes could be seen as 'incomplete processes of organization'.

These were just the beginnings of Schelling's philosophy of nature. In the meantime, strings had been pulled by his friends in Jena, and at the end of June 1798 he was offered a job there, as Professor of Philosophy. In the summer he travelled to Dresden, to meet up with August Wilhelm Schlegel and his wife, Caroline; with Friedrich Schlegel; and also with Novalis (Friedrich von Hardenberg), whom Schelling had first met in December 1797. In the spring of that year the Schlegel brothers had published the first volume of their *Athenäum*, the manifesto of the early Romantic movement. And here in Dresden Schelling enjoyed everything which was remarkable about the place – 'the Gallery, where the divine paintings of Raphael and Correggio are kept, the collection from antiquity, in which the old world continues in living statues', 'the whole wide and magnificent district around Dresden' – and also the 'marvellous company of good and happy men'.[8] It was the start of the friendship with the Romantics which has led to the description of him as their philosopher. He stayed there until October, with August Wilhelm Schlegel and Caroline. In October both August Wilhelm and Schelling went to lecture in Jena. Schelling lectured on his philosophy of nature and transcendental philosophy, and later on the philosophy of art. He published, in 1799, what he was writing for his lectures – a *First Sketch of a System of Naturphilosophie*, and then an *Introduction* to it. In 1800, he published his *System of Transcendental Idealism*. In the autumn of 1799 Friedrich Schlegel and Dorothea Veit and the Tieck family arrived in Jena. And so was gathered together, in the last year of the century and the first of the next, the group now known as the Jena Romantics. Late in 1799 Schelling wrote a poem, in rhyming couplets, a response, supposedly, to the 'religious paroxysms' of Tieck and Novalis.[9] Friedrich Schlegel had a mind to put it into *Athenäum*. But it was not published there. Schelling published part of it later, in his own journal, the *Zeitschrift für spekulative Physik*.[10] It captured the spirit of his enterprise, by describing the odyssey of the *Proteus of Nature*:

> Don't know how I could be in terror of the world,
> Because I know it inside and out,
> It is just an inert and tame animal,
> Which threatens neither you nor me,
> Which must submit itself to laws,
> And lie quietly at my feet.
> Yet there is a gigantic spirit trapped within,
> But it is petrified with all its senses,
> It cannot get out of its tight shell,
> Nor break its iron prison,
> Although it often stirs its wings,
> Stretches itself forcibly and moves,
> In dead and living things
> Strives mightily after consciousness.
> Thence comes the quality of things,
> Because it wells up and strives within,
> The force by which metals sprout,
> Trees in the spring shoot up,
> It seeks in all nooks and crannies
> To twist and turn itself out into the light.
> It spares itself nothing of the toil,
> Springs up now into the height,
> Extends its members and organs,
> And then again shortens and contracts itself,
> And hopes by turning and twisting
> To find the right form and shape.
> And so fighting with hand and foot
> Against the adverse element
> It learns in small to win a space,
> In which it first comes to awareness,
> Closed up inside a dwarf
> Of beautiful form and upright growth
> (Named in language the child of man)
> The gigantic spirit finds itself.

From iron slumber, from lengthy dream
Awoken it barely recognizes itself.
Is very much amazed about itself
With great eyes greets and takes the measure of itself,
Wants straight away with all its senses
To melt once again into the great Nature,
But once torn away,
Can no longer flow back again,
Stands narrow and small all its life
Alone in its own great world.
In anxious dreams it fears indeed
Lest the giant take courage and rear up,
And like the old God Saturn
Devour his children in wrath.
Does not know itself to be [the God],
Is forgetful completely of its own descent,
Torments itself with spectres.
Could thus say to itself:
I am the God, whom [Nature] cherishes in her bosom,
The spirit which moves in all,
From the first strife of dark forces
To the outpouring of the first juices of life,
Where force suffuses force, and matter matter;
The first flower, the first bud swells,
To the first ray of the newborn light
Which breaks through the night like a second Creation,
And from the thousand eyes of the world
Illuminates the heavens like day the night,
It is One force, One interplay and weaving,
One drive and impulsion to ever higher life.[11]

It was here in Jena that Schelling first had an effect on those who investigated nature, by telling them how to look at it. He was very successful in doing so, and not only there, but later among the *Naturforscher* and doctors in southern Germany – so successful, indeed, that D. H. F. Link, a most respected *Naturforscher*, felt compelled to write a corrective to *Naturphilosophie*, in 1806: in recent times, he complained, the *Naturforscher* had been advised to forget everything he had learnt, in order to be admitted into the Temple of *Naturphilosophie* where the origin of things would be revealed to him. But it was in fact only after the rise of the new speculative science that entrance to the Temple would prove unattractive.[12]

•

## NOTES

1   Johann Christoph Schwab, in *Preisschriften über die Frage: Welche Fortschritte hat die Metaphysik seit Leibnitzens und Wolffs Zeiten in Deutschland gemacht? Von Johann Christoph Schwab, Karl Leonhard*

*Reinhold und Johann Heinrich Abicht*, ed. by the Royal Prussian Academy of Sciences (Berlin, 1796; repr. Darmstadt, 1971), p. 4.

2   The following account is taken from Karl Otmar Freiherr von Aretin, *Heiliges Römisches Reich 1776–1806. Reichsverfassung und Staatssouveränität*, 2 vols. (Wiesbaden, 1967), I, pp. 241–371, ch. 4, and T. C. W. Blanning, *The Origins of the French Revolutionary Wars* (London and New York, 1986), pp. 69–204, chs. 3, 4 and 5.

3   The translation is taken from Blanning, *The Origins of the French Revolutionary Wars*, p. 178.

4   Quoted in Paul Hocks and Peter Schmidt, *Literarische und politische Zeitschriften 1789–1875* (Stuttgart, 1975), p. 105.

5   *Johann Gottlieb Fichte. Briefwechsel 1793–1795*, ed. Reinhard Lauth and Hans Jacob (Stuttgart–Bad Cannstatt, 1970), p. 298. The book about the revolution is his *Beitrag zur Berichtigung der Urtheile des Publikums über die französische Revolution*, which appeared in 1793.

6   See his dissertation *Antiquissimi de prima malorum humanorum origine philosophematis genes. III. Explicandi tentamen criticum et philosophicum* and the article 'Ueber Mythen, historische Sagen und Philosopheme der ältesten Welt', published in the journal *Memorabilien*, edited by H. E. G. Paulus. Both have been reprinted in later editions of Schelling's works.

7   'Allgemeine Uebersicht der neuesten philosophischen Litteratur', *Philosophisches Journal*, 6 (1797; reprinted Hildesheim, 1969), 188–9 (my translation). For the meaning of *Bildung* and its variants, particularly for the play, here so important for Schelling, between *symbol* and *form*, see Jacob Grimm and Wilhelm Grimm, *Deutsches Wörterbuch*, vol. I (Leipzig, 1854) for *anbilden*; vol. II (Leipzig, 1860) for *Bild*, *bilden* and *Bildung*.

8   *Aus Schellings Leben. In Briefen*, ed. G. L. Plitt, 3 vols. (Leipzig, 1869–70), I, p. 240.

9   Rudolf Haym, *Die Romantische Schule. Ein Beitrag zur Geschichte des deutschen Geistes* (Berlin, 1870; reprinted Darmstadt, 1977), p. 552.

10  *Zeitschrift für spekulative Physik*, 2 vols. (Jena and Leipzig, 1800–1801; reprinted Hildesheim, 1969), I, pp. 152–5.

11  Translation mine.

12  For another account of Schelling's turn to nature, see Wolfgang Wieland, 'Die Anfänge der Philosophie Schellings und die Frage nach der Natur', in *Natur und Geschichte. Karl Löwith zum 70. Geburtstag* (Stuttgart–Berlin–Köln–Mainz, 1967), pp. 406–440.

## FURTHER READING

### WORKS OF SCHELLING

*The Unconditional in Human Knowledge. Four Early Essays (1794–1796)* by F. W. J. Schelling, translation and commentary by Fritz Marti (Lewisburg and London, 1980)

*Ideas for a Philosophy of Nature as Introduction to the Study of that Science* (1797) by F. W. J. Schelling, trans. E. E. Harris and Peter Heath; introduction by Robert Stern (Cambridge, 1988)

*Introduction to the Outlines of a System of Natural Philosophy; or, On the Idea of Speculative Physics and the Internal Organization of a System of this Science (1799)*, translated from the German of Schelling by Tom Davidson, in *The Journal of Speculative Philosophy*, 1: 4 (1867), 193–220

*System of Transcendental Idealism (1800)* by F. W. J. Schelling, translated by Peter Heath; introduction by Michael Vater (Charlottesville, 1978) .

*On University Studies* by F. W. J. Schelling, translated by E. S. Morgan; ed. and with an introduction by Norbert Guterman (Athens, Ohio, 1966)

## FURTHER LITERATURE IN ENGLISH

Blanning, T. C. W., *The French Revolution in Germany. Occupation and Resistance in the Rhineland 1792–1802* (Oxford, 1983)

Butler, E. M., *The Tyranny of Greece over Germany* (Cambridge, 1935)

*Crabb Robinson in Germany 1800–1805. Extracts from his Correspondence*, ed. Edith J. Morley (Oxford, 1929)

Engelhardt, Dietrich von, 'Romanticism in Germany', in R. Porter and M. Teich (eds.), *Romanticism in National Context* (Cambridge, 1988), pp. 109–33

Esposito, Joseph L., *Schelling's Idealism and Philosophy of Nature* (Lewisburg and London, 1977)

*German Aesthetic and Literary Criticism: Kant, Fichte, Schelling, Schopenhauer, Hegel*, ed. David Simpson (Cambridge, 1984); in the same series: *The Romantic Ironists and Goethe*, ed. Kathleen M. Wheeler (Cambridge, 1984); and *Winckelmann, Lessing, Hamann, Herder, Schiller, Goethe*, ed. H. B. Nisbet (Cambridge, 1985)

Gower, Barry, 'Speculation in Physics: The History and Practice of *Naturphilosophie*', *Studies in the History and Philosophy of Science*, 3 (1973), 301–56

Lenoir, Timothy, 'Generational Factors in the Origin of *Romantische Naturphilosophie*', *Journal of the History of Biology*, 11 (1978), 57–100

Nauen, Franz Gabriel, *Revolution, Idealism and Human Freedom: Schelling, Hölderlin and Hegel and the Crisis of Early German Idealism* (The Hague, 1971)

Schiller, Friedrich von, *On the Aesthetic Education of Man, in a Series of Letters*, ed. and trans. E. M. Wilkinson and L. A. Willoughby (Oxford, 1967)

Snelders, H. A. M., 'Romanticism and *Naturphilosophie* and the Inorganic Natural Sciences 1797–1840: An Introductory Survey', *Studies in Romanticism*, 9 (1970), 193–215

Wetzels, Walter D., 'Aspects of Natural Science in German Romanticism', *Studies in Romanticism*, 10 (1971), 44–59.

# 3

# Romantic philosophy and the organization of the disciplines: the founding of the Humboldt University of Berlin

## ELINOR S. SHAFFER

From the turn of the nineteenth century a groundswell of criticism of the old universities made itself heard, and the many responses resolved themselves into an increasingly radical call for 'die Neuschaffung der Universität aus dem Geist des deutschen Idealismus' ('the new creation of the university out of the spirit of German idealism'). The crisis point was reached when in 1806 Napoleon, having defeated the Germans at Jena, ordered that the University of Halle be suspended. The new University was founded in Berlin in 1809, and first entitled Frederick William University, after the reigning king of Prussia, Frederick William III, who accepted the recommendations made by Wilhelm von Humboldt, then in the government of the liberal statesman Karl vom Stein and in charge of the reform of the entire educational system.[1] It was indeed a new foundation of the universities on the basis of contemporary German philosophy, the transcendentalism of Kant and the revisions of it in idealism, heralded by Fichte's *Wissenschaftslehre* (*Science of Knowledge*) (1794) and Schelling's first decade of publication, and still to unfold fully into the work of Hegel and the later Schelling. As Frederick William said: 'Der Staat muss durch geistige Kräfte ersetzen, was er an physischen verloren hat' ('The state must replace through intellectual power what it has lost in material power'). Without this philosophical development the movement we know as Romanticism is inconceivable. The relationship of Romanticism to the sciences ('science' both in the general sense of *Wissenschaft*, which may be translated as science, knowledge, scholarship or discipline, depending on context, and the more limited sense, usually made explicit, of 'natural science') can be seen most clearly and succinctly in the discussion of the nature and relations of the disciplines to one another that preceded the founding of the Humboldt University. This discussion started from first principles, and extended from the nature of *Bildung* (education or culture) and the idea of the university to the practical details of its conduct.

The five founding documents of the new university, written by men with the closest possible intellectual and personal links to the new philosophy and to the Romantic movement, comprise: F. W. J. Schelling, *Vorlesungen über die Methode des akademischen Studiums* (*Lectures on the Method of Academic Studies*) (given as lectures in 1802 in Jena and published the following year);[2] Friedrich Schleiermacher, *Gelegentliche Gedanken über Universitäten in deutschem Sinne*, with an 'Anhang über eine neu zu errichtende Universität' (*Occasional Thoughts on the German Idea of a University*, with an 'Appendix on a New University') (1808); Johann Gottlieb Fichte, *Deduzierte Plan einer zu Berlin zu errichtenden höhern Lehranstalt* (*Plan for an Institution of Higher Learning in Berlin* (1808)), which should be read in conjunction with his *Ideen für die innere Organisation der Universität* (*Ideas for the Internal Organization of the University*) (1807); Henrik Steffens, *Über die Universitäten* (1808–9) (a series of lectures 'On the Universities'); and Wilhelm von Humboldt himself, *Über die innere und äussere Organisation der höheren wissenschaftlichen Anstalten zu Berlin* (*On the Spirit and the Organizational Framework of Intellectual Institutions in Berlin*) (1809). As Humboldt did not employ Steffens's proposals in drawing up his plan, we shall omit consideration of them here; Fichte and Schleiermacher did not have access to each other's proposals until after the foundation. Fichte and Schleiermacher were among the first appointments to the new university (along with the historian Barthold Georg Niebuhr and the jurist Friedrich Karl von Savigny); Schelling was called to Berlin only in 1843.

To speak only of these documents is of course highly artificial, as an immense activity of philosophic and pedagogic comment prepared the way for the educational reform carried out by Humboldt. Rousseau and Pestalozzi, Schiller, especially his inaugural lecture at Jena on 'The Nature and Value of Universal History' (1792), and F. A. Wolf (the Homeric scholar), provided a stimulus to new educational theory and policy. Most essential of all was Kant's *Streit der Fakultäten* (*Conflict of the Faculties*) (1798), a seminal document in its day and one that has in recent times received much attention of a kind that makes it particularly important to be clear about its actual occasion and practical political consequences.[3] Kant's paper, a sharply polemical and well-considered response to the ban the king (Frederick William II) had in 1794 placed on his giving any further public lectures on religion (after the publication of *Religion within the Limits of Reason Alone*), made a powerful case for academic freedom. The strong continuity between the Enlightenment and Romanticism can be seen in the contribution of Kant to the reform programme of Humboldt. He argues that the 'lower faculty', the Faculty of Philosophy, should be considered the superior of the traditional 'higher faculties' of Theology, Law and Medicine, because it alone could exercise rational critique of the others; that it formed a kind of 'permanent opposition party, the left wing of the parliament of thinkers' ('eine

Oppositionspartei, die linke Seite des Parlaments der Gelahrtheit'); and that it was in the best interests of the state to allow it to exercise its function without interference.[4] Here Kant was putting into effect his notion of the *Klerisei* (the clerisy, in Coleridge's translation), a band of secular intellectuals who within the church of reason would be analogous to the clergy within the established churches.[5] The primacy of the Faculty of Philosophy and the independence of the university from state control and censorship became leading principles of all the founders; the sense of mission and the urgent need to create a new intellectual community informed them all. Moreover, each of the authors of the major documents wrote a good deal else that is relevant to the subject, which in turn is only fully understandable in relation to the larger corpus of his philosophic writings. Fichte in particular addressed himself to these matters in a variety of publications, for example his influential popular lectures, including his *Bestimmung des Menschen* (*The Vocation of Man*) (1800) and his five lectures on the *Bestimmung des Gelehrten* (*The Vocation of the Scholar*) (1805);[6] early translated into English, his writings were a force in the discussion of the modern university, from Coleridge (who read nearly all the founding documents and much of the literature around them) and the founders of London University in the 1820s to Newman's *Idea of a University* (1852), and were propagated in the United States even more effectively than elsewhere through Emerson. Further, each of these statements is itself a complex and rich, often an extensive document, or, as in the case of Humboldt's admirably succinct memorandum, a summary of a wide and deep rethinking of the cultural role of education. Furthermore, the relations among these writers form a topic in itself, even if confined to their communications on the nature of the university, the academies and the research institutes.[7] What follows must be understood, then, as a schematic account of the general features of a programme held in common by the reformers, while taking account of the considerable differences of emphasis.

The new philosophy carried out a new ordering and linking of the disciplines in the terms of the producing self or consciousness coming to understand its own productions. This task was set by Fichte's *Wissenschaftslehre* (1794), and was carried out most fully by Schelling and later Hegel (whose major publications followed the founding of the University). This task required the redefinition and reformulation of existing disciplines. It also led to the identification of new disciplines that were needed to elaborate certain phases of the self, and to others that were to bridge, unify or ground the whole cycle. Thus, for example, dialectics, arising as a response to the Kantian crux of the antinomies, is a discipline formulated, though in different ways, by all the thinkers of this school. Schelling's *Naturphilosophie* reorganizes the natural sciences according to the order and mode in which consciousness comes to regard a range of its products as objective. Art, again arising out of a major Kantian contention, namely that the ideas of reason can be realized only in art, is of unique significance in all their systems, creating in effect a new discipline, Romantic aesthetics. Humboldt's 'theory of human education' (*Menschenbildung*) is concerned with how the individual self is to be culti-vated in the light of the full potential of human consciousness as analysed by the new school both in the abstract and in its historical development. All of the documents display the urgency with which high philosophical speculation was seen as requiring a complete overhauling of the practical sphere. The indissoluble connection between theory and practice was a central plank in their philosophy and was built into the structures of the new university.

## SCHELLING AND THE UNITY OF KNOWLEDGE

Schelling's *Lectures on the Method of Academic Studies*, although they do not explicitly discuss the founding of a new university, give a very good sense of the Romantic idealist conception of knowledge and the relations among the disciplines which was to underlie the practical arrangements of the faculties and institutes in the Humboldt University. A remarkably precocious thinker, he was at the time of these lectures barely thirty, yet had published major works on natural philosophy and transcendental idealism, and would in the following year give his lectures on aesthetics, a central statement of Romanticism.

In his opening lecture Schelling puts eloquently his view of the ultimate unity of all knowledge: this tenet of idealist philosophy animates all the advocates of the new university and affects its practical arrangements. He addresses himself to the new philosophy's ambitious programme 'zu erklären, wie das Ich sich selbst als empfindend anschaue' ('to explain, how the self could perceive itself in the act of perceiving'), and to construct a new *Wissenssystem*, a system based on the relation of the self's modes of self-perception regarded and formulated as formal bodies of knowledge, as disciplines. In his second lecture, 'The Scientific and Moral Functions of Universities', Schelling puts the practical question: 'What should be done within the . . . actual structure of our universities today, in order that the unity of the whole may re-emerge amid the widespread specialization?' (p. 21) 'The usual view of the universities', he points out, 'is that they should produce servants of the state, perfect instruments for its purposes' (p. 23, n. 6). Kant had argued that the Faculty of Philosophy, through its function of exerting criticism, might even improve the quality of the state appointees.[8] Schelling contends that even such instruments of the state should be formed by science (*Wissenschaft*), and to achieve this, university teachers must be given 'intellec-tual freedom' (p. 22). This is essential if universities are to be 'real scientific institutions':

Incontestably, the state has the authority to suppress the universities or to transform them into industrial training schools; however, it cannot intend the universities to be real scientific institutions without desiring to further the life of ideas and the freest scientific development . . . Science ceases to be science the moment it is degraded to a *mere* means, rather than furthered for its own sake. (p. 23)

The relation of the university to the state, that is, the safeguarding of the freedom of scientific investigation despite the undoubted power and tendency of the state to repress it, is one of the repeated themes of all these documents.

The conflict that was to rend Oxford half a century later, between the function of teaching young men and the function of advancing knowledge, was decisively resolved for the Humboldt University in Schelling's terms. Even if universities were 'no more than institutions for the transmitting of knowledge',

to transmit knowledge intelligently one must be able to understand the discoveries made by others . . . Many of them are of a kind whose inner essence can be grasped only by a kindred genius through a rediscovery in the literal sense of the word. A teacher who merely transmits will often give a radically false version of what he learned . . . A man incapable of reconstructing the totality of his science for himself, or reformulating it from his own inner, living vision, will never go beyond mere historical exposition of the science. (pp. 26–7)

This view, that the teacher must himself be a researcher, is implied by the Romantic epistemological tenet that creative or innovative genius has a special communicative capacity and function, and that the learner can be said to understand only if he passes through the same train of experience as the communicator; thus, a teacher who has not carried out research is by definition incapable of teaching its methods or its results. 'Results' can themselves be understood only as the end product of the process of inquiry. The process of inquiry is, moreover, more valuable than the results; it is the process that the educated man learns, and can apply in whatever sphere he finds himself. Thus 'character-training' is a by-product of the process of inquiry properly understood. The indissoluble link between pure research and teaching was strongly asserted by all the founding documents, and was maintained in Humboldt's practical arrangements.

Further, there is no antithesis between knowledge and action: good science will also best serve practical ends. The university should have a wider function in the community:

If our universities have not yet begun to further culture in the universal sense, in addition to serving as nurseries of knowledge, it is because, even with respect to knowledge, they remain at a low level. (p. 28)

Low-level science cannot serve as the cultural leavening agent which is essential also for technological or 'applied' advances.

The debate about whether the ancient languages should continue to dominate the curriculum, at the expense of the modern languages, and later of the natural sciences, was to continue throughout the century. Schelling defended the study of the classics, but in new terms that were to be of profound significance in nineteenth-century Germany. In his third lecture, 'Prerequisites to University Studies', he argued that university professors are not mere teachers of language, they are philologists: 'the philologist stands with the artist and the philosopher at the highest peak . . . His task is the historical construction of works of art and science; he must understand their history and expound it vividly' (p. 39). Humboldt, a linguistic and philological scholar of the first rank, pressed this shift from the old 'general grammar' to the new historical philology, and saw to it that the new university, when it opened its doors in 1810, had both August Boeckh and Franz Bopp on its small but distinguished faculty. Schelling's defence of the study of classics is closely related to the new Romantic hermeneutics and aesthetics: 'Nothing forms the intellect so effectively as learning to recognize the living spirit of a language dead to us.' As Schleiermacher put the point, it is the effort to reconstruct imaginatively the alien world of the past that is most powerfully exemplified in the study of a language of a past civilization. This power of the imagination is precisely what the new *Bildung* would train and exercise. This study of language bears no resemblance whatever to mere rote learning. Moreover, and here we are close to the heart of Schelling's own doctrine of the ultimate unity of approaches through the study of nature on the one hand and the study of mind on the other which he had recently explored in his major work *System des transcendentalen Idealismus* (*The System of Transcendental Idealism*) (1800), there is no opposition between the study of classical languages and the study of science:

> To be able to do this [to recognize the living spirit of a dead language] is no whit different from what the natural philosopher does when he addresses himself to nature. Nature is like some very ancient author whose message is written in hieroglyphics on colossal pages, as the Artist says in Goethe's poem. (p. 40)

After his three opening lectures, Schelling proceeded to devote eleven further lectures to particular disciplines, beginning with mathematics and ending with art. To philosophy, as the focal discipline, he devoted his four central lectures (lectures 6 to 10). He contrasts 'the deeply implanted impulse and the desire to investigate the essence of things' to 'the understanding which non-philosophy calls sound common sense [which] wants the truth in hard cash' (p. 62).

The state considers philosophy to be dangerous, and tries to suppress it; but the state requires a proper philosophic grounding if it is to survive and flourish. This argument was a powerful one – it was tantamount to a threat of the overthrow of the state – in a time that had witnessed the French Revolution near at hand. It was widely held that revolutions in knowledge prepared the ground for political revolutions. A repressive state will in its turn be repressed: it must reform itself, or go under.

No one can undertake science without first having derived a notion of truth from philosophy; it is philosophy that awakens the interest in the 'most living and diverse cognition of the particular' (p. 57, n. 12). But philosophy must be balanced on the one hand by 'the richness of classical culture' and on the other by 'genuine experience based on the observation of nature'. Philosophy is seen as the prior, the central and the unifying subject, but one that requires both classical studies and the sciences. The argument that the state depends for its life on intellectual endeavour as much as intellectual endeavour depends on the state is stressed in a variety of ways by all these thinkers.

The three lectures devoted to natural science (11, 12, and 13) reinforce the link with philosophy. The first, 'On Natural Science in General' ('Über die Naturwissenschaft im Allgemeinen'), makes the essential point that nature symbolizes the ideas of cognition, and therefore can only be understood as a system, a whole. Any attempts, whether theoretical or experimental, which do not take this into account can solve only local puzzles. Schelling is in no way opposed to experimental science, but experiment must be placed within a 'construction' of nature. It is precisely this orientation that characterizes a 'method' rather than blind trial-and-error.[9]

His lectures 'On the Study of Physics and Chemistry' and 'On the Study of Medicine and the Organic Doctrine of Nature' introduce conceptions already more fully worked out in his earlier writings on *Naturphilosophie*. Because ideas of mind operative in the explanation of nature were the object of investigation, mechanical explanation must be subordinated to and rewritten in terms of the organic principle. His attempt here to show how the traditional Faculty of Medicine could be recast in terms of the philosophical idea of organism foreshadows the powerful effect of his thought on the life sciences, and the struggle of early disciples of *Naturphilosophie* such as M. J. Schleiden to break away from it after the discovery of the cell in 1828.

Philosophy is so vital for Schelling that it is not to be hived off as a separate discipline or faculty, but is essential to all the disciplines. It is objectified in different ways in the three positive sciences of theology, medicine and law, but is fully visible only in art. Thus philosophy belongs pre-eminently to the Faculty of Arts. Moreover, its special relation to the arts gives it a unique claim to freedom from the state (p. 8). It is to 'the science of art' that he devoted his final lecture.

In Schelling's philosophy, art was essential to the manifestation of the ideas of the mind; as Kant had argued in the Third Critique, the ideas of reflection, which included ideas such as God, the soul and freedom, which the old metaphysics had sought to demonstrate, could only be evidenced obliquely, through the products of human creativity; they had no objective existence.

Schelling had attempted to argue that there was a philosophical faculty which allowed the mind to view its own products, both as subjective and as objective; but even if this were so, they could be seen most vividly and perfectly in works of art, which exhibited the constructions of mind as empirical objects.[10] Moreover, this philosophical faculty was restricted to the few, whereas art could reach many. Thus the circle of communication of self-knowledge which was the model for the new university was seen in little in the work of art.

Schelling's parting shot, in the last paragraph of his final lecture, is a sharp reminder to the state that its quality will be judged by its contributions to public and private art:

In closing, let me say that it is a disgrace in those who have a direct or indirect part in governing the state not to have familiarity with or receptivity to art. Nothing honors princes and those in authority more than respect and encouragement to artists. It is a sad and shameful spectacle when those who have the means to promote art's finest flowering waste their money on tasteless, barbarous, vulgar displays. Even though the public at large may find it hard to grasp that art is a necessary, integral part of a state founded on Ideas, we should at least recall the example of antiquity, when festivals, public monuments, dramatic performances, and other communal activities together made up a single, universal, objective, and living work of art. (p. 151)

## FICHTE: THE ENCYCLOPAEDIC PRINCIPLE

Fichte's essay on the 'internal organization' of the university stresses the possibility of a new beginning, correcting the faults of the existing universities. In so far as the university merely reproduces what already exists in books (or oft-repeated lectures) it is wholly superfluous. The professorial monologue must give way to a dialogue. The teacher must not just tell the pupils what to do, he must demonstrate how he has done it. Fichte argues that there should be no 'provincial universities', but all should be universities of the Prussian monarchy, and should compete with one another in the same object. He calls for the establishment of a journal of the universities, through which all may communicate freely with one another on their progress. Students should travel to another part of the realm to university than that in which they were brought up, so that nationalism and cosmopolitanism may combine, and narrow patriotism be put aside (pp. 281–4).

Fichte in the *Vocation of the Scholar* (1805) centered on the figure of the scholar rather than on the relations of the disciplines, and founded that approach which we are now perhaps more familiar with from Carlyle's *Heroes and Hero-worship*, Kierkegaard's secular 'saints' and Emerson's *The American Scholar*, all powerfully influenced by Fichte. He treats the scholar at different phases of his development, which includes his phase as student, during which he should have social as well as academic contact with his teachers. The dignity conferred on the profession in Fichte's writings gave a powerful impetus to the 'professorial system' which to outside eyes became a German hallmark.

In his 'Plan for an Institution of Higher Learning at Berlin' he wrote eloquently of the university as 'a school for the art of the scientific use of the understanding' ('eine Schule der Kunst des wissenschaftlichen Verstandesgebrauches'), and of teachers and students as 'wissenschaftliche Künstler' ('scientific artists').[11] This is a recasting of Schiller's famous distinction between the 'Brotgelehrte', those who studied only to get a qualification to earn a living, and the 'philosophische Köpfe', the true students, the philosophical minds.

Fichte underlines Kant's and Schelling's stress on the Faculty of Philosophy: all students should receive a philosophical foundation in the first year of their course, and this should take the form of an encyclopaedic overview of the disciplines and their interconnections ('der gesamte wissenschaftliche Stoff, in seiner organischen Einheit' (p. 53)). The notion of the 'encyclopaedia' as a systematic, unifying organization of knowledge was a leading idea of the Enlightenment that found new forms in the philosophers Fichte and Hegel and in Romantic writers like Novalis (Friedrich von Hardenberg) and Coleridge. In Fichte's proposal, which combined the traditional role of the Faculty of Philosophy as 'preparatory' to the vocational Faculties of Theology, Law, and Medicine with the critical and unifying functions accorded it by Kant and Schelling, we can see how this critical overhauling of knowledge led to the 'liberal arts' foundation courses familiar in American universities.

If the teacher must be able to give such a philosophical overview, he must also be able to present research of his own, using 'data that no one before him has known so fully as he' (p. 56). Fichte stated the maxim for teachers that all the founders shared: never repeat what is already in the books. The teacher must reinterpret what is in the books; he must find new knowledge; and he must bring it to life through his own personal presence and experience. His new knowledge is to be understood as a contribution to human culture as a whole.

Able students must be enabled to attend the university, regardless of their origin; their period of study should be supported financially, because anxieties about subsistence oppress the spirit, and demean the character (p. 42). Financial support is implied by the position of dignity which is accorded the student in the new university.

On science he outlines a plan pregnant with the conflicts of the future. The sciences of nature, divided into *Naturgeschichte* (natural history) and its theoretical part, the *Naturlehre*, belong to the encyclopaedia, together with history. *Geschichte* (history) is the changing element, *Naturgeschichte* the permanent (p. 26). But natural science as *Naturwissenschaft* appears again, in the course of discussion of the three old faculties, in which natural science had traditionally been part of the Faculty of Medicine. This twofold entry, placing and nomenclature mirrors the conflict then coming to a head between older forms of science and new. The presentation in the encyclopaedic scheme maintains the old 'natural history', which still bore traces of the idea of the 'great chain of being', according to which individuals changed but the species was permanent, so that even while 'development theory' gained ground it retained the form of classification rather than evolution.[12] But Fichte also argues for the independent development of the 'natural sciences', which until now had too often been seen only as 'Magd der Heilkunde', handmaid to medicine (p. 62), as philosophy had been handmaid to theology. Fichte's proposal is for an independent institute of medicine. He clearly sees the dangers, and warns that even an independent institute should have a plan based on the encyclopaedia of the disciplines, and operate according to the maxim: never teach what is already in the book. Equally, the natural sciences should not disregard medicine, which may be able to apply its findings (p. 62). Thus Fichte tries to have it all ways: to free the developing natural sciences from the domination of the Faculty of Medicine, and to free medicine for its own purposes, while tying both to the encyclopaedic organization of the university through the Faculty of Philosophy. That these elements were likely to fly apart was already apparent, as indeed came about with the beginning of the founding of the technical universities in the 1840s.

## SCHLEIERMACHER AND HERMENEUTIC DIALOGUE

Schleiermacher's essay 'Occasional Thoughts on the German Idea of a University' presents a model for self-determination of all the parts of the university, which he wished to defend as 'German' pluralism as opposed to the Napoleonic model of bureaucratic centralization.

In its examination of the circle of the disciplines, the role of the lecturer and the student, the relation of the university to the specialized institutes and to the Academy of Sciences, and to its urban setting, his essay, of all the documents, remains closest to the life and conduct of the university. We glimpse the inspired teacher he is reputed to have been. Where Fichte emphasized the scholar (and the scholar-to-be), Schleiermacher emphasizes the teacher and the student, working together in mutual freedom. His lifelong concern with the nature of mutual understanding of speakers and hearers, with dialogue as a formal and a personal mode, and with the methods of interpretation, gives the stamp of a uniquely engaged personality to the doctrine shared by all these thinkers, the reciprocity of creative inquiry. Hermeneutics and dialectics were closely linked in the Socratic tradition he reinterpreted as the model for the circle of question and answer that led to new knowledge. The formation of a

new community which could reinterpret the past was vital both to his theory
and to his life practice. It was Schleiermacher's development of the notion of
the hermeneutic circle that Dilthey later took as the distinguishing mark of all
inquiry into the human (as opposed to the natural) sciences; but Schleiermacher, the founder of general hermeneutics, did not himself drive this wedge
between the natural and the human sciences.

For him too the Faculty of Philosophy forms the centre of the university. It
most fully mirrors the organization of all knowledge, and best serves the
university's function of awakening the idea of knowledge itself, 'das Lernen
des Lernens', learning to learn. At the heart of it is the 'Gesprächs-
konstellation' ('dialogic constellation') of the researching teacher and his students, which partakes of the Socratic academy, the craft guild or workshop of
artist and disciple, and the creative intimacy between individuals on which all
communication depends.

In the 'Appendix' he explores the implications of Berlin as the proposed
venue for the new university. His two major concerns are the
professionalization of university teaching and the welfare of the student body.
Both may be summed up under the heading of 'academic freedom', guaranteed in practical terms. He stresses that the new university teachers must be
full-time, and they must be paid sufficiently well to free them from adminis-
trative or other business to devote themselves to their profession (p. 632).
Their appropriate status should be that of the upper echelons of the civil
service. While being sited in the capital made transfers between the Institutes
and the university, and the civil service and the university, feasible, these must
not be casual. Moreover, the university must have its own independent
funding established at the outset, and not be dependent on the whim of the
state, nor on the income of the parents of the students, nor on the benefactions
of the rich (though these might have a role to play) (p. 632). He envisages a
community contribution towards the university in the form of lower prices
and concessions for public events, restaurants, rents and other services. For
the student body, he stresses the importance of 'student freedom' (p. 634).
This includes the control of their own finances; although some may abuse it, it
is essential if 'character formation and intellectual progress are to advance
together', and the student is to emerge 'ein gemachter Mann', a 'whole man'
(p. 634). If the university is '*in loco parentis*' (not a phrase employed by the
German reformers), it is an enlightened parent of a young adult member of
society with excellent prospects who must be granted intellectual, social and
financial independence in order to find his own way. The university, students
and teachers together, as well as the specialized institutes, are part of the
'wissenschaftliche Republik', the republic of learning (an extension of
Klopstock's famous essay of 1771 on the *Gelehrtenrepublik*) (p. 643); good
teachers will give their best students access to their social circle (p. 637). The

2    Statue of Wilhelm von Humboldt outside the Humboldt University.

city must contribute to student protection by keeping an alert police eye on the facilities for gambling and prostitution. In short, this is a sketch of the urban university, with practical proposals for taking advantage of its opportunities, and counteracting its disadvantages. A strong sense of a communal enterprise breathes through these pages.

## HUMBOLDT AND THE THEORY OF HUMAN EDUCATION

Finally, we return to Humboldt himself. His brief memorandum 'On the Spirit and the Organizational Framework of Intellectual Institutions in Berlin' (1809) sums up admirably the main points arrived at by his academic predecessors, and shows at its best the relationship between the higher bureaucracy and the professoriate that Schleiermacher had adumbrated. So pithy is it that one is tempted to quote it verbatim; a commentary on each of its sentences would yield up the lengthier disquisitions of Fichte, Schelling and Schleiermacher. Behind it lies his own lifelong attempt to found a new discipline, which as early as 1793 he identified and named the 'Theory of Human Education'.[13] It was a form of cultural anthropology, in which 'the formation of human character in its diversity of forms', including considerations of 'gender, age, temperament, nation, and period' could be studied.[14] This discipline would combine an overview of the modes of acquisition of human knowledge with the principles of its inculcation. It would cut across and organize existing materials, in, for example, pedagogy and law, supply a view of religion in relation to education and give substance to the title 'philosophical history of mankind' suggested by the work of Herder and Lessing.[15] In this formulation of his interests in terms of a 'new discipline' that would complete 'the encyclopaedia of our disciplines' we see both his wish to vie with the new disciplines that as a young man he saw as primary, Kant's transcendental philosophy and F. A. Wolf's classical studies (*Altertumswissenschaft*), and to establish a wider foundation on which advances in individual subjects could be ranged and against which they could be assessed. The theory of human education would provide a critique of advances in individual subjects from the standpoint of the criteria of general and personal culture. The influence of Goethe on Humboldt, as on his brother Alexander, was great, and Goethe's view (expressed most dramatically in his critique of Newton's optics, and his elaboration of his own physiological optics)[16] that the very process of conceiving and carrying out any experimental work must be considered in its relation to the development of the researcher's faculties, as well as to the objective results obtained, undoubtedly played a role in his humane pedagogy. His interest in his brother's scientific work, and their mutual influence on each other, was great; Alexander von Humboldt's last work, *Kosmos*, is a monument to the aesthetic and organic conception of the sciences represented by Schelling, Goethe and Steffens. Humboldt followed Kant and Schiller in pointing to the 'harmonious and balanced play of all the faculties in the completion of a particular goal' as essential to *Bildung*, and in locating this in its most highly developed form in the experience of art.[17] In the formulations of all these theorists it was individual self-culture which was seen as the vital factor in the progress of human culture as a whole. As Humboldt put it:

Every effort in behalf of the progress of the human race that does not take its start from the education of the individual will simply be fruitless and chimerical; whereas if the individual is provided for, the influence on the progress of the whole follows of itself and without being expressly directed to it.[18]

For Humboldt this was linked with the individual's need for political freedom.[19] Humboldt's own particular academic interests developed within this general framework: his special study of general linguistics and of a variety of particular world languages that had never been codified, especially those of Central America, were an essential part of the philosophical theory of the education of mankind. He belonged to a group of thinkers who saw that language was itself a social institution.[20] His professional concern with the primacy of speech in human culture, like Schleiermacher's with the conditions of communicated understanding, made him alert to the need to create precisely the right institutional setting. His theoretical statements are marked by the strong practical bent that was to make him the principal architect of the new educational institutions.

Humboldt's memorandum begins with the pregnant statement: 'The idea of disciplined intellectual activity, embodied in institutions, is the most valuable element of the moral culture of the nation' (p. 242). This being so, the university must be free to pursue its purpose, the unceasing process of inquiry into *Wissenschaft*.[21] Independence and freedom from distraction (*Einsamkeit*, solitude) are required to carry it out. At the same time, free inquiry is a collaboration. 'Both teacher and student have their justification in the common pursuit of knowledge' (p. 243). Humboldt stresses the difference between school and university: in the schools finished bodies of knowledge are presented; in the universities *Wissenschaft* does not 'consist of closed bodies of permanently settled truths', but of ceaseless pursuit of new knowledge.

He sums up the appropriate relationship between the state and the university uncompromisingly:

The state must always remain conscious of the fact that it never has and in principle never can, by its own action, bring about the fruitfulness of intellectual activity. It must indeed be aware that it can only have a prejudicial influence if it intervenes. The state must understand the intellectual work will go on infinitely better if it does not intrude. (p. 244)

The state's 'legitimate sphere of action' must be limited to supplying 'the organizational framework and the resources necessary for the practice of *Wissenschaft*' (p. 244). Moreover, the state should be aware that its manner of provision of the organizational framework and resources can be damaging to the essence of *Wissenschaft*. In general, the state must 'demand nothing simply for its own needs' (p. 246). By safeguarding the university's freedom the state could earn honour for itself as a 'Bildungsstaat'. Humboldt's aim was to make the university independent of the state, in particular through his attempt to prevail upon the king to grant the university a permanent endowment in landed property.[22] He did not gain the conferral of independent financial resources on the university through the grant of land (this far-sighted plan was to be realized only by the Land Grant Act that established the state universities in the United States in 1865); but he gained recognition of the 'human right' of education and of academic freedom.

In recent years, as the nature of the university has again come under scrutiny, the question of whether the reform envisaged in the plans for the Humboldt University succeeded or failed has been heatedly discussed. If the

preponderance of recent opinion has claimed its success, and seen the achievements of the whole German university system as growing out of that success,
those who most value the aims embodied in the founding documents stress
that the later development of the German university did not fulfil Humboldt's
vision.[23] He himself retired from public life in 1819 in indignation at the onset
of conservative government under Metternich, the vision of the unity of the
disciplines faltered in the march towards scientific specialization, a positivist
hierarchy of the sciences challenged the philosophic order of idealism and the
very numbers who flocked to the universities swamped the Socratic dialogue
between scholar and student. Nevertheless, 'the idea of the university'
remains.

## NOTES

1   Wilhelm von Humboldt was 'Chef der Sektion' for 'Kultus and Unterricht'
    (ecclesiastical and educational affairs) in the Prussian *Innenministerium*
    (Ministry of the Interior) from February 1809 to June 1810.
2   All five documents are to be found in Ernst Anrich (ed.), *Die Idee der
    deutschen Universität* (Darmstadt, 1956).
3   See, for example, J. Derrida, 'The Conflict of the Faculties', in his *Essays on
    Education and Politics* (Cambridge, Mass., 1989).
4   I. Kant, *Streit der Fakultäten* (1798), ed. Klaus Reich (Hamburg, 1959; repr.
    1975), p. 29.
5   E. S. Shaffer, 'Metaphysics of Culture: Kant and Coleridge's *Aids to
    Reflection*', *Journal of the History of Ideas* (Apr.–June 1970), 215–16. For the
    further development of the idea of the clerisy in England, see C. Turk,
    *Coleridge and Mill* (Aldershot, 1988), pp. 169–186; B. Knights, *Idea of the
    Clerisy* (Cambridge, 1978); R. Williams, *Culture and Society 1780–1950*
    (London, 1958); E. S. Shaffer, 'The Hermeneutic Community: Coleridge and
    Schleiermacher', in R. Gravil and M. Lefebure (eds.), *The Coleridge
    Connection* (London, 1989); N. Leask, *Coleridge and the Politics of
    Imagination* (London, 1988).
6   The English titles are sometimes misleading: Fichte's *Popular Works*, trans.
    William Smith, vol. I (London, 1848) contains *The Vocation of the Scholar*
    (1794); *The Nature of the Scholar* (1805); and *The Vocation of Man* (1800).
    *The Vocation of Man* was also translated, by Mrs Percy Sinnell, as *The
    Destination of Man* (London, 1846). The most relevant are the 1805 lectures.
       His *Addresses to the German Nation* (1807) contain his ideas on pre-
    university educational reform, see especially lectures 2, 3 and 9–11.
7   For a large selection of further documents, letters and speeches relating to the
    founding of the university, see Wilhelm Weischedel (ed.), *Idee und
    Wirklichkeit. Dokumente zur Geschichte der Friedrich-Wilhelm-Universität
    zu Berlin*, 2 vols. (Berlin, 1960), vol. II.
8   Kant, *Streit*, p. 22.
9   Schelling, *Sämmtliche Werke* (Stuttgart and Augsburg, 1856–61), V, p. 323.
10  For Schelling's complex development of Kant's argument in the *Critique of*

*Judgement* as it affected Romantic aesthetics, see Elinor S. Shaffer, 'The "Postulates in Philosophy" in the *Biographia Literaria*', *Comparative Literature Studies*, 7:3, 297–313.

11  Fichte, 'Deduzierte Plan einer zu Berlin zu errichtenden höhern Lehranstalt', in Weischedel (ed.), *Idee und Wirklichkeit*, p. 34.

12  W. Lepenies, *Das Ende der Naturgeschichte* (Munich, 1976), pp. 54–77.

13  K. Menze, *Die Bildungsreform Wilhelm von Humboldts* (Hanover, 1975), p. 19. Humboldt wrote a programmatic fragment on 'The Theory of Human Education' between 1793 and 1795; see the Royal Prussian Academy of Sciences' edition of his *Gesammelte Werke* (Berlin, 1841–52), I, pp. 282ff.

14  Letter to F. A. Wolf, 23 December 1796, *Gesammelte Werke*, V, pp. 175ff.

15  Letter to Christian Gottfried Körner, 27 October 1793, in *Ansichten über Asthetik und Literatur von Wilhelm von Humboldt. Seine Briefe an Christian Gottfried Körner*, ed. F. Jonas (Berlin, 1880).

16  For Goethe's concern with the negative effects on the researcher and the student of Newton's optical experiments see Dennis L. Sepper, pp. 189–98 in this volume, and the group of papers by Böhme, Sepper, Hegge and Zajonc in *Goethe and the Sciences: A Reappraisal*, ed. F. Amrine *et al.* (Dordrecht, 1987). For the linking of this position with the Romantics see Frederick Burwick, *The Damnation of Newton: Goethe's Color Theory and Romantic Perception* (Berlin and New York, 1986).

17  Menze, *Bildungsreform*, pp. 30–8: 'Die in der Kunst erschaffene Welt ist vollendete Bildungswelt . . .' ('The world created in art is the world of culture fulfilled . . .') (p. 35).

18  Humboldt, *Gesammelte Werke*, II, p. 15.

19  Klemens Menze, *Wilhelm von Humboldts Lehre und Bild vom Menschen* (Ratingen, 1965), pp. 54–93.

20  Hans Aarsleff, 'Wilhelm von Humboldt and the Linguistic Thought of the French *Idéologues*', in *From Locke to Saussure: Essays on the Study of Language and Intellectual History*, ed. H. Aarsleff (London, 1982), p. 350.

21  Edward Shils's translation, from which this text is taken (*Minerva*, 8 (1970), 242–50), renders *Wissenschaft* as 'pure science and scholarship'.

22  Paul R. Sweet, *Wilhelm von Humboldt. A Biography* (Athens, Ohio, 1980), II, p. 64.

23  Klemens Menze, 'Ist die Bildungsreform Wilhelm von Humboldts gescheitert?', *Collectanea Philologica*, ed. G. Heintz and Peter Schmitter (Baden-Baden, 1985), pp. 381–401; Kurt Müller-Vollmer, Introduction to W. von Humboldt, vol. II, *Politik und Geschichte*; Walter M. Simon, *The Failure of the Prussian Reform Movement* (Ithaca, N.Y., 1955).

## FURTHER READING

### PRIMARY SOURCES

All five founding documents of the Humboldt University are to be found in Ernst Anrich (ed.), *Die Idee der deutschen Universität. Die fünf Grundschriften aus der Zeit ihrer Neubegründung durch klassischen Idealismus und romantischen*

*Realismus* (Darmstadt, 1956). See also *Philosophies de l'Université. L'Idéalisme allemand et la question de l'Université. Textes de Schelling, Fichte, Schleiermacher, Humboldt, Hegel*, eds. Luc Ferry, J.-P. Pesron and Alain Renaut (Paris, 1979). Further documents, letters and speeches relating to the founding of the university are to be found in vol. II of *Idee und Wirklichkeit. Dokumente zur Geschichte der Friedrich-Wilhelm-Universität zu Berlin*, ed. Wilhelm Weischedel, 2 vols. (Berlin, 1960).

Immanuel Kant, *The Conflict of the Faculties*, English/German parallel text, ed., trans. and intro. by Mary J. Gregor (New York, 1979)

J. G. Fichte, *The Science of Knowledge*, trans. and ed. Peter Heath and John Lachs (Cambridge, 1970). Translations of all of Fichte's writings on education except the 'Addresses to the German Nation' are to be found in G. H. Turnbull, *The Educational Theory of J. G. Fichte* (Liverpool, 1926). Johann Gottfried Fichte, *Addresses to the German Nation*, trans. R. F. Jones and G. H. Turnbull (London, 1922), repr. with an introduction by George A. Kelly (New York, 1968).

F. W. J. Schelling, *On University Studies*, trans. E. S. Morgan, ed. with an introduction by Norbert Guterman (Athens, Ohio, 1966)

W. von Humboldt, 'On the Spirit and the Organizational Framework of Intellectual Institutions in Berlin', trans. Edward Shils, in *Minerva*, 8 (1970), 242–50. Other writings of Humboldt on educational policy are to be found in *W. von Humboldt*, vol. II, *Politik und Geschichte*, ed. and introduced by Kurt Müller-Vollmer (Fischer Taschenbuch Verlag, 1978). See also W. von Humboldt, *On Language: The Diversity of Human Language-Structure and its Influence on the Mental Development of Mankind*, introduced by Hans Aarsleff (Cambridge, 1988).

## SECONDARY SOURCES

Bruford, W. H., *The German Tradition of Self-Cultivation* (Cambridge, 1975)

Brockliss, L. W. B. (ed.), *History of Universities*, special issue: 'History and Change in the European University in the Age of the Liberal Revolution 1760–1830', vol. VII (1987–8)

Shaffer, Elinor S., 'The Hermeneutic Community: Coleridge and Schleiermacher', in Richard Gravil and Molly Lefebure (eds.), *The Coleridge Connection. Festschrift for Thomas McFarland* (London, 1989)

Sweet, Paul R., *Wilhelm von Humboldt. A Biography*, 2 vols. (Athens, Ohio, 1980)

Peck, Geoffrey M., 'Berlin and Constance: Two Models of Reform and their Hermeneutic and Pedagogical Relevance', *The German Quarterly*, 60:3 (Summer, 1987), 388–406

# 4

# Historical consciousness in the German *Romantic* Naturforschung

## DIETRICH VON ENGELHARDT

### (translated by Christine Salazar)

## PRELIMINARY OBSERVATIONS

History and metaphysics are brought into a close relationship by the *Naturforscher* (literally: investigators of nature) of German Romanticism; according to them this relationship is valid for nature as well as for the natural sciences. The history of nature and the history of the natural sciences are immanently connected. For the Romantic *Naturforscher*, however, history is always more than the gathering and rendering of facts – essentially history means the interpretation of the past in the light of ideas and the linking of it with the present and the future. The history of nature and the history of knowledge about nature have a common origin, have passed through a separate development and are heading towards a common future.

The concepts of history held by the Romantic *Naturforscher* are related to the historical ideas of the philosophers and historians of their period. From the eighteenth century Herder, Hamann and Justus Möser are especially influential. Those finding attention and acceptance in their own period are Franz von Baader, Schelling, Hegel, Wilhelm von Humboldt, Novalis, Schlegel and Adam Müller. Their own involvement in *Naturforschung* brings specific emphases. Conversely, the investigation of the organic realm and its regularities in the natural sciences influences the historical understanding of the philosophers, historians and jurists of the period. The Romantic *Naturforscher* are themselves investigators of history; they not only criticize the separation of disciplines theoretically, they also strive to overcome it by their own literary contributions in other fields. This contiguity is particularly apparent in the case of medicine: many medical men are at the same time *Naturforscher* or specialize in *Naturforschung*. The concept of history held by the Romantic *Naturforscher* coincides with the medico-historical concept of Romanticism.

It is a well-known fact that the influence of Schelling's metaphysical natural philosophy dominates the Romantics' understanding of nature, but other theological as well as philosophical currents of the past and present also exercise their influence. The reverberation of Hegel's ideas is relatively minor; apart from a series of *Naturforscher* influenced by this philosopher (J. Müller, G. F. Pohl, J. E. Purkinje, etc.), consequences of Hegelian historical philosophy can be detected mainly in Romantic historiography of medicine. The Romantic *Naturforscher*, however, remain sceptical and negative towards the speculative deduction of culture and nature as it is employed by the Idealists Schelling and Hegel; a deduction of that kind would transgress the boundaries of human consciousness. They equally maintain a distance from Kant's transcendental philosophy. Romantic *Naturforschung* cannot be identified with *Naturphilosophie* which is itself not homogeneous around 1800: it can be transcendental, metaphysical or sensualistic–empirical. Equally there are major divergences between the individual *Naturforscher*. Nevertheless, according to the view of the Romantic *Naturforscher*, physics and metaphysics should not be mutually exclusive, but interrelated. According to them – in the interest of a meaningful and beneficial investigation into nature – one must not deny their interrelation. On the same grounds, a convincing historiography of the natural sciences should only be possible through the correlation of facts and ideas.

## EVOLUTION OF NATURE

Romantic *Naturforschung* advocates the integration of the objective and subjective dimensions of historical consciousness, that is to say the integration of the temporalization of nature and the temporalization of knowledge about nature – a view opposed both to the parallelism of these dimensions of the Objective (nature) and the Subjective (knowledge) in the era of the Enlightenment, and to their separation during the Renaissance and the age of Positivism.[1] For the Renaissance there existed only knowledge about nature; nature herself was considered as unchanged as far as species of plants and animals were concerned, in accordance with belief in the biblical creation story. History of the natural sciences was a revival of antiquity and a starting point for an unlimited future. In the Positivism of the nineteenth century, on the other hand, the historicization of nature (as in Darwin) became more accentuated, accompanied by a growing loss of interest – on the part of the *Naturforscher* – in the history of the natural sciences. To be more precise, there was a separation of empirical *Naturforschung* and historiography of science.

The eighteenth century begins to perceive nature as subject to change. At the same time there appear numerous publications by *Naturforscher*, relating to the history of the sciences; comprehensive accounts are undertaken by A. G.

Kästner, Joseph Priestley, J. S. Bailly and T. O. Bergman. The cosmological, astronomical and geological studies of Steno, Descartes, Thomas Burnet, Leibniz, John Woodward and William Whiston suggest the temporalization of animated nature as well. Some species were supposed to have disappeared, other species to have come into being, and at the same time many species were supposed to have been conserved; to some extent even actual transformation was considered as a possibility. Essential contributions to this temporalization of the organic world are made by B. de Maillet (1720; 1748), P. L. de Maupertuis (1751), J. B. Robinet (1768), A. N. Duchesne (1769), Denis Diderot (1769), J.-L. G. Soulavie (1780; 1784), G. L. L. de Buffon (1778) and Erasmus Darwin (1794; 1796). The verdict of their own times on this historicization of nature, however, is not always positive among the established *Naturforscher*: the concept of the continuity of species, founded on the Bible, is still too powerful. In Germany, around the year 1800, J. F. Blumenbach, C. F. Kielmeyer, F. S. Voigt and other *Naturforscher* take up these ideas and essays in the temporalization of nature, and they also discuss their possible political and ethico-religious consequences.

As is generally known, a new phase of this development is associated with Lamarck, who, around 1800, attributes the mutation of species as a genealogical transformation to external influences and, at the same time, to an inherent tendency of nature towards increasing complexity. By way of semi-liquid bodies which mediate between the organic and the inorganic, the temporalization is extended to the inorganic world. According to Lamarck this external/internal concept of history should also be valid for the history of mankind and should also be compatible with religion.

The ideas of the Romantic *Naturforscher* on the history of nature are related to the varying attempts at historicization made in the eighteenth and at the beginning of the nineteenth century, that is, to the spectrum of ideas ranging from partial extinction and partial creation of species, on the one hand, and their extensive transformation, on the other, to explanations of causality and finality based on the interior of the organism or on dependence upon the environment. The extinction and new creation of species are not called in question by the Romantic *Naturforscher*; but the changes of nature and the internal relation of natural phenomena come to be based on an ideal systematization. Ideogenesis, not actual descent, is the view prevalent among the Romantic *Naturforscher*. Ideas, and their hierarchic arrangement, carry the development of natural forms; this development cannot be made to depend upon actual factors alone. The distance of the Romantic *Naturforscher* from Buffon and Lamarck is obvious; they are separated from Buffon by the idea of an internal correlation of development of the entire natural world, from Lamarck by his metaphysical foundation.

In 1801 Henrik Steffens (1773–1845) explicitly speaks of a 'theory of evolution', but in the sense of an idealist, 'internal history of nature'. The multiplicity of plants and animals is based on the dynamism of expanding and contracting forces; this dynamism is also the cause of the geological processes. Nature has 'passed gradually' through the various 'levels of animalization' so as to reach its perfection in mankind.[2] The principle of progression is again and again complemented by the principle of regression. It is on these grounds that Johann Wilhelm Ritter (1776–1810) criticizes Steffens for having disregarded involution for the sake of evolution; only the combination of evolution and involution could make the development of nature intelligible, otherwise one would 'only represent mute sequences, never anything living'.[3] According to Hans Christian Oersted (1778–1851), 'alternating expansions and contractions' of the natural forces would have made earlier creatures become extinct 'to make room for the present chain of beings, with man at its top';[4] this fundamental change is itself, in turn, liable to change. Lorenz Oken (1779–1851) explicitly rejects the idea of actual descent, as is suggested by this and similar statements: 'The expression that earth and metal evolved into coral is hardly to say that the earth as such really changed into coral, just like the above claim that it became metal, or that the air became sulphur . . . everything is to be taken in a philosophical sense.'[5] Dynamics and statistics are combined with each other; extinction and new formation are relative to external nature, the inner being of nature being supposed to remain untouched by this transformation. An 'actual generation and perishing'[6] is unthinkable, as is also stressed by Carl Gustav Carus (1789–1869).

However, some amongst the Romantic *Naturforscher* or those sympathizing with Romanticism, do not take sufficient account of the difference between material and ideal dimensions, a fact to which Hegel explicitly takes exception. Thus, in the case of the zoologist and palaeontologist Georg August Goldfuss (1782–1848), the border-line between real and ideal descent seems blurred: 'The animal remains, found in the various older and more recent geological strata, prove that the animal kingdom – regarded as a single organism – actually went through a metamorphosis similar to that of the foetus, its periods being contemporary with the formation periods of the globe.'[7]

In the case of the Romantic *Naturforscher*, the temporalization, or rather the historicization, of nature shows itself in principle as an evolutionary involution, or an involutionary evolution, usually understood as being ideal, more rarely real, predominantly attributed to final causes, although the important function of real factors is acknowledged. The Idealist *Naturphilosophen*, Schelling and Hegel, reject the notion of actual descent with even more conviction than the Romantic *Naturforscher*. Schelling sees in the organic natural phenomena the result of an alternation of productivity (*natura naturans*) and inhibition (*natura naturata*) of the original natural force, and he regards them as fixations or finite realizations existing *per se*, which cannot pass immediately to another level of fixation. So, according to him, one cannot think of actual descent: 'Thus the claim that the various organizations have actually been shaped by a gradual development from each other is the misunderstanding of an idea.'[8] Seen from this point of view, Charles Darwin's later causal factors – mutation and selection – could not suffice as an explanation for the change in nature; like Hegel's natural philosophy, Schelling's 'dynamic graduation' is a philosophical and not an empirical dynamic representation of the static natural history of the eighteenth century.

Hegel is equally explicit in his rejection of the notion of an 'actual descent' ('Realdeszendenz') in the signification the word had in his time.[9] For Hegel, development is the mark both of animate and inanimate nature, but this development is a speculative combination of concept and appearance, which has its philosophical foundation in ontology and at the same time has to hold good in the analysis of concrete natural phenomena. This is the dialectical perspective in which Hegel also formulates his criticism of Romantic *Naturforschung*, which according to him only had 'a dim image of the idea, the unity of the concept and of objectivity and of the fact that the idea is concrete'.[10]

## DEVELOPMENT OF THE NATURAL SCIENCES

For the Romantic *Naturforscher*, nature has a history, and history is at the same time also nature. History of science and history of nature acquire an internal relationship. Likewise, the Romantic history of medicine links the history of diseases with the history of medical theory and of therapy. History of medicine is the history of culture, and diseases have a cultural aspect. For the Romantic *Naturforscher*, the history of the natural sciences and of culture in its totality always remain related to nature – causally and phenomenally as well as teleologically. The evolution of nature aims at producing culture; all the differences between nature and culture cannot annul their basic identity. The development of the social and spiritual world is based, as Steffens explains, on an impulse identical with 'the formative drive [*Bildungstrieb*]' of nature as a whole.[11] In the appearances and goal of *Naturforschung*, too, this correlation of nature and culture can be observed.

The Romantic *Naturforscher* made many contributions to the history of the natural sciences, and also to history in general, to sociology and politics, to pedagogical issues and to questions concerning the political role of the university. Historical problems were treated by them, sometimes in separate studies, sometimes integrated in empirical investigations, and sometimes presented in university lectures. Finally, important clues to the understanding of history of the Romantic *Naturforscher* can be found in their extensive autobiographies, travel journals and correspondence.

For the Romantic *Naturforscher*, the history of the natural sciences is a return to origins, but on a higher level; a knowledge of history is an essential condition for further progress. In Romanticism there is a renewed attempt to bring the historiography of the natural sciences and empirical *Naturforschung* – the separation of which had become more and more accentuated during the Enlightenment – into an immanent and mutually stimulating relationship. The nineteenth century in turn will definitively establish the dichotomy between them which prevails to this day.

The history of the natural sciences should harmonize with the system of forms of cognition and with the hierarchy of the individual natural sciences. Sensuality, understanding and reason are considered as the essential dimensions of the human mind, and according to Carus the stages of scientific development correspond to them. The initial phase consisted in the instinctive observation of external nature; then followed the period of intelligence with insights into the changes and correlations of natural phenomena in the form of 'natural history' and its grounding in a 'science of nature'. The last phase is the era of recognition – mediated by reason – of the eternal ideas in the natural phenomena.[12] Dietrich Georg Kieser (1779–1862) emphasizes the correlation between the hierarchy of the natural sciences and the sequence of their formation in recent times: 'Wakened to life on a higher level since the twelfth century, the human mind now forms the different spheres of natural science in a regular development, rising from the lower to the higher.'[13]

Historiography of science, too, is determined by the unity of philosophical and metaphysical dimensions. In the concept of the history of science, as well as in the understanding of nature, Romanticism and empirical investigation do not represent alternatives to one another. The history of the natural sciences should have a philosophical base, something which had been neglected or deliberately rejected by the eighteenth-century *Naturforscher*.

For Romantic historiography of science subjective viewpoints are considered just as unsatisfactory as the pragmatic orientations prevalent in the historical writing of the Enlightenment. Besides, history cannot be reduced to a description of the past; the cognition of the internal necessity is the decisive element – in accordance with the Idealist understanding of Kant's, Schelling's and Hegel's history of philosophy: history of science is the history of ideas and never mere source critique. Without its Idealist orientation, history appears – according to Steffens – as 'an incongruous, disconnected multiplicity of facts, the depressing result of which seems to be an irregular alternation of regress and progress to no purpose or goal'.[14] However, the Romantic *Naturforscher*'s philosophical history of science takes up the annalist, biographical and pragmatical positions of the past and integrates them into its own ideal-based enterprise.

The development of the natural sciences is genetic, possesses an internal logic and depends on economic and social factors. Internal and external dimensions do not have to be mutually exclusive. Disruptions and delays of progress can occur, but basically the development cannot be brought to a halt. The fact that simultaneity of political and scientific development does not necessarily mean concrete interaction can, they supposed, be proved by, for example, the contemporaneity of the French Revolution with John Brown's medical system or Lavoisier's antiphlogistic chemistry.

In the chronological subdivision of scientific development they distinguished between the Origin, Antiquity, the Middle Ages and finally the current developments, and this development as a whole is seen as an enhanced return to the Origin. For the *Naturforscher* the history of science is regressive progress or progressive regress.

In the beginning, there prevailed a unity of being and consciousness, of science, belief and life. According to Ignaz Paul Vitalis Troxler (1780–1866), in the beginning of the history of science 'the ideal and the real were fused in one', the history of *Naturforschung* had begun with the dissolution or the loss of this unity and its final phase would overcome the 'strife between idealism and realism'[15] and lead the way back to the original identity.

Profound insights into nature are, they supposed, hidden in religious and poetic texts of the early ages of the history of mankind, and can be discovered by appropriate interpretation. The physicist and chemist J. Christoph Salomo Schweigger (1779–1857) published many writings from this perspective.[16] For him the mythology of Antiquity is the result of an extinct natural science. Castor and Pollux are symbols for negative and positive electricity; Proteus signifies water as being the basic constituent of the material world, the Phoenix the course of the sun. According to Schweigger, pagan religion forced man to use symbolic expressions and only the advent of Christianity overcame this tyranny over the knowledge of nature, and gave rise to truly free investigation.

The Romantic *Naturforscher* bring Antiquity and the Middle Ages into proximity in a way which had previously been rather uncommon. The concept of two ages – ancient times and a new age – is popular. The Middle Ages are judged ambivalently by the Romantic *Naturforscher*, and do not meet with general rejection, as was the case in the Enlightenment. Steffens considers this phase, too, as a necessary 'stage of formation for the whole human race'.[17] According to Oersted, chemistry in the Middle Ages was hidden by an 'impenetrable veil of mysticism',[18] but this did not justify the verdict that 'that period is a blank for us'.[19] At the same time, however, they warn against a glorification of the Middle Ages. No age is put in an absolute position above all others, every period is considered as at once eternal and yet relative in its appearance. Even alchemy and astrology are seen as being in accordance, in their basic idea, with the internal reason of development.

For the Romantic *Naturforscher* the Renaissance was the decisive break: Copernicus, Kepler, Galileo, Bacon, Descartes and Newton are the key figures of this new phase, which – according to Oersted – at first presented itself as a 'period of fermentation' or a 'period of contradiction and strife'.[20] The following centuries then rendered the possibility of 'harmony and correlation' between nature and the natural sciences visible and actual, or so Ritter hoped.[21] Among the external causes responsible for this development they list the invention of printing, the establishment of academies, the Crusades, the fall of Constantinople, the discovery of America and the Reformation. They honour outstanding *Naturforscher* of the new age for their specific achievements, but often their verdict is ambivalent. Newton's colour theory is subjected to criticism, but this attitude is by no means shared by all *Naturforscher*. Jakob Joseph von Görres (1776–1848) is 'totally disgusted' by Goethe's polemics against Newton; he claims that Goethe is mostly wrong as far as experiments are concerned and that his rejection of Newton's theory springs from a 'bad mood'[22] rather than from true understanding. In their own age they saw an expanded world-historical opportunity for the history of the natural sciences and its philosophical historiography.

With progress, a basic potentiality should become realized; it should be possible to regain the original unity on a higher level. Although external conditions do play their part and are worthy of the historian's interest, what really counts is the ideal level, and the historiography of science must follow its lead.

The knowledge gained from the history of science should not remain mere theory. Again and again the connection is made between the history of science and current investigations. History fosters the present and the present casts light on history. On the basis of a knowledge of its past development, Friedrich Joseph Schelver (1778–1832) holds botany to have the task of recognizing a natural order of plants for itself and in relation to the totality of nature: grasping these connections, a 'higher sense of nature' has already awakened from its slumber and at the same time shows the way back 'to those most ancient mysteries, which themselves could not find the expression'.[23] The Idealist foundation of the history of science allows a connection of the change in appearances with the persistence of ideas; the ideas have a temporal aspect; they become realized with the passage of time. Ritter is convinced that there is 'not history of physics, but history = physics = history'.[24] Despite all the differences, natural, scientific and cultural development are immanently interrelated.

## HISTORY OF THE FUTURE AS THE UNITY OF NATURAL EVOLUTION AND CULTURAL DEVELOPMENT

Romantic *Naturforschung* always relates the natural sciences and their history to nature, culture and their respective histories; they have sprung from nature and are moulded by their cultural environment; conversely, their further progress should serve nature and culture, ameliorate the political order, bring about social reforms and heighten man's sensory and moral education.

The concord of natural and cultural development is essentially complemented by the idea of a new and common history of nature and culture, which may be taken to have come into being with the metaphysical/Romantic era around 1800. In the view of the Romantic *Naturforscher*, developmental history is not limited to biology; the true transmutation reaches beyond the limits of nature, towards mankind and society and – by means of culture – it again influences nature. According to Ritter the evolution of nature lifts man up to an even higher world, the transcendental: 'Amidst the undifferentiated, however, we become the continuation of another evolution; we gravitate towards something more sublime, the thread of which is only a new age for us.'[25]

At the same time the entire movement is characterized by an explicit diversity of socio-political standpoints. The metaphysical exegesis of nature and the natural sciences, as well as their respective developments, does not lead to a unified image of society and its history. Lorenz Oken gets involved in the *Wartburgfest*, a rally for German freedom and unity; Karl Adam August von Eschenmayer (1768–1852) offers a justification of the constitutional reform in Württemberg; Gotthilf Heinrich von Schubert (1780–1860) deduces the class hierarchy from the order of the three kingdoms of nature in natural history. In general, however, in the view of the Romantic *Naturforscher*, the state and the individual should not represent opposites, but should be mutually complementary. The sciences, philosophy and the arts rank above the individual and the state, giving them their aims and receiving their tasks from them.

The maxim about man's natural and cultural fate is also valid for nature; her fate, too, is determined by herself as much as by man. The idea of unity of nature and mind not only leads to socio-cultural involvement, but also to an involvement on the part of nature: 'Not only does man need the earth for his life and activity, but the earth also needs man', writes Carus.[26] Ritter states: 'Integrating nature is his [man's] raison d'être.'[27] The idealization of nature and naturalization of culture should lead the history of the future beyond the schism of nature and culture. It should be possible to recover the original identity of nature and man, of being and consciousness, on a higher plane.

In this new era the exploitation and destruction of nature would become impossible, as would anti-natural anthropologies and anti-natural social projects. According to the Romantic *Naturforscher*, empirical knowledge, so greatly increased in recent times, with its consequent practical and technical applications, has made man's duties towards nature only too obvious. Man's relationship with nature, they say, must not limit itself to power and sovereignty (as in Bacon and Descartes). In the future the historically formed relationship of mankind and nature should rather combine the realization of the Other (natural teleology) with the realization of the Self (anthropocentricity). Monism goes deeper than any dualism. According to Novalis, mankind's mission is the 'education of the earth',[28] and in man nature will be able to become wholly mind. The contemporary situation, in nature as well as in culture, demonstrates this connection of fates. It should be possible to stop or avoid 'nature's mocking at the grave of history', which Johann Ferdinand Koreff (1783–1851), faced with the erosion of the Italian landscape caused by man, sees as a threatening possibility.[29] The essential goal of natural history, as well as of the history of mankind, is to comprehend nature, science and culture in their internal and external relations and to realize this relation.

We cannot describe the Romantic *Naturforscher*'s socio-political and historical ideas by using current opposites such as 'Restoration' and 'Revolution', 'Progressive' and 'Reactionary'. Origin and history, continuity and change are always considered together – in the world of nature as well as in that of culture. Development is always related to ideas, which can be realized in a chronological sequence and at the same time be fitted into a systematic order. Development in culture as in nature is not linear. This development is not merely heading towards the future, but rather turns back to the beginnings of nature and culture, and recovers a wholeness which has been lost in the specialization and empiricization of scientific and cultural development. From this perspective, the history of science obtains its global meaning: it is contemplation of the past, analysis of the present and prognosis of the future.

## FAILURE AND SIGNIFICANCE

The historical consciousness of the Romantic *Naturforscher* did not prevail. It was Buffon, Lamarck and Darwin who determined the ideas about evolution of the times following them, and history of science took the empirical and positivist path. The metaphysical contemplation of nature and history was abandoned in practice if not in theory. The progress of the natural sciences could do without metaphysics and history. The attacks of the nineteenth-century natural scientists against a metaphysical interpretation of nature were mirrored in the turning of the philosophy of their times towards theory of knowledge, anthropology and history of the discipline itself, accompanied by an explicit lack of interest in nature. Among the natural scientists, interest in the history of their respective disciplines decreased, and thenceforth the development of the natural sciences and their historiography continued along separate paths. An endangered nature and the opposition of the natural sciences to the humanities are characteristics of our present situation.

The historical consciousness of the natural sciences in the Romantic period not only merits the historian's attention, it is also of a lasting significance. Today again the concept of development has become a key-word for practice and theory. Nature, culture, the individual, society and science all develop, and there arises the question of concurrence and divergence of the types of development in these diverse realms of reality. In each separate sphere autonomy and heteronomy, causality and finality, self-regulation and external direction present themselves in a particular way.

In general, the development of science is nowadays not related to natural evolution, but nevertheless some connections have become increasingly clear. Nature has to be understood as a part of culture, and in the same way culture has a natural foundation. The development of culture threatens natural evolution, and achieving insight into the shared fate of nature and culture is an essential task for the future. However, along with the concurrences one also has to mark the divergences. Conceding ideography to nature (e.g. geological formation and biological evolution) is not to deny the absence of consciousness in the natural world. Establishing a contiguity with cultural evolution by using the principle of finality in biological evolution limits man, as a historical subject, in his pretensions to power over history.

With its synopsis of nature, science and culture, the Romantic *Naturforscher*'s historical consciousness (substantially influenced by the *Naturphilosophen* of the German Idealist period) can give valuable incentives both to present-day reflections on natural and cultural evolution and to an interdisciplinary concept of the historiography of the natural sciences.

### NOTES

1  D. von Engelhardt, *Historisches Bewusstsein in der Naturwissenschaft von der Aufklärung bis zum Positivismus* (Freiburg and Munich, 1979).

2  H. Steffens, *Beyträge zur innern Naturgeschichte der Erde* (Freiburg, 1801; repr. Amsterdam, 1973), p. 88; see also p. 256.

3  J. W. Ritter, *Fragmente aus dem Nachlasse eines jungen Physikers*, 2 vols. (Heidelberg, 1810; repr. Leipzig, 1984), I, p. 92.

4  H. C. Oersted, 'Betrachtungen über die Geschichte der Chemie', *Journal für die Chemie und Physik*, 3 (1807), 230.

5  L. Oken, *Abriss des Systems der Biologie* (Göttingen, 1805), p. 53.
6  C. G. Carus, 'Von den Naturreichen, ihrem Leben und ihrer Verwandtschaft', *Zeitschrift für Natur- und Heilkunde*, 1 (1820), 26f.
7  G. A. Goldfuss, *Grundriss der Zoologie* (Nuremberg, 1826), p. 33.
8  F. W. J. Schelling, 'Erster Entwurf eines Systems der Naturphilosophie', in vol. III of his *Sämtliche Werke* (Munich, 1927), p. 63.
9  G. W. F. Hegel, 'System der Philosophie. 2. Teil. Die Naturphilosophie', in vol. IX of his *Sämtliche Werke* (Stuttgart, 1958), § 249, p. 59.
10  G. W. F. Hegel, 'System der Philosophie. 1. Teil. Die Logik', in vol. VIII of his *Sämtliche Werke* (Stuttgart, 1964), § 231, p. 441.
11  H. Steffens, 'Über die Bedeutung eines freien Vereins für Wissenschaft und Kunst, 1817', in his *Alt und Neu*, vol. I (Breslau, 1821), p. 149.
12  C. G. Carus 'Unterschied zwischen descriptiver, vergleichender und philosophischer Anatomie, von der Entwicklung dieser einzelnen Methoden in verschiedenen Zeitaltern, und von der zweckmässigsten Aufeinanderfolge in Studium dieser verschiedenen Methoden', *Litterarische Annalen der gesammten Heilkunde*, 4 (1826), 1–30.
13  D. G. Kieser, 'Über die Bedeutung der Naturwissenschaften für das Leben der Zeit', *Amtlicher Bericht der 14. Versammlung Deutscher Naturforscher und Ärzte* (1837), p. 45.
14  H. Steffens, 'Über die Bedeutung', p. 151.
15  I. P. V. Troxler, *Elemente der Biosophie* (Leipzig, 1808), pp. XI–XII.
16  See J. C. S. Schweigger, *Ueber die älteste Physik und den Ursprung des Heidenthums aus einer missverstandenen Naturweisheit* (Nuremberg, 1821); *Einleitung in die Mythologie auf dem Standpunkt der Naturwissenschaft* (Halle, 1836); *Ueber das Elektron der Alten und den fortdauernden Einfluss der Mysterien des Alterthums auf die gegenwärtige Zeit* (Greifswald, 1848).
17  H. Steffens, *Zur Geschichte der heutigen Physik* (Breslau, 1829), p. 28.
18  H. C. Oersted, 'Betrachtungen über die Geschichte der Chemie', *Journal für die Chemie und Physik*, 3 (1807), p. 199.
19  *Ibid.*, p. 201.
20  *Ibid.*
21  J. W. Ritter, *Die Physik als Kunst. Ein Versuch, die Tendenz der Physik aus ihrer Geschichte zu deuten* (Munich, 1806), p. 3.
22  Görres to D. Runge, 7 July 1817, in his *Ausgewählte Werke und Briefe*, vol. II (Kempten and Munich, 1911), p. 193.
23  F. J. Schelver, 'Die Aufgabe der höheren Botanik', *Verhandlungen der Kaiserlichen Leopoldinisch–carolinischen Akademie der Naturforscher*, 2 (1821), 616.
24  Ritter, *Fragmente*, I, p. 104.
25  *Ibid.*, p. 91.
26  Carus, 'Von den Naturreichen', p. 72.
27  Ritter, *Die Physik als Kunst*, p. 3.
28  Novalis, 'Blüthenstaub, 1798', in his *Schriften* (Darmstadt, 1965), II, p. 427.
29  J. F. Koreff, 'Ueber die in einigen Gegenden Italiens herrschende böse Luft', *Magazin für die gesamte Heilkunde*, 9 (1821), 152f.

## FURTHER READING

On the *Naturforschung* of Romanticism, see R. Ayrault, 'En vue d'une philosophie de la nature', in his *La Genèse du romantisme allemand*, vol. IV (Paris, 1976), pp. 11–167; C. Bernoulli and H. Kern (eds.), *Romantische Naturphilosophie* (Jena, 1926); D. von Engelhardt, 'Bibliographie der Sekundarliteratur zur romantischen Naturforschung und Medizin 1950–1975', in R. Brinkmann (ed.), *Romantik in Deutschland* (Stuttgart, 1978), pp. 307–30; D. von Engelhardt, 'Romantische Naturforschung', in his *Historisches Bewusstsein in der Naturwissenschaft von der Aufklärung bis zum Positivismus* (Freiburg and Munich, 1979), pp. 103–57; A. Faivre, 'La Philosophie de la nature dans le romantisme allemand', in Y. Belaval (ed.), *Histoire de la philosophie*, vol. III (Paris, 1974), pp. 14–45; A. Gode-von Aesch, *Natural Science in German Romanticism* (New York, 1941; repr. New York, 1966). A. Grassl, *Die Romantik, ein Gegenpol der Technik* (Bonn, 1954); H. A. M. Snelders, 'Romanticism and *Naturphilosophie* and the Inorganic Natural Sciences 1797–1840; *Studies in Romanticism*, 9 (1970), 193–215.

On the general historical concept of Romanticism, see K. Borries, *Die Romantik und ihre Geschichte*; H. H. F. Flöter, 'Die Begründung der Geschichtlichkeit der Geschichte in der Philosophie des deutschen Idealismus', unpublished University of Halle doctoral dissertation, 1936; H. Lübbe, *Die Transzendentalphilosophie und das Problem der Geschichte. Untersuchungen zur Genesis der Geschichtsphilosophie (Kant–Fichte–Schelling)*, Habilitationsschrift Erlangen, 1956; K. Löwith, *Weltgeschichte und Heilsgeschehen* (4th edn, Stuttgart, 1961); F. Meinecke, *Die Entstehung des Historismus* (Munich, 1936; 4th edn, 1959); L. Renthe-Fink, *Geschichtlichkeit. Ihr terminologischer und begrifflicher Ursprung bei Hegel, Haym, Dilthey und Yorck* (Göttingen, 1968); W. Schulz, *Philosophie in der veränderten Welt* (Pfullingen, 1973; 5th edn, 1984).

On the historical concept of Romantic medicine, see D. von Engelhardt, 'Historisches Bewusstsein in der Medizin der Romantik', in E. Seidler and H. Schott (eds.), *Bausteine zur Medizingeschichte, Festschrift für Heinrich Schipperges zum 65. Geburtstag* (Stuttgart, 1984), pp. 25–35; E. Harms, 'The Early Historians of Psychiatry', *American Journal of Psychiatry*, 113 (1957), 749–52; E. Heischkel-Artelt, 'Die Geschichte der Medizingeschichtsschreibung', in W. Artelt (ed.), *Einführung in die Medizinhistorik* (Stuttgart, 1949), pp. 202–37; G. B. Risse, 'Historicism in Medical History: Heinrich Damerow's "Philosophical" Historiography in Romantic Germany', *Bulletin of the History of Medicine*, 43 (1969), 201–11; H. von Seemen, *Zur Kenntnis der Medizinhistorie in der deutschen Romantik* (Leipzig, 1926); O. Temkin, 'German Concepts of Ontogeny and History around 1800', *Bulletin of the History of Medicine*, 24 (1950), 227–46.

On Schelling's philosophy of the organic, see W. Bonsiegen, 'Zu Hegels Auseinandersetzung mit Schellings Naturphilosophie in der "Phänomenologie des Geistes"', in L. Hasler (ed.), *Schelling. Seine Bedeutung für die Philosophie der Natur und der Geschichte* (Stuttgart–Bad Cannstatt, 1981), pp. 167–72; D. von Engelhardt, 'Die organische Natur und die Lebenswissenschaften in Schellings Naturphilosophie', in R. Heckmann et al. (eds.), *Natur und Subjekt. Zur Auseinandersetzung mit der Naturphilosophie des jungen Schelling* (Stuttgart–Bad Cannstatt, 1985), pp. 39–57; W. Förster, 'Die Entwicklungsidee in der deutschen Naturphilosophie am Ausgang des 18. und zu Beginn des 19. Jahrhunderts', in G. Stiehler (ed.), *Veränderung und Entwicklung. Studien zur vormarxistischen Dialektik* (Berlin, 1974), pp. 171–210; E. Mende, 'Die Entwicklungsgeschichte der

Faktoren Irritabilität und Sensibilität in deren Einfluss auf Schellings "Prinzip" als Ursache des Lebens', *Philosophia Naturalis*, 17 (1979), 327–48; W. Szilasi, 'Schellings Beitrag zur Philosophie des Lebens', in his *Philosophie und Naturwissenschaft* (Bern, 1961).
On Hegel's philosophy of the organic, see H. Boehme, 'Das Leben als Idee. Die Idee des Lebens in Hegels Wissenschaft der Logik', in *Die Logik des Wissens und das Problem der Erziehung* (Hamburg, 1981), pp. 154–63; O. Breidbach, 'Das Organische in Hegels Denken', unpublished University of Würzburg doctoral dissertation, 1982; D. von Engelhardt, 'Die biologischen Wissenschaften in Hegels Naturphilosophie', in R. P. Horstmann and M. J. Petry (eds.), *Hegels Philosophie der Natur* (Stuttgart, 1986), pp. 121–37; J. N. Findlay, 'The Hegelian Treatment of Biology and Life', in R. S. Cohen and M. W. Wartofsky (eds.), *Hegel and the Sciences* (Dordrecht and Boston, 1984), pp. 87–100; K. Nadler, 'G. W. Fr. Hegel und C. G. Carus. Zum Verhältnis idealistischer und romantischer Natur–philosophie', *Sudhoffs Archiv*, 31 (1938), 164–88; M. Riedel 'Grundzüge einer Theorie des Lebendigen bei Hegel und Marx', *Zeitschrift für philosophische Forschung*, 19 (1965), 577–600; H. Querner, 'Die Stufenfolge der Organismen in Hegels Philosophie', *Hegel–Studien*, Supplement 11 (Bonn, 1974), pp. 153–64.
Publications by Romantic *Naturforscher* on the history of science include J. Dollinger, *Von den Fortschritten, welche die Physiologie seit Haller gemacht hat* (Munich, 1824); H. C. Oersted, 'Übersicht der neuesten Fortschritte in der Physik', *Europa*, 1:2 (1803), 20–48; H. C. Oersted, *Materialen zu einer Chemie des neunzehenten Jahrhunderts*, pt 1 (Regensburg, 1803); H. C. Oersted, 'Betrachtungen über die Geschichte der Chemie', *Beiträge zur vergleichenden Zoologie, Anatomie und Physiologie*, 1 (1806), 103–22; J. W. Ritter, *Die Physik als Kunst. Ein Versuch, die Tendenz der Physik aus ihrer Geschichte zu deuten* (Munich, 1806); J. W. Ritter, 'Versuch einer Geschichte der chemischen Theorie in den letzten Jahrhunderten', *Journal für die Chemie, Physik und Mineralogie*, 7 (1808), 1–66; J. C. S. Schweigger, *Ueber die älteste Physik und der Ursprung des Heidenthums aus einer missverstandenen Naturweisheit* (Nuremberg, 1821), *Einleitung in die Mythologie auf dem Standpunkt der Naturwissenschaft* (Halle, 1836) and *Ueber das Elektron der Alten und den fortdauernden Einfluss der Mysterien des Altertums auf die gegenwärtige Zeit* (Greifswald, 1848); and H. Steffens, *Zur Geschichte der heutigen Physik* (Breslau, 1829).

# 5

## Theology and the sciences in the German Romantic period

### FREDERICK GREGORY

Among historians of science there has been a definite trend of late to portray the relationship between science and religion in the nineteenth century in terms other than the traditional ones of warfare. By focusing on the social groups that corresponded with the various intellectual positions assumed in the nineteenth century, it has been possible for historians to depict the claims of science as mere ideological property whose primary importance lies on the social plane as opposed to the cognitive.

In this study, the relationship between science and religion is also not depicted as one of confrontation. In the German Romantic period, however, it is not necessary to employ social categories to avoid outdated military metaphors. The intellectual attempts of Romantic thinkers to determine the significance of nature and of our knowledge of nature for religion contained no hint that the relationship between the two was antagonistic. To the extent that the so-called 'conflict history'[1] of John Draper and Andrew Dickson White has proven unacceptable, the assessment of the Romantics bears renewed scrutiny.

I have chosen two individuals from the Romantic period whose approaches to the issue before us reflect the two basic intellectual alternatives that were available to members of the age. While there were of course many variations possible within either of these options, the two directions I have in mind were represented first by those who took their cue from the work of Kant, and second by the followers of Schelling. A key point of difference between the two groups emerged in their respective positions on the nature and role of intuition. Those who preferred Schelling accepted an intellectual intuition in place of Kant's sense intuition, thereby moving beyond what they perceived to be the unnecessary restrictions of Kant's system where the knowledge of nature was concerned.[2]

The choice of representatives of these two directions is not difficult. No treatment of theology in the Romantic period could possibly ignore Friedrich

Schleiermacher, who was the dominant theological mind of the first three decades of the century. Although Schleiermacher's sympathy for Schelling's *Naturphilosophie* has long been acknowledged, relatively little attention has been paid until recently to the central place the concept of nature played in Schleiermacher's thought. As for the Kantians of the time, there is one who made it a central concern of his work to define the boundary between science and religion. He was Jakob Friedrich Fries, who in the English-speaking world is known through the *Dialogues on Morality and Religion*, which only appeared in English translation in 1982.[3] Although Fries and Schleiermacher both received their early education in the same school of the pietistic 'Herrnhuter' sect, they came to very different perceptions of the relationship between science and religion.

Prior to Schleiermacher's epoch-making *Speeches on Religion* of 1799, the German theological tradition divided itself into two fundamental styles. The distinguishing characteristic of these styles stemmed from the stance taken by each on the trustworthiness of reason in religious matters. Orthodox rationalists, as Karl Barth calls them,[4] were convinced that much could be known through reason, and that even those revealed truths that were beyond reason were not therefore necessarily contrary to reason. Pietists, on the other hand, avoided reliance on a reasoned systematic theology, trusting their personal experience of 'rebirth' and their retreat into religious communities separate from the world to provide a foundation for their faith.

While the rationalistic theology of the early eighteenth century did not question the fundamental tenets of Christianity simply because it was rationalistic, the same could not be said of rationalistic philosophy, especially in the latter half of the century. Eighteenth-century philosophers, for example, *did* challenge the religion of their day. They were not content with a rational analysis of such fundamentals as miracles and God's superintendence of nature and history; rather, they questioned them. Indeed, rationalistic philosophy directly criticized the validity of the traditional arguments for God's existence that eighteenth-century rationalistic theology had endorsed, convincing pietistic theologians more than ever that reason alone was no sure ground on which to construct a sound theology.

Critical rationalism invaded theology toward the end of the eighteenth century in the thought of the theologians known as 'neologists'. Contrary to orthodox rationalists, these men were not willing to argue that the content of revelation had to be compatible with reason. In the words of Karl Aner, they 'pushed this content to one side as contrary to reason'.[5] With the neologists the content of the Christian religion lost its capacity for a distinctly protestant outlook. John Dillenberger and Claude Welch note that for neologists whatever truth remained 'was not really different from that which thoughtful men were saying outside the church . . . In these men Protestantism had become a victim of the Enlightenment.'[6]

This, then, was the context in which Schleiermacher's famous speeches must be understood. Writing from within the ranks of the theological community, Schleiermacher articulated a constructive re-interpretation of the very meaning of religion. Schleiermacher's positive re-evaluation of the essentials of Christianity contrasted sharply both with the traditional theological approaches of the eighteenth century and with the destructive tendency of those Schleiermacher called religion's 'cultured despisers'.[7] Schleiermacher's thought was a defence of the Christian faith against philosophers and even against the neologists, whose modernistic approach to Christianity became enormously popular in the nineteenth century under the name of theological rationalism.

But if Schleiermacher's work represented a defence of the Christian faith, it was vastly different from the traditional apologetics of the eighteenth century. Schleiermacher realized that the debates between the older orthodox rationalists and pietists were, in the face of the new rationalism, largely irrelevant. No longer could the theological debate focus, as it had in the past, on how the fundamentals were to be apprehended and articulated. Now in need of re-interpretation, indeed a constructive, living and vibrant re-interpretation, was the very foundation and nature of religion itself.

Schleiermacher rejected the notion, dear to rationalists old and new, that religion consisted of metaphysical analyses that produced a set of beliefs. He also denied Kant's conclusion that because religion could not have its foundation in knowledge, it had to be identified with the human capacity for moral sense. According to Schleiermacher religion referred primarily to that specific component of human consciousness that was neither cognitive nor volitional. Religion was born of feeling (*Gefühl*), which was prior both to cognition and volition. Cognition and volition both have reference to a mediated external object, but in religion, said Schleiermacher, we encounter a primary human experience.

Here was a view of religion that changed the rules governing the theological debates. For example, according to Schleiermacher it was no longer necessary to forsake Christianity simply because one did not believe in miracles; on the contrary, Christianity truly understood did not rely on a traditional grasp of miracles at all. To Schleiermacher God's presence *was* evident in nature, but it was evident far more in everyday natural events than it could ever be in miracles.[8] God, frequently synonymous with the Infinite in Schleiermacher's *Speeches on Religion* of 1799, was apprehended primarily in human feeling about the universe, not in knowledge of it.

The question before us, however, is of precisely what relevance the knowledge of nature was to religion for Schleiermacher? Some have maintained that Schleiermacher abandoned all claim to any religious knowledge. Hegel, in particular, castigated Schleiermacher's basing of religion on feeling. 'If religion in man is based solely on feeling', Hegel wrote in his Preface to H. F. W.

Hinrichs's *Religion in Its Internal Relation to Science* (1822), then 'it is correct that this has no further determination than to be a feeling of dependence, and so a dog would be the best Christian, because it has this feeling most intensely'.[9]

It is true that Schleiermacher rejected the possibility of a concept (*Begriff*) of God, and it is also clear that there are numerous places in Schleiermacher's writings in which he appears to deny that religious intuition can have anything to do with reflection on the construction of a system of thought. Any attempt to organize individual religious intuitions was arbitrary and indefinite.[10] Hegel's criticism of Schleiermacher centred on this point. In Hegel's *Believing and Knowing* of 1802 Hegel said that Schleiermacher's position was too subjective. Near the end of his discussion of F. H. Jacobi's philosophy Hegel identified Schleiermacher's *Speeches* as the highest point to which 'the Jacobian principle' had developed. But, Hegel continued, the individual's intuition of the universe, being the central component of Schleiermacher's approach, was too particular a basis on which to establish the religious community. In Schleiermacher the community was to cultivate individual intuitions, with the result that 'an infinite number of communities and particularities are validated'.[11]

But, as Richard Brandt has persuasively shown, 'Hegel's view of Schleiermacher's system was an abbreviated and distorted one.'[12] Brandt has argued at length that Schleiermacher *did* believe that religious intuitions furnish objective knowledge of the nature of the universe; indeed, as is evident from the subtitle, Brandt wrote his study to establish this implication of Schleiermacher's system. In denying that responses to the merely beautiful and sublime in nature could be religious intuitions, Schleiermacher was deliberately setting himself apart from neo-Kantian thinkers like Fries.[13]

New light is shed on Schleiermacher's conception of the relationship between natural science and religion when one extends one's inquiries beyond the philosophical question about the possibility of religious knowledge to the general place of nature in Schleiermacher's thought. Schleiermacher's best-known biographer, Wilhelm Dilthey, writes that 'it is impossible to represent Schleiermacher's system without starting with the philosophy of nature and its historical value, for Schleiermacher's system is founded on the truth of this philosophy of nature'.[14] Dilthey's claim is supported by the fact that the concept of nature appears more than 820 times in his two-volume *Christian Faith* alone.[15] Much prior to these volumes Schleiermacher vehemently rejected what Hasler has called Kantian dualism and Fichtean solipsism because in both the objective conditions affecting the actions that could change society were blocked out.[16] What he missed in these philosophers he found·in the *Naturphilosophie* of the young Schelling. Here was a philosophy of the real as well as the ideal, here was support for Schleiermacher's own conviction that both the realms of the physical and the ethical shared an ontological foundation.

Recent scholarship on the relationship of Goethe and Hegel to the natural sciences confirms that Schleiermacher was not alone in his need to find an approach to nature which included an ethical dimension. In what is called Goethe's and Hegel's 'ecological' understanding of natural science one recognizes the same concern as that felt by Schleiermacher to move beyond the implications of natural science for epistemology to a genuine ontology of nature where the ethical and the natural could not be separated.[17] For Schleiermacher, however, it was Schelling who had articulated most effectively how nature should be treated. Like Heinrich Steffens, who also built a bridge from Schelling's *Naturphilosophie* to ethics in the first decade of the century,[18] Schleiermacher responded eagerly to Schelling's representation of the physical realm as an evolving process within human history whose structure was affected by the accompanying self-realization of humanity. Because of this mutual participation of the ideal in the real and of the real in the ideal ('das Ineinander von Realem und Idealem') Schleiermacher felt he had found a solid justification for viewing moral action as a continuous dialectical shaping of nature through reason.[19] Natural law and moral law agreed in their productive and formative functions; both captured reality's creative and evolutionary dimensions in such a way that they expressed what was normative for a given stage of development. As in Schelling, the prototype of natural law for Schleiermacher was not rigid mechanism, but developing organism.[20]

In the end, then, the relationship between natural science and religion for Schleiermacher was hardly one of mutual hostility, but one of necessary mutual reinforcement. Researchers were free to pursue naturalistic as opposed to supernaturalistic explanations of natural phenomena, even when the phenomena appeared to be miraculous, because the harmony between natural science and religion occurred at a level of religious knowledge deeper than that of the mediated external object. Schelling's intellectual intuition provided the ultimate guarantee of the common ground of these two expressions of human experience.

Initially, Jakob Fries was also attracted to Schelling's *Naturphilosophie*. After training in philosophy at Jena, Fries successfully landed a post in Heidelberg. Here he was responsible for teaching mathematics and philosophy, though physics later was added to his duties. The reason he was able to handle such a variety of responsibilities was not only because of his training in natural science and mathematics, but also because in his published works he had sketched out a philosophical position in which natural science and mathematics occupied a central place.

Like Schelling, Fries was interested in the relation between natural science and philosophy; indeed, Fries was initially strongly attracted to Schelling's philosophy of nature. By 1805, however, he had become a decided opponent of Schelling's *Naturphilosophie*. Elsewhere I have examined the relation between Fries and Schelling.[21] Suffice it here to say that for Fries Schelling had abandoned the sound principles of critical philosophy in order to escape into the mystical world of non-sensual intellectual intuition.

But Fries was not only a philosopher of natural science, he was also vitally interested in religion. As a loyal Kantian, he was concerned to delineate the limitations of reason in order to clarify the place of the non-rational in human experience. To do this Fries developed a philosophy of natural science, to be sure along Kantian lines, yet far beyond what Kant himself had carried out. But if Fries was out of the mainstream of Romanticism because of his persistent hold onto the critical principles of Kant when the mood of the day called for the overthrow of the Kantian rift between phenomena and noumena, he was nevertheless a child of his time in his unique understanding of the role of *feeling* in religious experience. Friesian Kantianism is without question *Romantic* Kantianism. This is nowhere more clearly evident than in Fries's book of 1805, entitled *Knowledge, Belief, and Aesthetic Sense,*[22] in which Fries attempts to set forth clearly the relationship between natural science and religion.

As one might surmise from the title, the goal of Fries's effort is to distinguish three separate modes of conviction; hence, it is the acquiring and possession of conviction that for Fries is the primary human experience. Since ultimately we all adopt our views because we *feel* certain of them, it is within a context in which feeling provides final consent that Fries will distinguish different modes by which we acquire and possess conviction.

The first means by which we acquire and possess conviction is through knowing (*Wissen*). There are two different kinds of knowledge involved in the process of knowing, immediate knowledge and mediated knowledge, and the difference between the two must be appreciated if the nature of scientific knowledge is to be understood.

Immediate knowledge comes for Fries in two varieties. The first, intuition (*Anschauung*), exists in the form either of the intuitively understood axioms of mathematics or of the intuitive apprehension of sense perception.[23] The two are related through Kant's great discovery that space and time are but forms of our sensuality, from which emerge the fundamental relations of mathematics.[24]

The second variety of immediate knowledge is a bit more difficult to describe. It lies in our unconsciousness, and we become aware of its presence only through reflection, that is, only through the use of judgement.[25] In the 1808 *New Critique of Reason* Fries puts it this way:

When I say, for example, every substance persists, every change has a cause . . . I acknowledge laws of nature . . . But these very laws, of which I again become conscious in the judgement, must lie in my reason as immediate knowledge, only I need the judgement in order to become conscious of them.[26]

So much for immediate knowledge. What of mediate knowledge? When we use judgements and inferences in combination with immediate knowledge from any of its sources to gain new knowledge, the knowledge gained is *mediate*, not *immediate* knowledge. It must draw on some external immediate knowledge in order to come into existence, and its certainty arises not from itself, but from the immediate knowledge on which it relies.

This mediated knowledge is scientific knowledge, for natural science, through the use of judgements and inferences, draws on the immediate knowledge of the senses, mathematics and the categories of the understanding to acquire new knowledge. Should anyone inquire about the certainty of a scientific assertion, any justification will, according to Fries, ultimately have to hark back to some form of immediate knowledge. Should someone claim: 'But I will prove it to you scientifically!', even then they cannot escape the realm of the mediate, for the essence of proof is to justify one assertion on the grounds of another. But in the process there are always premises, or immediate knowledge in Friesian terms, that are left unquestioned. Hence proof, writes Meinhard Hasselblatt, 'is basically only a postponement of the real justification'.[27]

When the immediate knowledge of the senses is ordered according to mathematical laws and the categories of the understanding, as happens in natural science, the result is knowledge that is both necessary and mediate. Indeed, a genuinely scientific description ties natural phenomena together with strict necessity. For Fries the world of *Wissen* is the same one that earlier had been made famous by Laplace and the French mechanists. Natural science is pure mechanism, and the world it describes is strictly determined with no exceptions:

Whoever knew the state of each and every thing in the world in a single moment of the past, present, or future could calculate from it the whole being of the world and the complete course of events – how motion follows from motion, how life comes from life, and how thought develops from thought for all the past, present, and future.[28]

It is important to understand how knowing originates within our mental capacities, for Fries believed, as he put it, that 'we succeed in defending the rights of belief [*Glaube*] mainly by showing how knowing arises only subjectively in reason'.[29] Like Kant, Fries wanted 'to win a place for belief', and he wanted to do it by humbling *Wissen*.[30] For one thing, proof has been substantially damaged in the Friesian analysis. Proof, he claims, is only a mediate process that cannot of itself guarantee anything since the immediate knowledge on which it depends results from either the purely subjective structure of our mental capacities or the non-necessary nature of empirical information.

And yet a real gain for belief does come out of this analysis of knowing, for among the abilities that the understanding possesses of itself, independently of intuition, is that of negation.[31] Reason (*Vernunft*) has become aware of its own limited nature through its analysis of knowing. But having become conscious of these limitations, it now, through the logical process of negation, can at least think away the limits. If reason shows itself to be tied to phenomena in knowing, then it at least can formulate a category that is the negation of phenomena. Fries is careful to point out that such a process does not produce any *positive* notion of what lies beyond the limitations of knowing. The origin of believing in contrast to knowing 'must remain thoroughly negative, really having no positive content other than that of the negation of all negations, or the denial of limitations':[32]

Believing therefore originates immediately out of the innermost essence of reason, providing us with its own view of the reality of things, the knowledge of its own higher world, the world of ideas.[33]

But how does believing express itself? Since it does not have concepts and judgements at its disposal, for they belong to the understanding and require the empirical world for their formation, it can be expressed only as we think away the limitations imposed by the understanding through 'ideas of that which is beyond all limits of the finite'.[34] Whatever these ideas are, then, we should not expect them to resemble anything we encounter in the world of natural science.

If the world of knowing is confined to space and time, the way to negate this restriction is to think space and time away. The negation of these limitations can be captured by the idea of an eternal realm outside space and time. A further characteristic of the phenomenal world, which is of course tied up with its being restricted in time and space, is its rigid necessity. There are two ways in which necessity can be imagined as cancelled. One is to imagine a realm in which freedom, not necessity, reigns; another is to think of the absence of necessity in the presence of purpose. Both of these characteristics are in fact encountered in our consciousness – we have a consciousness of our freedom, and we know ourselves to be purposeful beings.

While our experience testifies to the presence of a world that stands in marked contrast to the finite world of knowing, Fries is careful to point out that it is not within our power to prove the reality of this world. What we can do is to negate the limiting notion of existence in time in the idea of immortality, and negate the limitation of necessity in the idea of an autonomous free will and a highest purpose. These ideas, products of pure reason, lie deep in the 'belief of reason'.[35] Belief can express nothing positive or concrete about them, since they are born of negation.[36]

If this exposition of Fries has been successful, it should now be clear why a proper understanding of Friesian belief preserves its proximity to knowledge; in fact, Fries is clear that the two are closely related.[37] Both result from a characteristic activity of a specialized aspect of reason: knowledge from the subjectively restricted understanding as it acts in conjunction with sense, and believing from the logical capacity of the understanding to negate, at least formally, those same subjective restrictions. At the risk of distortion, one could invoke by way of analogy the classic circumstance encountered in medieval cosmology. When once the cosmos was rationally determined to be finite, the question could immediately be asked, 'But what then lies outside it?' Nor is this analogy out of place here, for Fries constantly associates knowing with the finite realm, and believing with the infinite, or, as he prefers to say, the eternal.

For Fries there is one more mode of apprehending human experience and possessing conviction, *Ahnen*, and it is fundamentally different from both knowing and believing. In the forthcoming translation of Fries's *Wissen, Glaube und Ahndung*, *Ahnen* is translated, for reasons that will become clear below, as 'aesthetic sense'.[38]

First of all, *Ahnen* refers to a feeling – it has no relation to processes born of reason like knowing or believing. As indicated above, feeling is the ultimate source of our conviction, it provides the ultimate assent. *Ahnen*, of course, shares this primacy, and therefore it is not just a third category tacked on to Fries's analysis. But *Ahnen* is also a focused feeling: it is the feeling that the eternal is reflected, albeit in an imperfect and restricted manner, in the finite.

Since we cannot, according to Fries, have any concept at all of the realm of the eternal from within the world of knowing, and since we cannot obtain any positive notion of the eternal from the world of belief, it being merely the logical negation of knowing's limitations, how can we ever possibly recognize the eternal in the finite?

Whatever the answer, we certainly cannot expect that the means by which we will recognize the eternal in the finite will remotely resemble those of knowing and believing. We will not meet any concepts or ideas. *Ahnen*, after all, is feeling; it does not belong to reason, but stands behind it. But what feelings?

Fries argues that in the feelings of the beauty and sublimity of nature we encounter the eternal *in the finite*. Why should we respond emotionally as we look out from the tops of Swiss mountains? Why is it that flowers are beautiful? Fries would reply that it is only possible if there is purpose, specifically, purpose in itself. Beauty and sublimity are ends in themselves. There must be a realm of purpose for beauty and sublimity to be possible. But the eternal is exactly that – a realm of purpose, the world of believing that was born of reason itself. We are one source of purpose, but in *Ahnen* we encounter it and recognize it *in nature*. In our encounter with a purpose whose source is outside us, we give life to our belief. Herein lies real religion, and not in the imposed purpose of our misplaced conceptual knowing of organism. 'Knowing', writes Fries, 'which is the infinite maternal principle, eternally marries believing, the eternal paternal principle, and from them has been produced the third, *Ahnen*':[39]

*Ahnen* is nothing without believing. It apprehends in nature the reality of believing, but belief cannot be expressed without knowing, and then only in opposition to knowing... A positive notion of the eternal is possible for us through its relation to the finite, but we can lay hold of it only in feeling through the beauty and sublimity of nature.[40]

In the end Fries was true to his Romantic Age, for the only expression we can hope to give to *Ahnen* is through poetry and the arts.

Were religion to be associated only with faith and natural science solely with knowledge, then the relationship between science and religion would be one of mutual exclusion. The two realms would be totally separate and independent. Indeed Fries allowed total freedom to the man of science to gain whatever knowledge possible. He even rejected Kant's denial that complete scientific knowledge of life and organism was theoretically possible.

Yet Fries was neither content to permit natural science and religion to remain completely isolated from each other, nor willing to resort to Schelling's intellectual intuition to bring them together. His solution was, however, equally indicative of the Romantic period as was Schelling's: through aesthetic sense of the infinite in the finite the realms of religion and natural science intersected.

Although the development of theology outside of Germany in the nineteenth century frequently ignored or was unaware of the Romantic interpretations of the relationship between religion and natural science, the challenge taken up by Schleiermacher and Fries would re-emerge in the twentieth century. These two thinkers faced squarely a fundamental problem that was lost in the warfare between science and religion in the nineteenth century. Only in this century has it become clear again that science cannot avoid the Why, nor religion the How. The Romantics, however, began with this assumption.

## NOTES

1   Otto Zöckler, *Geschichte der Beziehungen zwischen Theologie und Naturwissenschaft mit besonderer Rücksicht auf Schöpfungsgeschichte*, 2 vols. (Gütersloh, 1877 and 1879), I, pp. 1–2.

2   For a treatment of the antagonism between the Kantians and the
    *Naturphilosophen* on this point, see my 'Kant's Influence on Natural
    Scientists in the German Romantic Period', in Robert Visser *et al.* (eds.),
    *New Trends in the History of Science* (Amsterdam, 1989).
3   Jakob Friedrich Fries, *Dialogues on Morality and Religion*, ed. D. Z. Phillips
    (London, 1982).
4   Karl Barth, *Protestant Theology in the Nineteenth Century. Its Background
    and History* (London, 1972), p. 163.
5   Quoted by Barth, *Protestant Theology*, p. 164, from Karl Aner, *Die
    Theologie der Lessingzeit* (Halle, 1929), p. 245.
6   John Dillenberger and Claude Welch, *Protestant Christianity* (New York,
    1954), pp. 155–6.
7   Friedrich Schleiermacher, *Über die Religion. Reden an die Gebildeten unter
    ihren Verächtern* (Berlin, 1799).
8   Schleiermacher's position on miracles is set forth most clearly in his lectures
    on the life of Jesus, which were reconstructed from student notes in 1864 and
    published as *Das Leben Jesu*. A translation appeared in 1975 as *The Life of
    Jesus*, ed. Jack C. Verheyden (Philadelphia, 1975). See pp. 415–16, 432f., 455–
    6, 479–80. See also Ueli Hasler, *Beherrschte Natur. Die Anpassung der
    Theologie an die bürgerliche Naturauffassung im 19. Jahrhundert* (Bern,
    1982), p. 138, and Martin Redeker, *Schleiermacher: His Life and Thought*,
    trans. John Wallhausser (Philadelphia, 1973), pp. 123–4.
9   Quoted by Richard Brandt, *The Philosophy of Schleiermacher: The
    Development of His Theory of Scientific and Religious Knowledge* (New
    York, 1941), p. 325.
10  See Punjer's variorum edition of the *Reden* (Braunschweig, 1879), pp. 60–4.
11  'Glauben und Wissen oder die Reflexionsphilosophie der Subjektivität, in der
    Vollständigkeit ihrer Formen, als Kantische, Jakobische, und Fichtesche
    Philosophie', pp. 313–414, in G. W. F. Hegel, *Gesammelte Werke*, ed. H.
    Büchner and O. Poggeler (Hamburg, 1968), IV, pp. 385–6. For similar
    criticism of Schleiermacher's *Der Christliche Glaube*, see Brandt, *Philosophy
    of Schleiermacher*, pp. 322ff., where the author discusses Hegel's 1822 Preface
    to Hermann F. W. Hinrich's *Die Religion in innerem Verhältnisse zur
    Wissenschaft*.
12  Brandt, *Philosophy of Schleiermacher*, p. 325.
13  Fries, who is dealt with below, explicitly identified the apprehension of
    nature's beauty and sublimity as pure religious experience.
14  Wihelm Dilthey, *Leben Schleiermachers*, ed. M. Redeker, 2 vols. (Berlin,
    1966), II, pt 1, p. 451.
15  Hasler, *Beherrschte Natur*, p. 26.
16  *Ibid.*, pp. 41, 70.
17  See F. Amrine, F. J. Zucker and H. Wheeler (eds.), *Goethe and the Sciences:
    A Reappraisal* (Dordrecht, 1987), pp. 143–5, 170–1, 341; Michael John Petry
    (ed.), *Hegel und die Naturwissenschaften* (Stuttgart–Bad Cannstatt, 1987),
    pp. 8, 24–8, 430.
18  Perty (ed.), *Hegel*, pp. 72–3; see especially note 16.
19  *Ibid.*, pp. 72, 75–8. For an analysis of the role evolution played in
    Schleiermacher's social theory, see Heino Falcke, *Theologie und Philosophie
    der Evolution: Grundaspekte der Gesellschaftslehre F. Schleiermachers*
    (Zürich, 1977).

20 Hasler, *Beherrschte Natur*, pp. 80–1.
21 F. Gregory, 'Die Kritik von J. F. Fries an Schellings Naturphilosophie', *Sudhoffs Archiv*, 67 (1983), 145–57.
22 J. F. Fries, *Sämtliche Schriften*, ed. Lutz Geldsetzer and Gert König, 26 vols. (Aalen, 1968), III, pp. 413–755. References to this work will be cited hereafter as *WGA*, and the translations from it are mine. (*Ahndung* is an old form of the modern *Ahnung*.)
23 Fries is willing to speak of intuitive knowledge only on these levels of mathematics and sense perception; indeed, he vehemently opposes all other forms of intuition, for example Schelling's intellectual intuition, because no governing framework, such as that of space and time, can be specified for mystical intuitions. Kant would have agreed, for he said that 'without sensibility we cannot have intuition' (*Critique of Pure Reason*, trans. Norman Kemp Smith (New York, 1965), A68). For Kant's rejection of the possibility of intellectual intuition for the understanding, see *Critique*, Bxi-xii, B307, A255.
24 For a discussion of Fries's philosophy of mathematics, see my 'Neo-Kantian Foundations of Mathematics in the German Romantic Period', *Historia Mathematica*, 10 (1983), 184–201.
25 *WGA*, p. 22.
26 *Neue Kritik der Vernunft* in *Sämtliche Schriften*, IV, pp. 341–2 (my translation).
27 Meinhard Hasselblatt, *Jakob Friedrich Fries: Seine Philosophie und seine Persönlichkeit* (Munich, 1924), p. 23. See *WGA*, p. 26.
28 *WGA*, p. 81.
29 *Ibid.*, p. 118.
30 *Ibid.*, p. 119; Kant, *Critique of Pure Reason*, Bxxx.
31 *WGA*, p. 101.
32 *Ibid.*, pp. 122, 129.
33 *Ibid.*, p., 123.
34 *Ibid.*, pp. 134–5. Ideas, of course, belong solely to reason (*Vernunft*), as opposed to concepts, which belong to the understanding. See pp. 76, 126.
35 *Ibid.*, pp. 62, 69, 128–30. Fries uses the phrase 'belief of reason' in one of his discussions of intellectual intuition. See p. 57.
36 *Ibid.*, p. 136.
37 'Believing is more a kind of apprehending that is more closely related to knowing than to feeling' (*WGA*, p. 236).
38 *Knowledge, Belief, and Aesthetic Sense*, ed. Frederick Gregory, trans. Kent Richter (Cologne, 1989).
39 *WGA*, p. 56.
40 *Ibid.*, pp. 250–1.

## FURTHER READING

Barth, Karl, *Protestant Theology in the Nineteenth Century. Its Background and History* (London, 1972)
Brandt, Richard, *The Philosophy of Schleiermacher: The Development of His Theory of Scientific and Religious Knowledge* (New York, 1941)

Fries, Jakob, *Knowledge, Belief, and Aesthetic Sense*, ed. Frederick Gregory (Cologne, 1989)

Hasler, Ueli, *Beherrschte Natur. Die Anpassung der Theologie an die bürgerliche Naturauffassung im 19. Jahrhundert* (Bern, 1982)

Lovejoy, Arthur, *The Great Chain of Being. A Study of the History of an Idea* (New York, 1960)

Otto, Rudolf, 'Science of Religion', part 2, in *Religious Essays* (London, 1937)

Schleiermacher, Friedrich, *On Religion: Speeches to its Cultured Despisers* (New York, 1958)

Welch, Claude, *Protestant Thought in the Nineteenth Century*, vol. I: 1799–1870 (New Haven, 1972)

# 6

## Genius in Romantic natural philosophy

### SIMON SCHAFFER

'So the Experimenter is also simply the Genius.'
(Novalis, *Allgemeine Brouillon*)

'Natural Philosophy is the genius that has regulated my fate.'
(Victor Frankenstein, in Mary Shelley, *Frankenstein*)

What 'ingenuity' was for Baroque experimenters, 'genius' was for Romantic natural philosophers. The craft skill of seventeenth-century natural philosophers was deemed praiseworthy. Experimenters commonly addressed each other as 'ingenious', naturally clever and inventive.[1] Romantic sciences were pursued by men of 'genius'. Only after the mid-eighteenth century did this term come to be opposed to 'ingenuity' and its cognates. Goethe recalled the usages of the 1770s: 'In the common parlance of the day genius was ascribed to the poet alone. But now another world seemed all at once to emerge; genius was looked for in the physician, in the general, in the statesman, and, before long, in all men who thought to make themselves eminent either in theory or in practice'.[2] Both terms were highly contested. Critics denied experimenters' claims that 'ingenuity', machine-minding, could be proper philosophy. The proliferation of genius in the late eighteenth century was also viewed with derision by learned servants of the *ancien régime*. In 1786 the Westphalian conservative civil servant Justus Möser, exponent of the school of *Sturm und Drang*, described the 'mania for genius' as an 'epidemic'. Goethe, who collaborated with and much admired Möser as a German version of Benjamin Franklin, was equally scathing. '*Genius* became a universal watchword', and so corrupted; 'it seemed almost necessary to banish it entirely from the German language'.[3]

These conflicts were marks of major social crises. Savants sought standing as members of the academic Faculties and learned societies of Baroque polite culture. Their role was defined through their relations with artisans and instrument-makers, professors and clerics. In the Revolutionary epoch, natural philosophers sought their place on a social map in the process of rapid change. Natural philosophy was pursued in the academies, in scientific societies and by the public lecturers. Each was under attack. The corporate universities were viewed as bastions of the society of orders, challenged by

82

specialist research schools in innovative areas such as jurisprudence, philology, chemistry and mathematics. New journals and extramural networks appeared. The Paris Academy of Sciences, pattern for all metropolitan scientific societies, was abolished in 1794. Between 1790 and 1815 eight German universities were closed. Student numbers reached an historic low.[4] The public performers, including electrical demonstrators and purveyors of 'animal magnetism', were assailed either as aids of revolution or as frivolous charlatans of corrupted public taste. The concept of 'genius' was a central term in these debates. Its etymology was complex. The Latin term betokened the personal deity which watched over an individual's fate. In 1790 Kant argued that since genius could not 'indicate scientifically how it brings about its products', the word came from 'that peculiar guiding and guardian spirit given to a man at birth from whose suggestion these original ideas proceed'.[5] Its senses had proliferated to include the specific character of an epoch or a culture, and that of personal skill. It captured key themes of Enlightenment philosophy, such as the problem of historical and social relativism (in which sense it was used by J. G. Herder in his philosophy of history) and that of the relationship between creative power and rule-bound instruction, in which sense philosophers of taste and judgement, notably Edmund Burke and Immanuel Kant, debated its formation.[6] Genius began to be understood not as a peculiar capacity possessed by a creative artist, but as the power which possessed him. Imagery of divine fire and excessive power was used to analyse this phenomenon. A natural philosophy was developed to analyse active powers. This philosophy appealed to the active powers of the genius to legitimate its natural knowledge. This form of natural philosophy is often distinguished by the German label *Naturphilosophie*. Such labelling can obscure the complex and important connections between the changes in the culture of the sciences which took place in the German lands from the 1790s and the structure and impact of eighteenth-century experimental philosophies. In this chapter, the term 'natural philosophy' is used to describe the basic structures of inquiry into nature which dominated European culture until the early nineteenth century. 'Genius' was a term which was vital in the changes which these structures underwent during this period. Many of those protagonists whose views are discussed in this chapter moved from enthusiastic approval of the French Revolution to conservative, apologetic and in some cases radically reactionary political and philosophical accounts of the intelligentsia and the state. These careers were vitally important in the formation of the view of the scientist as genius. Two aspects of such connections between the genius and natural philosophy are described. First, the relationship between the power of genius and the power of the rules of taste and training is placed in the context of Revolutionary debates on the place of the natural philosopher. Second, the natural philosophies which aimed at working with the power of genius in mind and matter are examined as a fundamental aspect of Romantic culture.

These issues of pedagogy and natural philosophy correspond to important conjunctures in Romantic natural science. In the 1790s Burke used his political aesthetics to argue that spurious 'genius' of natural philosophers had fomented subversion through intellectual conspiracy. Kant argued that natural philosophers were not entitled to the label of 'genius'. After the Revolution natural philosophers used the terms of 'genius' to portray strategies for understanding and controlling power. The research and teaching work of the professoriate was described in terms of the dangers of genius and the virtues of disciplined pedagogy. J. G. Fichte's *Wissenschaftslehre* (1794), published soon after his appointment to the Jena philosophy chair, was announced as 'the first system of freedom', in emulation of the Revolution. In the prefaces Fichte added in 1797, he attacked those 'who see nothing more in the sciences than a comfortable way of earning a living . . . to whom no means are shameful to suppress the destroyer of their trade'.[7] some advocated renewed faculty discipline as a solution to this legitimacy crisis, such as the Prussian reformer Ludwig Jacob, who argued in 1798 that 'an orderly, upright man with a well-ordered erudition and a gift for communicating it' would make a better professor than 'a genius who has offensive morals and who does not think it worth the labour to employ diligence on lectures for his students'.[8] F. W. J. Schelling responded in his 1802 Jena lectures on university studies that 'a teacher who merely transmits will often give a radically false version of what he has learned'. Much of the philosophy to be developed in the academy was of a kind 'whose inner essence can be grasped only by a kindred genius through a rediscovery in the literal sense of the word'.[9] The professoriate must be filled with those who could recapitulate the process of the emergence of knowledge through discovery, and thus transcend the constraints of rule-bound scholasticism.

Genius seemed threatening to established academic rules and their powers, especially where natural philosophy was pursued as public theatre in the salons of the *ancien régime*, by mesmerists and physiognomists, electrical philosophers and pneumatic performers. Fichte satirized this feature of oppressive pre-Revolutionary society. In 1793 the Jena lawyer August Rehberg, ally of Möser, published an indictment of the Revolution, arguing that peasants were intellectually incapable of defending or comprehending political rule. Fichte immediately responded with his anonymous 'rectification of public judgement', arguing that conservative hostility to popular right on the basis of popular suffering was due to the fact that 'our epoch is in general much more sensitive to the needs of thought than those of nature'. Defenders of aristocratic power were consumed by the demands of spurious intellectual need: 'you have modern furniture, but you still lack a picture gallery. Perhaps you have this, but you still lack a cabinet of curiosities'.[10] Fichte used a conventional attack on the corruption of public taste in a radical way. The demands of nature must be analysed and understood. The needs to which polite civil society responded were refined, imaginary, the result of an intellectual disease. Philosophers had the task of substituting natural power for artificial taste, the understanding of the genius for the sensibility of the mere consumer, and thus becoming spokesmen for nature's citizens, the people.

Fichte's attempt to connect the work of philosophical genius, natural power and popular right was a representative articulation of Revolutionary idealism. In 1793 the French Republic's Committee of Public Instruction declared that 'true genius is almost always *sans culotte*'. The implication was that genius had been collectivized. 'Jacobinism' was soon made the label for any foreign allies of this idealism. The Jacobin Georg Forster, professor of natural history and then Rhineland delegate to the National Convention in 1793, compared the Revolutionary movement to an insect swarm: 'those beings whom one sees at the head of our Revolution are not elevated like demigods above their fellow citizens, and one sees none of them before whose genius the souls of others bow down'.[11] The attack on Jacobinism used aesthetic criteria, and it turned its gaze on the natural philosophical authority which the Revolutionaries claimed. Burke, author of a *Philosophical Enquiry into the Sublime and the Beautiful* (1756), translated into German in 1773, was the most prominent of these critics. His *Enquiry* was part of a British debate of which key texts included Edward Young's *Conjectures on Original Composition* (1759), Joshua Reynolds's *Discourses on Art*, delivered at the Royal Academy from 1769, and the influential *Essay on Genius*, composed in 1758 and published in 1774 by Alexander Gerard, divinity professor at Aberdeen. Reynolds told his students that there was a difference between imitation and genius, but that genius never transcended rule and discipline: 'Genius begins not where rules, abstractedly taken, end; but where known vulgar and trite rules have no longer any place.' With 'a kind of scientifick sense' these rules could be communicated so as to direct attention away from 'nicety and research' towards 'the true art of seeing nature' as a whole. 'The great business of study is to form a mind . . . to which all nature is then laid open.' Gerard shared Reynolds's concern with the process of training in skilful association of detailed technique and phenomena, and his subsequent claim that 'genius' represented the faculty closest to the design of natural power, through which this association most energetically expressed itself. 'Genius' was just the faculty of mind which made it most natural: 'this faculty bears a greater resemblance to nature in its operations, than to the less perfect energies of art'.[12]

The nature which Gerard envisaged was organic, and the power in nature to which genius was closely allied was that of vegetable growth. Edward

Young, above all, made the contrast between simple educational formation in technique and the rules of genius into the contrast between mechanical motion and organic life. The Königsberg visionary Johann Georg Hamann joked to Kant in 1759 that his 'conversion' from the materialism of the Enlightenment to Lutheran pietism had transformed him into 'a genius. And that role is one I was born for.' Hamann was deeply affected by German translations of Young's *Conjectures* in 1761, and argued in his *Aesthetica in Nuce* (*Aesthetics in a Nutshell*) (1762), the fundamental text for subsequent German Romantic reflections on creativity and aesthetics, that 'the earliest nourishment came from the plant realm'. Citing Francis Bacon, Hamann argued for the relation of botany and the 'poetic art': 'In Nature we have nothing but a confusion of poems . . . the scholar's task is to collect them; that of the philosopher, to expound them; the poet's humble part is to imitate them or even more audaciously to bring them into order.'[13] These contrasts of mechanism and vitalism, the connection between genius and natural power and the opposition of instruction and discovery were the dominant means through which the calling of the natural philosopher was analysed.

Burke and his admirers held that there were bands of self-styled enlightened philosophers whose sinister associations masked silent plots to subvert established order. There was nothing to choose between the radical savants and the enthusiast mob. New enlightenment was but old illumination. Astrology, mesmerism, alchemy, the Eleusinian mysteries, electrotherapy and prophecy all became linked to the radical cause. Such movements suggested targets for conservative comment: millenarians and prophets who foresaw an imminent change in the civil and moral order; radical physicians who appealed to a theory of genius as the result of purely mechanical operations in the mind in order to change humanity; and rationalist metaphysicians who applied the principles of their philosophy to the reconstruction of the state. Future prospects were ludicrously glorious and based on a risible natural philosophy: the anti-Jacobin wits claimed that the radicals hoped to raise man 'to a rank in which he would be, as it were, *all* MIND; would enjoy unclouded perspicacity and perpetual vitality; feed on *oxygene*, and never die, but *by his own consent*'.[14]

The apostles of reason often seemed to be tools of unreason. In 1796 Burke picked on natural philosophers whose 'dispositions' made them 'worse than indifferent about those feelings and habitudes which are the support of the moral world. Ambition is come upon them suddenly; they are intoxicated with it.' Sometimes, as in the satires directed against the radical physician Thomas Beddoes's use of gaseous medicines in Bristol in the 1790s and the 'pneumatic revelries' of his friends, the 'intoxication' became literal. When, in 1799, the young Humphry Davy supposed that pneumatic chemistry as pursued by Beddoes and himself, 'in its connection with the laws of life, [would] become the most sublime of all sciences', he intended to point out the aesthetics and significance of his chosen vocation.[15] The mental habits of these men were viewed as the cause of their threatening policy. Burke said that 'the heart of a thoroughbred metaphysician comes nearer to the cold malignity of a wicked spirit than the frailty and passion of man. It is like that of the Principle of Evil himself, incorporeal, pure, unmixed, dephlegmated, defecated evil.' The intellectuals' overestimate of their own mental capacity was accompanied by the bestialization of humanity: 'They find themselves possessed of faculties which enable them to speculate and to discover; and they find that the operation of those faculties is quite unlike the things which they contemplate by their means.' The consequence was reformist arrogance and corrupted politics: 'they feel a satisfaction in this distinction'.[16]

These distinguished minds were seen as 'Illuminati'. The Masonic conspiracies spread through Europe provided the appropriate model with which to analyse the Revolution. 'The detestable doctrines of Illuminatism have been preached among us', claimed the Edinburgh natural philosophy professor John Robison in his *Proofs of a Conspiracy against all the Religions and Governments of Europe* (1794).[17] Their revolution followed from a false natural philosophy. Burke compared bad policy with bad cosmology: 'to destroy any power growing wild from the rank productive force of the human mind, is tantamount, in the moral world, to the destruction of the apparently active properties of bodies in the material'. The radicals were compared with wily impresarios, cunning magicians, dissolute gamblers. Burke sought a corrective aesthetics of the state: 'to make us love our country, our country ought to be lovely'. The conspirators could not hope to understand or to use simplistic mechanical philosophy to make models of these aesthetic judgements. The Jacobin theatre of politics was a world of illusion and crude spectacle. Their philosophical supporters were no better than wizards, making use of 'poisonous weeds and wild incantations'. It was scarcely surprising that deluded natural philosophers found such allies congenial. Since the state was a 'body politic' a failure to understand the 'true genius and character' of any natural body would lead to a failure to understand proper politics. Burke argued that 'the deceitful dreams and visions of the equality and rights of men' would end in 'an ignoble oligarchy' formed of coteries of rootless men.[18] The habit of association was a natural consequence of the reformers' intellectual position. Robison listed the 'Corresponding – Affiliated – Provincial – Rescript – Convention – Reading Societies' as British manifestations of this habit. Their light was derived from the wrong use of the inquiring mind. 'We see that it is a natural source of disturbance and revolution', wrote Robison; '*Illumination* turns out to be worse than darkness.'[19]

Popular usage, spurred on by the conspiracy theories of writers such as Robison and the abbé Barruel, sanctioned these connections between genius,

illumination and revolutionary magical science. A Lutheran minister, Peter
Will, supplied London readers with translations of the spiritualist works of
the Swiss physiognomist and pastor Johann Lavater. Lavater declared that
'each man is a man of genius in his large or small sphere. He has a certain circle
in which he can act with inconceivable force.'[20] His career inspired Goethe's
remarks on the epidemic of 'genius': 'every talent which rests on a decided
natural gift seems to have something of magic about it'. Goethe argued that
Lavater's real gift for divination of private character became transformed into
a radical spiritualist programme, akin to the mesmerism Lavater admired,
which promised his devotees a delusory 'general distribution of mental
gifts'.[21] Political campaigners who aimed at power through new techniques
for inspecting the mind also advocated radical change in the social order.
Radical performers in the 1790s sought to emulate Lavater's amazing tri-
umphs. Romantic theorists after 1800, notably August Schlegel, would seek
an idealized science of physiognomy as the basis of a new epistemology and a
new critical theory.[22] In popular literature these ideals were especially obvious. Will also translated works such as the naturalist and journalist Karl
Grosse's *Der Genius* (1791–5), given the striking English title *Horrid Mysteries*. This book inspired the young Ludwig Tieck with a fit of divine madness; it
was the novel which Thomas Love Peacock imagined Shelley kept under his
pillow.[23] Grosse's story was an account of the sinister workings of 'genius' as
Illuminatism, an international philosophical conspiracy which could perceive
the darkest purposes 'in order to correct them'. Commentators such as Will
and Adolf von Knigge explained the connection between subversion and
genius. Through illumination 'a spirit of philosophical investigation began'
and 'many great geniuses were raised from mental lethargy'. But since 'only
inferior geniuses' could be led by such powers and 'those that are endowed
with superior gifts are spoiled, degenerate, are misguided, or rule over the rest
at the expense of their fellow associates', the Revolution inevitably ensued.[24]
Goethe summed up: 'at this time genius was thought to manifest itself only by
overstepping existing laws'. Critical philosophy set the task of establishing the
relationship between rule and the excessive power of creative genius so as to
construct a proper account of the work of science. Peacock's satire hit the
mark when he envisaged Shelley reflecting upon the doctrines of 'the sublime
Kant' and the 'Illuminati', who selected 'wisdom and genius from the great
wilderness of society'. Citing Kant's third *Kritik* (1790), Goethe wrote that
'the time was yet distant when it could be affirmed that genius is that power of
man which by its deeds and actions gives laws and rules'.[25]

   In the third *Kritik* Kant analysed the natural philosopher's role in terms of
the concept of 'genius' as excessive natural power. The work was addressed to
the central issue of the relationship between the purposive in the organic
realm, spuriously attributed to some innate natural power, and the judgement

of art, spuriously based on a notion of unconstrained creativity. The book was composed as a direct result of Kant's fight with Forster in 1786–7 on the origin of races, involving the rejection of Forster's vital materialist account of spontaneous generation of living forms.[26] Where Burke attacked Jacobin materialists by pointing out their unwarranted claims to genius and the mechanization of mental power, Kant attacked Forster by connecting his account of power with that of erroneous classical aesthetics. He admired Burke's aesthetic system but convicted it of reliance on the empirical physiology of bodily fibres, so impotent against enthusiasm's diseases.[27] In the 1760s and 1770s, developing his account of genius and taste, Kant targeted figures such as Hamann, Lavater and Emmanuel Swedenborg as representatives of this *Schwärmerei* (fanaticism). Mystical language and deluded aesthetics characterized this wave of passionate philosophizing about the powers of mind and body. Only a rectified account of the philosophical mind would cure this disease. In a celebrated 1766 attack on Swedenborg, *Träume eines Geistersehers* (*Dreams of a Spirit Seer*), and in subsequent writings on the spirit realm, Kant argued repeatedly that there were no means in rational natural philosophy by which the immaterial could be experienced as part of sensible nature. He told the philosopher Moses Mendelssohn in 1766, with respect to his attack on Swedenborg, that 'the total extermination of all these chimerical insights would be less shameful than the dream science itself, with its confounded contagion'.[28] 'Dreamy' claims that such means existed and had been realized in theatrical performances were characteristic of the natural philosophies of radical mesmerists and naturalists alike.

Between 1770 and 1790 Kant refined his account of the genius of the natural philosophical performer. In the 1770s he allowed that Newton's original cosmological work, for example, was an act of genius. By 1790 he denied this: 'we can readily learn all that Newton has set forth . . . however great a mind was required to discover it'. The replicability of natural philosophical work distinguished it from the achievements of excessive creative power. In his 'analytic of the sublime', reworking received views on genius and the power of rules, Kant argued that '*genius* is the talent (or natural gift) which gives the rule to art'.[29] The canon of works of genius could lead to the exemplification of a natural rule which could then become part of a curriculum. But genius itself could not be imitated. Hence proper science could not be the work of genius. This conclusion may be compared with that of the radical divine and natural philosopher Joseph Priestley. Priestley and Kant agreed that Newton's natural philosophy was not the work of genius. But Priestley argued in 1767 that 'an opinion of the greater equality of mankind in point of genius would be of real service in the present age'. Newton only seemed a genius because he had laid out his work deductively. In contrast, Kant was no empiricist and no leveller. He implied that the deductive structure of New-

ton's work was just what differentiated his work from genius.[30] The implication was that proper natural philosophy was necessarily part of pedagogy;
inimitable showmanship, and excessive power, were alien to the spirit of the
rule-governed sciences.

Kant's contrast between the genius of art and the authority of natural
philosophy was both compelling and hotly debated. In the closing passages of
his *System des transcendentalen Idealismus* (*System of Transcendental Idealism*) (1800) Schelling argued that 'only what art brings forth is simply and
solely possible through genius'. In the sciences discovery was possible through
genius but never necessary. Kepler's discovery of planetary mechanics was an
act of genius; Newton's was not. The signs of such a creative act were
moments where a discoverer developed claims whose meaning was clarified
only subsequently, just as Newton revealed what must have been present
'unconsciously' with Kepler.[31] This assessment of the relative 'genius' of
Kepler and Newton became a Romantic commonplace, expressed forcefully,
for example, by Schelling's disciple S. T. Coleridge in his *Biographia Literaria*
(1817) and elsewhere. Coleridge's commentary on Schelling's views helped re-
emphasize the novelty of the claim that 'there is a philosophic, no less than a
poetic genius', and, even more importantly, that this genius would be able to
divine the basis of mental power through its historical inquiry into the basis of
natural power.[32] Scientific genius emerged from unconsciousness to consciousness through the sciences' history. This made Romantic aesthetics of
the sublime the key to the status of genius in natural philosophy.

For Schelling, 'the obscure concept of *genius*' was identified with unexpected goals of inquiry and work, attributed by the agent to the overwhelming
power of destiny lying above and beyond it. He referred here to Kant's
account of the sublime in the third *Kritik*. Kant had argued 'of nature regarded
as might' that such an overwhelming power was not due to objects in nature
but to the act by which the witness became conscious of his superiority to
nature. 'Clouds piled up in the sky, moving with lightning flashes and thunder
peals; volcanoes in all their violence of destruction', could 'raise the energies
of the soul above their accustomed height and discover in us a faculty of
resistance of a quite different kind'. This 'faculty of resistance' became a
proper object of study.[33] Jacobin natural philosophers interpreted mental
powers, including spiritual communication and inspired vision, in terms of
the natural powers of meteorological catastrophe and mundane spectacle.
'The "inborn spirits" that held sway in nature and produced the indestructible
qualities have themselves become bits of matter which can be caught and
imprisoned in test tubes and retorts', Schelling sneered. Romantics sought to
interpret such natural phenomena through the powers whose effects they
elicited in the sensitive inquiring mind of the natural philosopher. Schelling
lectured in 1802 on the comparison between such natural philosophy and the

tragic struggle against fate: 'the mind's arduous efforts to gain insight into primordial nature and the eternal inner essence of its phenomena is no less sublime a spectacle'.[34] Natural philosophy, possessed of historical self-consciousness and directed at the action of the natural philosopher's own mind, could now become the work of genius.

This project was typified by the work of the group which gathered around Schelling in Jena from 1798, soon after the publication of his essay *Ueber die Weltseele (On the World Soul)*. It included Novalis (Friedrich von Hardenberg), Friedrich Schlegel, Henrik Steffens, Ludwig Tieck and the young natural philosopher Johann Wilhelm Ritter. When Ritter left Jena for Munich in 1805 a new nucleus formed including Schelling and the spiritualist philosopher Franz von Baader, who had reacted to Schelling's 1798 essay as 'the first enjoyable sign of the resurrection of natural philosophy'.[35] The coherence of these groups has been exaggerated, but the aim to treat 'physics as an art', the title of a lecture Ritter delivered in Munich in 1806, was fundamental.[36] Novalis promoted the project of an historicist and sublime natural philosophy in opposition to French materialism and the principles of Jacobinism. In his extraordinary manuscript *Die Christenheit oder Europa (Christianity or Europe)* (1799), composed after his reading of Burke, Novalis composed an eschatological history in which the successive Fall through Reformation, Enlightenment and Revolution prefaced a new golden age. Using 'the magic rod of analogy', rather than the spurious magics of Jacobins and mystics, Novalis invited the Germans to participate in 'a superior epoch of civilisation'.[37] His strategy was to compile a new *Encyclopédie* to rival that of Diderot. Novalis analysed the theory of 'physics as art' through the question of whether 'few men have genius in experimentation'. His conclusion was that to experiment was precisely to act as a genius: 'the true lover of nature plainly describes himself, through his skill in making, simplifying, combining, analysing, romanticising and popularising experiments . . . and in his artistically detailed description and representation of the observations'.[38]

Novalis made the routines of experimental natural philosophy into powerful resources for a sublime physics. Important practical techniques were those of the Wernerian mining school at Freiberg, where he had studied fom 1797 to 1799, and of galvanic animal electricity, which obsessed natural philosophers after 1791.[39] These were the central concerns of Ritter, whose posthumous *Fragmente* argued that 'Earth history is human history' and that 'man's anatomy and that of the body of the Earth and that of the greater human body are a unity'.[40] Experimentation as an investigation of the self was the key technique of Ritter's strategy. In his late work on the electrostatic figures initially produced by Georg Lichtenberg and Ernst Chladni, Ritter argued that his aim was 'to rediscover or else to find the original or *natural* script by means of electricity'. Spark tracks were treated as signs to be interpreted as part of human history. All these trials showed that 'we possess an inner sense – as yet not developed – for knowing the world. It does not see, nor hear, but it knows.'[41] In his early work (1796) with Alexander von Humboldt on galvanism, Ritter's concern was to identify the relationship between the polarity of the elements of a galvanic cell with his own physiological responses. Galvanic discharges were shown to have varying effects on his own eyes, ears and tongue, according to the polarity of the discharge. Humboldt joined in, applying electrodes to his back to raise painful blisters and using his own body as a pole through which to excite the motions of frogs.[42] During this set of trials he submitted to Schiller's journal the poem *Der Rhodische Genius* (*The Genius of Rhodes*), in which he metaphorically described the morbidity of chemical affinity and the separating 'genius' of vital force. Here 'genius' became identified with the power which the naturalist could recover and render visible and, simultaneously, with the power of his own experimental perception.[43]

This active investigation of natural power through such auto-experimentation is perhaps the most characteristic aspect of Romantic natural philosophy. The excessive power made visible in galvanism and in pneumatics could be connected with the powers of the experimenter's own presence. In Bristol, Humphry Davy administered nitrous oxide to himself and his fellows and carefully recorded the ecstatic results: 'I existed in a world of newly connected and newly modified ideas. I theorised, I imagined that I made discoveries . . . Nothing exists but thoughts! The universe is composed of impressions, ideas, pleasures and pains!' These researches prompted his remark that 'the laws of gravitation, as well as the chemical laws, will be considered as subservient to one grand end, perception.'[44] Whereas Davy would turn from introspective pneumatics to public lecturing on voltaic chemistry after his move to London in 1800, in Munich, Ritter, Schelling and von Baader worked hard on divination experiments with hand-held pendulums and rods. Davy made Ritter one of his chief butts.[45] But both men's trials were the subject of satire and scepticism from established natural philosophers, and in both cases the protagonists theorized their work in terms of the 'genius' which drove them to investigate these sublime phenomena. In his *Consolations in Travel*, Davy had his 'Unknown' persona teach that 'it suits the indolence of those minds which never attempt anything . . . to refer to accident that which belongs to genius'. His poem 'The Sons of Genius', submitted to Robert Southey's *Annual Anthology* in 1799 is just such an apologetic text. Davy's favoured theme of the flight of young birds appears here, as it does in his late verses on a soaring eagle and its young of the 1820s. Kant interpreted taste as 'the discipline of genius; it clips its wings'. Davy's aerial images, in which the soaring bird gazed back upon its offspring, were significant because they embodied the image of disciplined, emulative genius as the characteristic means by which natural philosophy was to be transmitted, yet they also allowed a creative reworking of the natural philosopher's vocation.[46]

Furthermore, both Davy and Coleridge, his self-appointed interpreter and conscience, tried hard to draw the line between true genius and the false fanaticism of the radical and deluded vulgar. In his *Biographia Literaria*, for example, Coleridge used exactly the same image as that which Forster had used to describe the Jacobin leadership: 'the swarming of bees'. But where Forster argued that the egalitarianism of the Revolutionary leadership was a virtue, Coleridge pointed out that 'the German word for fanaticism . . . is derived from the swarming of bees, namely . . . Schwärmerey', and used this etymology to explain his model of the 'self-sufficing power of absolute *Genius*', which stood over against such vapourings. Davy tried out the same politically loaded distinction. In his notebooks he recorded that 'the man of genius always feels more power than he is able to develop', and that hence the outstanding feature of genius was 'a contempt for public opinion'. But the historical sense was crucial for these disciples of intellectual privilege. Davy distinguished between contempt for popular fashion and respect for precedent and antiquity. It was precisely the power of sympathy with the course of history which allowed the true genius to liberate himself from public culture. This imaginative retrospection was a necessity throughout the Revolutionary period, especially for youthful radicals seeking their career, and more elderly savants who then needed to account for their own past excesses. Thus 'genius' allowed retrospection full, creative, rein. Historicism and heroism became central features of scientific pedagogy and public display. As professors within the reformed academies, and as performers on the new stages of nineteenth-century science, such savants used the principles of reflection upon the power of genius in nature and culture to authoritative effect. They aimed at new rules to replace those of classical taste and education. Coleridge reacted to Davy's work in electrochemistry as a foreshadowing of the time 'when all human Knowledge will be Science and Metaphysics the only science': it was Coleridge, famously, whose refusal of the title 'philosopher' to members of what he considered the plebeian British Association for the Advancement of Science in 1833 prompted the coinage of the very term 'scientist'.[47]

The impact of natural philosophical 'genius' was significant precisely because of its meaning for the vocation of the natural philosopher. It was this 'genius' which Mary Shelley evoked (1818) in her account of Victor Frankenstein's mortal choice of calling – at Ingolstadt, home of the Illuminati, her protagonist stood poised between the paths offered by a chemistry professor spouting Davy and the textual lore of Paracelsianism.[48] No doubt the ideology which Romantic sociologists developed in the aftermath of the Revolution was of immense significance for the construction of the scientific career.

Patterns of work such as the heroic privilege of discovery, the use of a disciplinary history as a means of legitimation of the division of labour in the sciences and the integration of laboratory teaching and lecture performance were all aspects of Romantic natural philosophy and its aftermath, the emergence of organized natural science.

The histories of the sciences which were composed by textbook writers in natural science in the later nineteenth century did not normally describe this connection between Romantic culture and the emergence of the scientist's role. They rejected Coleridge's claim that through this account of the 'genius' in the sciences the 'science of nature becomes finally natural philosophy, the one of the two poles of fundamental science'. However, an example from the career of one of the textbooks' heroes, Hans Christian Oersted, shows how the disciplinary histories and stories about scientific discovery emerged. In 1806 as a newly appointed professor at Copenhagen after his collaboration with Ritter and propagandizing for him in Paris, Oersted delivered a telling set of lectures on the history of chemistry. His argument was that the polar law of alternating states, displayed through the genius of Abraham Gottlob Werner in geology and through the genius of Ritter in electrochemistry, also applied to the historical development of the sciences. 'There are times when a multitude of great geniuses step forward at once', and these are followed by a period 'when the great ideas of the previous time are explained, arranged and determined'. Genius appeared in the sciences through 'the dread of that universal death which most powerfully stimulated the slumbering creative force'. He claimed that 'the development of the Earth was the same as that of the human mind', and that 'the development of science is itself part of natural science'. As he inaugurated his own work on 'electrical conflict', Oersted told his students that the laws of 'material' and 'spiritual' nature were identical.[49] Such professorial appeals made cultural politics an integral part of the nineteenth-century transformation of natural philosophy into natural sciences.

## NOTES

1   S. Shapin and S. Schaffer, *Leviathan and the Air-Pump: Hobbes, Boyle and the Experimental Life* (Princeton, 1985), pp. 129–31.
2   J. W. von Goethe, *Poetry and Truth*, ed. K. Breul, 2 vols. (London, 1913), II, pp. 285–6.
3   Goethe, *Poetry and Truth*, pp. 138, 286; J. Möser, *Patriotisches Archiv für Deutschland*, 4 (1786), 397–408, reprinted in H. Brunschwig, *Enlightenment and Romanticism in Eighteenth-Century Prussia*, trans. F. Jellinek (Chicago, 1974), p. 214.
4   J. L. McClellan, *Science Reorganized: Scientific Societies in the Eighteenth*

*Century* (New York, 1985), pp. 253–9; R. S. Turner, 'The Growth of Professorial Research in Prussia, 1818 to 1848: Causes and Context', *Historical Studies in the Physical Sciences*, 3 (1971), 137–82; J. Conrad, *The German Universities* (Glasgow, 1885), pp. 282–3; C. McClelland, *State, Society and University in Germany 1700–1914* (Cambridge, 1980); K. J. Caneva, 'Conceptual and Generational Change in German Physics: The Case of Electricity 1800–1846' (unpublished Princeton University Ph.D thesis, 1975), pp. 470–3.

5   I. Kant, *Critique of Judgement*, ed. J. H. Bernard (New York, 1951), § 46, p. 151.

6   J. Ernst, *Der Geniebegriff der Stürmer und Dränger und der Frühromantiker* (Zurich, 1916); O. Schlapp, *Kants Lehre von Genie und die Entstehung der Kritik der Urteilskraft* (Göttingen, 1901).

7   Fichte to Baggesen, April/May 1795, in J. G. Fichte, *Briefwechsel*, ed. H. Schulz, 2 vols. (Leipzig, 1925), I, p. 449; J. G. Fichte, *The Science of Knowledge*, ed. P. Heath and J. Lachs (Cambridge, 1970), p. 5.

8   L. H. Jacob, *Ueber die Universitäten in Deutschland* (Berlin, 1798), p. 255, translated in R. S. Turner, 'University Reformers and Professorial Scholarship in Germany 1760–1806', in L. Stone (ed.), *The University in Society*, 2 vols. (Princeton, 1974), II, p. 517.

9   F. W. J. Schelling, *On University Studies*, ed. N. Guterman (Athens, Ohio, 1966), p. 26.

10   U. Vogel, *Konservative Kritik an der bürgerlichen Revolution: A. W. Rehberg* (Darmstadt, 1972); J. S. Fichte, *Sämtliche Werke*, ed. J. G. Fichte, 8 vols. (Berlin, 1845–6), VI, pp. 184–5.

11   R. Darnton, *The Literary Underground of the Old Régime* (Cambridge, Mass., 1982), p. 39; G. Forster, 'Pariser Umrisse' ('Parisian Sketches') (21 November 1793), in his *Werke*, ed. G. Steiner, 4 vols. (Frankfurt, 1970); compare Udo Kürten, 'From Theory to Practice: Georg Forster and the French Revolution', in Francis Barker, *et al.*, (eds.), *1789: Reading Writing Revolution* (Colchester, 1982), pp. 209–17.

12   Sir Joshua Reynolds, *Discourses on Art*, ed. R. R. Wark (London, 1966), pp. 89, 179–80; A. Gerard, *Essay on Genius* (1774), ed. B. Fabian (Munich, 1966), pp. 60–4; M. H. Abrams, *The Mirror and the Lamp: Romantic Theory and the Critical Tradition* (Oxford, 1953), pp. 198–213.

13   J. G. Hamann, 'Aesthetica in Nuce' (1762), in his *Sämtliche Werke*, ed. J. Nadler, 6 vols. (Vienna, 1949–57), II, pp. 198–9; for Hamann on Edward Young, see p. 200; Hamann to Kant, 27 July 1759, in *Kant: Philosophical Correspondence 1759–99*, ed. A. Zweig (Chicago, 1967), p. 35. Compare R. G. Smith, *J. G. Hamann: a Study in Christian Existence* (London, 1960), p. 197; F. C. Beiser, *The Fate of Reason: German Philosophy from Kant to Fichte* (Cambridge, Mass., 1987), pp. 33–7.

14   C. Garrett, *Respectable Folly: Millenarians and the French Revolution in France and Britain* (Baltimore, 1975): *Poetry of the Anti-Jacobin*, ed. C. Edmonds (London, 1890), p. 147.

15   E. Burke, *Letter to a Noble Lord* (London, 1796), in R. Hoffman and P. Levack (eds.), *Burke's Politics* (New York, 1949), pp. 532–4; D. Stansfield, *Thomas Beddoes* (Dordrecht, 1984), ch. 7; Humphry Davy, 'An Essay on Heat, Light and the Combinations of Light' (1799), in his *Collected Works*, ed. J. Davy, 9 vols. (London, 1839–40), II, p. 86.

16   Burke, *Letter to a Noble Lord*, p. 534; Burke, *Reflections on the Revolution in France* (London, 1910), p. 6.
17   J. Robison, *Proofs of a Conspiracy against all the Religions and Governments of Europe* (3rd edn, London, 1798), p. 481.
18   Burke, *Reflections*, pp. 93, 178, 190, 223.
19   Robison, *Proofs*, pp. 431, 479.
20   J. C. Lavater, *Essays on Physiognomy* (1775–8) (7th edn, London, 1850), p. 415.
21   Goethe, *Poetry and Truth*, pp. 284, 286; Lavater, *Essays*, p. civ.
22   Brunschwig, *Enlightenment and Romanticism*, p. 187; A. Gode-von Aesch, *Natural Science in German Romanticism* (New York, 1941), pp. 223–5.
23   K. Grosse, *Horrid Mysteries* (1791–5), ed. D. P. Varma (London, 1968), pp. xiii–xiv; T. L. Peacock, *Nightmare Abbey* (1818), ed. R. Wright (Harmondsworth, 1969), p. 47.
24   Grosse, *Horrid Mysteries*, pp. xv–xvii.
25   Goethe, *Poetry and Truth*, p. 286; Peacock, *Nightmare Abbey*, p. 47.
26   E. Riedel, 'Historizismus und Kritizismus', *Kant-Studien*, 72 (1981), 41–57; Beiser, *Fate of Reason*, pp. 153–8.
27   Kant, *Critique of Judgement*, § 29, pp. 118–19.
28   Kant, *Dreams of a Spirit Seer illustrated by Dreams of Metaphysics*, trans. E. F. Goerwitz (New York, 1900); Kant to Mendelssohn, 8 April 1766, in *Philosophical Correspondence*, p. 55; Kant, *Religion within the Limits of Reason Alone*, trans. H. H. Hudson and T. M. Greene (New York, 1960), p. 162; R. E. Butts, *Kant and the Double Government Methodology* (Dordrecht, 1984), pp. 70–98, 291–7.
29   Kant, *Critique of Judgement*, § 47, pp. 150–1.
30   J. Priestley, *The History and Present State of Electricity* (London, 1767), p. 574; *The History and Present State of Electricity* (3rd edn, London, 1775), II, pp. 167–9.
31   F. W. J. Schelling, *Sämtliche Werke*, ed. K. F. A. Schelling, 14 vols. (Stuttgart, 1856–61), III, p. 623, translated in *System of Transcendental Idealism*, ed. P. L. Heath (Charlottesville, Va., 1978), p. 227.
32   S. T. Coleridge, *Table Talk and Omniana* (Oxford, 1917) [8 October 1830]; Coleridge, *Biographia Literaria* (1817), ed. J. Engell and W. J. Bate, 2 vols. (Princeton, 1983), I, pp. 92, 299.
33   Schelling, *Sämtliche Werke*, III, pp. 613–14, and *System*, p. 222; Kant, *Critique of Judgement*, § 28, p. 100.
34   Schelling, *On University Studies*, pp. 124, 132.
35   W. D. Wetzels, *Johann Wilhelm Ritter, Physik im Wirkungsfeld der deutschen Romantik* (Berlin and New York, 1973), pp. 15–17, 30–1, 48–9, and 'Aspects of Natural Science in German Romanticism', *Studies in Romanticism*, 10 (1971), 44–59; F. X. von Baader, *Sämtliche Werke*, ed. F. Hoffman, 16 vols. (Leipzig, 1851–60), III, p. 249.
36   J. W. Ritter, *Die Physik als Kunst* (Munich, 1806), esp. pp. 57–9. See Erich Worbs, 'Novalis und der schlesische Physiker Johann Wilhelm Ritter', *Aurora*, 23 (1963), 85–92; B. Gower, 'Speculation in Physics: the History and Practice of Naturphilosophie', *Studies in History and Philosophy of Science*, 3 (1973), 327–39.
37   *Novalis Schriften*, ed. P. Kluckhohn and R. Samuel, 4 vols. (Stuttgart, 1960–75), III, pp. 518–19.

38  Novalis *Schriften*, III, p. 256 (my translation); L. E. Wagner, *The Scientific Interests of Friedrich von Hardenberg* (Ann Arbor, 1937); J. Neubauer, *Bifocal Vision: Novalis' Philosophy of Nature and Disease* (Chapel Hill, 1971), ch. 4.

39  O. Wagenbreth, 'Werner–Schüler als Geologen und Bergleute', in *Abraham Gottlob Werner Gedenkschrift* (Leipzig, 1967), pp. 163–78; A. Ospovat, 'Romanticism and German Geology', *Eighteenth-Century Life*, 7 (1982), 105–17; A. Hermann, *Die Begründung der Elektrochemie und die Entdeckung der ultravioletten Strahlen von J. W. Ritter* (Frankfurt-on-Main, 1968).

40  J. W. Ritter, *Fragmente aus dem Nachlasse eines jungen Physikers*, 2 vols. (Heidelberg, 1810), II, p. 37.

41  Ritter, *Fragmente*, II, pp. 204, 230 (the discussion of Lichtenberg's figures is on pp. 225–69); J. W. Ritter to H. C. Oersted, 31 March 1809, in M. C. Harding, ed., *Correspondence de H. C. Oersted*, 2 vols. (Copenhagen, 1920), II, pp. 223–4. Compare Wetzels, *Ritter*, pp. 87–90; D. von Engelhardt, 'Mesmer in der Naturforschung und Medizin der Romantik', in H. Schott (ed.), *Franz Anton Mesmer und die Geschichte des Mesmerismus* (Stuttgart, 1985), pp. 97–8.

42  F. A. von Humboldt, *Versuche über die gereizte Muskel und Nervenfasser* (Berlin, 1797); K. E. Rothschuh, 'Alexander von Humboldt und die Physiologie zeiner Zeit', *Sudhoffs Archiv*, 43 (1959), 97–113.

43  Gode-von Aesch, *Natural Science*, p. 193n.

44  H. Davy, 'On Heat, Light, and the Combinations of Light', in T. Beddoes (ed.), *Contributions to Physical and Medical Knowledge, principally from the West of England* (Bristol, 1799), reprinted in H. Davy, *Collected Works*, ed. J. Davy, 9 vols. (London, 1839–40), II, p. 85; *Researches Chemical and Philosophical chiefly concerning Nitrous Oxide* (London, 1800), reprinted in Davy, *Works*, III, pp. 289–90.

45  Wetzels, *Ritter*, pp. 52–3; L. W. Gilbert, 'Einige kritische Aufsätze über die in München wieder erneuerten Versuche', *Annalen der Physik*, 26 (1807), 369–449; Ritter to Oersted, 20 April 1807, in Harding (ed.), *Correspondence*, II, pp. 192–206.

46  Davy, *Consolations in Travel* (1830), in Davy, *Works*, IX, pp. 354–5; 'Sons of Genius' (c. 1795) and 'The Eagles' (c. 1822), in 'Memoirs of the Life of Sir Humphry Davy', in Davy, *Works*, I, pp. 23–7, 279. See D. M. Knight, 'The Scientist as Sage', *Studies in Romanticism*, 6 (1967), 65–88; T. H. Levere, 'Humphry Davy, "The Sons of Genius" and the Idea of Glory', in Sophie Forgan (ed.), *Science and the Sons of Genius: Studies on Humphry Davy* (London, 1980), pp. 33–58. Compare Kant, *Critique of Judgement*, § 50, p. 163.

47  Coleridge, *Biographia Literaria*, I, p. 30; Davy, *Works*, I, p. 151; S. T. Coleridge to Dorothy Wordsworth, November 1808, in *Coleridge, Collected Letters*, ed. E. L. Griggs, 6 vols. (Oxford, 1956–71), III, p. 38; T. H. Levere, *Poetry Realized in Nature: Samuel Taylor Coleridge and Early Nineteenth-Century Science* (Cambridge, 1981), p. 73.

48  M. Shelley, *Frankenstein* (1818), ed. M. Hindle (Harmondsworth, 1985), pp. 83, 95.

49  Coleridge, *Biographia Literaria*, I, p. 257; H. C. Oersted, *The Soul in Nature* (London, 1852), pp. 321–4; R. C. Stauffer, 'Speculation and Experiment in the Background of Oersted's Discovery of Electromagnetism', *Isis*, 48 (1957),

33–50; L. P. Williams, 'Kant, *Naturphilosophie* and Scientific Method', in R. N. Giere and R. S. Westfall (eds.), *Foundations of Scientific Method: The Nineteenth Century* (Bloomington and London, 1973), pp. 3–22.


# FURTHER READING

Abrams, M. H., *The Mirror and the Lamp: Romantic Theory and the Critical Tradition* (Oxford, 1953), esp. ch. 8, on British and German critical theory of originality in composition
Becker, G., *The Mad Genius Controversy* (Beverly Hills, 1978)
Brunschwig, H., *Enlightenment and Romanticism in Eighteenth Century Prussia*, trans. F. Jellinek (Chicago, 1974), ch. 11, 'The Interpretation of Life as a Miracle', and ch. 12, 'The Revolt of the Intellectuals'
Engell, J., *The Creative Imagination* (Cambridge, Mass., 1981) and de Porte, M. V., *Nightmares and Hobbyhorses* (San Marino, 1974) for studies of enthusiasm and imagination in the arts
Ernst, J., *Der Geniebegriff der Stürmer und Dränger und der Frühromantiker* (Zurich, 1916), the basic study of the genius concept in this period
Grappin, P., *La Théorie du génie dans le préclassicisme allemand* (Paris, 1952), which covers the debates on creativity up to and including Hamann, Herder and Goethe
Honour, H., *Romanticism* (Harmondsworth, 1979), ch. 7, 'The Artist's Life'
Knight, D. M., 'The Scientist as Sage', *Studies in Romanticism*, 6 (1967), 65–88, on Davy and Oersted
Yeo, R. R., 'Genius, Method and Morality: Early Victorian Images of Sir Isaac Newton', *Science in Context*, 2 (1988), for a detailed study of the uses of genius in scientific biography.

# PART II

SCIENCES OF THE ORGANIC

# Doctors contra *clysters and feudalism:* *the consequences of a Romantic* *revolution*

NELLY TSOUYOPOULOS

## THE TRADITION

> But always the filthiness, that lies a little deeper . . .
> T. S. Eliot, *Four Quartets.*

Who has not experienced in his life that unutterable feeling of satisfaction which is due to a relaxed condition of the bowels? Poetry and philosophy do not ordinarily pay much attention to this phenomenon, although it is so fundamental to human character. But physicians understand better the secret undercurrents of life and have always been sensitive to the necessary cleansing of an organism of its impurities! For more than two centuries, the 'natural cloaca' of the body was the most important object of medical treatment and the point of inspiration in therapeutic advance.

That is why, at the turn of the eighteenth to the nineteenth century, the so-called 'antigastric' method was the most famous and most widely accepted method in European, especially in German, therapeutics:

I would never underestimate the great benefits of the *antigastric* method . . . I believe that the cleansing of the 'main path' must be the first and most important step of medical practitioners and that this 'natural cloaca' is indeed in most cases the convenient path for artificial cleansing. I only want to warn of abuse which is now usual.[1]

These are the words of the young doctor Christoph Wilhelm Hufeland (1762–1836), written at the beginning of his career, the words of a physician who soon became one of the most famous and representative personalities of the medical profession.

The antigastric method lacked neither medical tradition nor theoretical foundation. It was derived from a general pathological principle, according to which diseases are caused by impurities which are produced in the fluids of the body. This pathological principle was believed to be the consequence of the medical theory known as the 'humoral system' which originated in Hippocratic and Galenic medicine. An essential point of the old humoral system was the assumption that the circulating fluids (humours) of the body, their quality, quantity and mixture, are basic for life, a measure for all physiological and pathological functions.

But in the West this system was much criticized and, after the seventeenth century, it was abandoned and replaced by other systems, supposed to meet the new standards of scientific investigation, systems like atomism and mechanism. But while the systems were changing and rapidly succeeding one another, the pathological principle of the 'dangerous impurities' remained stable for all systems, even those contradicting one another. The antigastric method corresponded to this principle of pathology: emptying and cleansing the body of impurities, freeing the organism of its own poisonous products, avoiding putridity. The practice of medicine mainly consisted of a rich variety of cleansing techniques such as laxatives, vomitives, purgatives, blood-letting, blood-leeches, cupping glasses, vesicatories (plaster for raising blisters on the skin), blood-cleaning decoctions, stool-suppositories and – of course – clysters (liquid enemas administered by squirting into the rectum), for which the inventive fantasy seemed unlimited (see fig. 3).

The idea of the human body as producing uncleanness, leading to putridity and death, was such a general conviction that the 'cleansing techniques' were obligatory treatment for every disease. This monotonous, coercive proceeding is the most astonishing characteristic of the antigastric therapeutics. Variety of diseases existed in name only: tuberculosis, or mania, hypochondria or smallpox, all were, in fact, reduced to impurity which had to be cleansed. The technique used and the frequency of application were the only criteria which separated a bad practitioner from a good one. Some examples of this therapy should make this clear.

## A case of mania

A young man, following a severe illness of bilious fever, was suffering from headaches, sleeplessness and depression. On 24 January 1796 this young man was suddenly taken with mania, so that he had to be restrained. The family asked the *Wundarzt* of the small town, where the patient was living, to visit him. The healer came and prescribed blood-letting at once and an emetic for the following day. But as the patient did not show any signs of recovery, the family sent for the physician from Weimar (a journey of five hours), who was the director of the lunatic-asylum of the town. This man at first gave his advice without visiting the ill person. He ordered one blood-letting, various emetics and a mustard-plaster to be applied to the legs as well.

But as the treatment did not help, the physician came to visit the patient himself:

3   The 'steam-clyster', invented by the court physician in Hanau, Johann
    Kämpf. From *Neues Magazin für Ärzte*, ed. E. G. Baldinger (Leipzig, 1779).

On the second of February, I visited him myself and, indeed, I found him in a terrible
state . . . He was bound by the hands and feet and he was speaking continuously, mainly
about geometry . . . I told his brother to forbid him to speak and as the patient did not
obey I told him to give him a strong box on the ear. This occurred and the afflicted
person was silent for five minutes . . .

The physician ordered the *Wundarzt*, who was also present, to administer the
following treatment to the patient: 'some clysters with mild soap and water,
which had indeed the expected efficacy . . . I also ordered five or six blood
cupping glasses around the neck and then I left.' This treatment appeared to be
successful, for the patient was very quiet and the *Wundarzt* seemed satisfied
when he informed the physician in Weimar:

The clysters seem to have had excellent results . . . The patient looks a little tired,
obviously from the last mixture of the clysters which made him run often for
relaxation. I hope we have defeated this disease![2]

Nevertheless clysters did not always result in making the patients quiet, as
they did in the case of the young maniac.

### *A case of acute disease: the story of a strange fever*

A girl, eleven years old, who had always been weak and pale, was suddenly ill.
On 12 January 1795 the child complained of chills and headache and she went

to bed earlier than usual. During the night, she felt very feverish. On the second day, the symptoms were the same, and the family called the doctor. He came and immediately ordered a purgative (sal ammoniac with rhubarb tincture), a clyster and foot-bath. By the third day the fever had diminished but the child began to vomit. The doctor again ordered a purgative, a clyster and a vomitive. On the fourth day, the fever was again high and the child vomited again and again. The doctor ordered another purgative which she immediately regurgitated and after that he ordered 'a vomitive of six grains of ipecacuana and after the first effective result, I repeated the same dose'.

The same symptoms occurred on the following days and the same treatment was ordered, mainly purgatives and clysters.

On the eighth day the situation was at its worst:

Sadness reigned the whole day. Because she had no stool, I gave her Epsom-salts repeatedly, in small doses . . . At one o'clock she had convulsions of the face and the right arm and leg . . . then, for a quarter of an hour she lay quiet, without critical symptoms . . . But then, all of a sudden, convulsions broke out again, she was deadly pale, foam came out of her mouth and she was dead![3]

Even the *Lebenskraft*, the metaphysical principle of German vitalism, was connected with the impurities of the fluids in explaining several diseases, for example hypochondria:

The most important cause of hypochondria is the deficiency of the *Lebenskraft* . . . in the digestive organs. Deficiency of the *Lebenskraft* of the liver causes a corrupt bilious secretion. This bilious impurity is, for a great number of hypochondriacs, the cause of their troubles.[4]

The enormous anxiety over impurities was the result of considering the body as a lifeless machine which could, at any time, be destroyed by its own waste products.

## THE REVOLUTION

The dance along the artery
The circulation of the lymph
Are figured in the drift of stars

T. S. Eliot, *Four Quartets*

During the last decade of the eighteenth century (*circa* 1798) a widespread movement towards a fundamental reform of medicine arose in Germany. The movement found an enthusiastic reception very quickly even in circles outside the medical profession, especially among the Romantic poets and philosophers like Novalis (Friedrich von Hardenberg) and Schelling. Finally, the interactions between the new medical ideas and the *Naturphilosophie* of Schelling gave the movement its theoretical principles.

Most important, but usually overlooked, are the practical aspects of the movement. In fact, it was part of the general movement against the feudal structure of sociopolitical life which arose in Germany at the turn of the eighteenth to the nineteenth century. The medical movement was also an effort to abolish the feudal structure of the medical system. The rise of the movement, its special aims, its quick success at the beginning, its suppression and degeneration after 1815, are phenomena reflecting the simultaneous sociopolitical developments.

This fact is not surprising if we recognize that the medical system was one of the most firmly established institutions of the feudal system in Germany. The main characteristic of the system was the existence of several classes of healers. According to feudal medical law (already in the medical edict of *Churfürst* Friedrich Wilhelm in 1685), medical healers were divided into two main classes. One class consisted of the so-called *medici*. These were doctors who had a university degree. To the second class belonged practitioners who had not attended university but some other practical school. These were not obliged to have completed their schooling but mainly to have practical experience and to be examined by an authority on the subject for which they wanted to have a licence. This class of practitioners had several subdivisions: *Wundarzt* first class or town-*Wundarzt*; *Wundarzt* second class or country-*chirurgus*; *Bademeister*; and barber.

All these were professionals. They were licenced healers and as such protected by law from competition. Feudal elements of the system are first of all the criterion of division of these healers: it was not specialization or some other medical consideration but the necessity to divide the power of the profession between the universities and the guilds. According to the corporations law the *Wundarzt*, the *Bademeister* and the barber were subject to the power and the rights of the guilds. This fact had many consequences for the medical care of the time. According to the law, the medical competence of the *medici* was that of 'internal cure', whereas the 'external cure' should stay exclusively in the hands of the other practitioners. The terminology 'internal cure', 'external cure', which was introduced by the medical edict of 1725 (enacted by the king of Prussia, Friedrich Wilhelm I) did not correspond to a specialization in general medicine or surgery – as it is usually explained – and it was, in fact, only effective in cases in which the rights of the one class had to be limited or protected with regard to the other class. If a doctor of the university, for example, who formally had the right to practise anywhere but only for 'internal cure', decided to settle in a small or middle-sized town, where a *Bademeister* or barber was established, he had to be careful not to enter the domain of competence of the other practitioners.

The feudal element consisted not in the existence of healers with varying educational backgrounds and degrees (a kind of hierarchy according to education, degree and function exists even today in the modern hospital) but in the fact that different kinds of healers with different educational backgrounds and different degrees were destined for the different social classes. A very important criterion of social differences in feudalism is the size and kind of the town or village where one has to live. This was also the most important criterion for the quality of medical care (more decisive than the financial situation of the patient). For example, the poor of the great towns (with over 10,000 inhabitants) could generally have very good medical care. The poorest of these towns were taken care of by the community. The community physician had to be a doctor of medicine. In most cases he was one of the best, because the position was considered to be honourable. The hospitals of these towns were designated for the journeymen and the house servants who belonged to the poorest of the citizens. The treating physicians of the hospitals were also doctors of the universities and educated surgeons. In Bamberg, for example, the hospital, built in the 1790s, was considered to be one of the best of its time. The poor patients were always treated by famous doctors, such as Adalbert Friedrich Marcus, Andreas Röschlaub and Franz von Walther. Marcus was at the same time the physician-in-ordinary of the sovereign prince-bishop of Ethal.

In the middle-sized and small towns, on the other hand, where the guilds were very powerful, doctors had no chance. These were the domain of the *Wundarzt*, who usually was a *Bademeister*. The countrymen also were in the hands of the guilds, medically speaking. The numerous villages in the country constituted the domain of the country-chirurgus, who was usually a barber. In their domains these medical persons were healers for all cases: general medicine, surgery, dentistry and obstetrics, with no risk of being accused of quackery. According to the law, they had the right to treat all diseases (also the 'internal cure' if 'no other medical person could be reached in a distance of one hour').[5]

The new medical movement demanded a fundamental change in the established medical system: uniform medical care for all people. This should begin with uniform education of all physicians and the abolition of the several classes of healers. For this purpose, first of all a reform of the education at the universities was necessary. Young doctors were not able to compete with the *Bademeister* in the small towns and villages, not only because of the power of the guilds but also because they had no practical experience during their studies at the universities. This situation had to be changed, which was possible only if the whole system of medicine changed. 'Single reforms of medicine are good for nothing. The real evil goes very deep and if we succeed in combatting it everything will change fundamentally. Medical practice will then be quite different from what it is today.'[6]

The theoretical knowledge, which the student learned at university, had to become really related to practical medicine, and this knowledge had to have an impact on practical medicine. For this purpose the doctrine of the Scottish doctor John Brown was the right support at the right moment. Contrary to all other systems, John Brown's doctrine promised to change pathology and therapy. More precisely, it promised to change the 'unchangeable' principle of 'impurities of the body' and, of course, the 'antigastric' method of therapy. This was a revolution! If a new doctrine was able within a short time to change traditional medical treatment, on which the existence of the medical professions (doctors as well as barbers) was based, if it changed the 'unchangeable' pathological principle, about which the most famous of the profession had written so many books, and for which the academies had awarded so many prizes; if this change really took place, that was truly a revolution!

Much has already been written about the theoretical basis of the movement[7]; so I shall give only a brief outline of the main principles necessary in order to understand the consequences of their realization. The theory originally came from Brown, but in Germany it was substantially changed. Its final form was a combination of Brown's doctrine, as interpreted by Röschlaub, with the theory of the German physiologist Blumenbach about the *Bildungstrieb* (formative drive), as interpreted by the philosopher Schelling. In medical terms it meant a combination of the theory of stimulation (excitability) with the theory of metabolism and regeneration. In other words, life is the ability of the organism to respond to stimuli of the environment with activities like assimilation and regeneration (self-reproduction). It is obvious, then, that the level of stimulation is responsible for health *and* disease, rather than the purity or impurity of the fluids. Consequently, medical treatment had to help the organism to regulate its defence system rather than cleanse it of poisonous matters, thus weakening its natural defence mechanisms and regulating powers.

Central to the new therapeutic orientation is the consideration of whether the organism is too weak to meet the challenge of the stimuli, and thus needs support, or if its reaction to the stimuli is already so strong that it can harm itself, and it therefore needs to be calmed.

These two main situations of lost balance are known as 'sthenia' and 'asthenia', according to Brown's terminology, but they were changed in their meaning through the interpretation of Röschlaub and Schelling: each of these situations, 'sthenia' and 'asthenia', expresses now a synthesis of active and passive vital forces which are in false proportion to one another, whereas the normal state is an ideal synthesis, a 'correct' proportion of activity and passivity determined by the rhythm of self-reproduction (regeneration) of each individual organism.

The terms 'sthenia' and 'asthenia' are very often misunderstood and interpreted by scholars as nosological simplification, that is to say, as an effort of Brownianism to reduce all diseases to only two. This is, however, not the case. 'Sthenia' and 'asthenia' are not intended to simplify or reduce the numerous nosological forms of disease. This can be seen very clearly if one examines the statistical tables of the hospital of Bamberg or of the Medico-Clinical Institute in Würzburg (fig. 2), the two main centres of Brownianism.[8]

In any case the concepts of 'sthenia' and 'asthenia' were not intended to abolish the old nosologies, but rather the traditional pathology and therapy. Of course, very few medical professionals were able to understand the new, complicated theory (just as with modern 'stress theory'). Its medical and scientific implications were understood much later. The majority of the profession understood in any case its practical consequences: the end of 'humoral pathology' and antigastric therapeutics. As such it was accepted or attacked and rejected.

The immediate and most obvious result of the movement was a sudden critical attitude of many doctors towards the cleansing techniques, which led to their actual limitation in therapy:

Everybody got tired of the prate of the so-called systems . . . and everybody was soon convinced that the 'old motley', as it existed at that time, was good for nothing . . . the *professores* as well as the *barbae tonsores*. They no longer wanted to begin a therapy with vomitives, laxatives and blood-letting. We began to consider the organism as something to be respected [and] to investigate exactly all harmful influences on it and to study its internal and external activities.[9]

With these words Adalbert Friedrich Marcus recalled the beginning of the reform movement, in which he, as one of the first partisans and supporters, had participated.

The change of general pathology and treatment is well reflected in the table issued by the 'Institute for Ill Journeymen' in Würzburg between the years 1786 and 1802. The 'Institute' was a clinic for the members of the guilds of artists and artisans and stood under the protection of the ecclesiastic prince-bishop Franz Ludwig of Ethal, who also was the founder of the hospital in Bamberg. Georg Adelmann, a physician of the 'Institute', edited the tables of the diseases in 1803. Adelmann commented on the tables of the diseases and spoke about the changes which had occurred during this period:

The so-called 'impurities of the first path' played, according to the tables, a very important role during the first years of the 'Institute', especially during the years 1786–1795, when the court physician Ehlen was doctor of the Institute . . . A change occurred when the doctors Thomann and Christian von Siebold were employed as directors of the Institute, during the years 1796–7 . . . The expression 'impurities of the first path' appeared for the last time in the year 1798. After this time Prof. Georg Nicolas Thomann alone was the director of the Institute . . . Prof. Thomann had never considered the 'impurities of the first path' as the cause of a disease.[10]

According to the theory of the 'impurities', it was usual to make additional use of the 'cleansing method' as a prophylactic cure. Such cures, especially when

| Nomen | Locus natalis | Vitae genus | Aetat. |
|---|---|---|---|
| Joseph. Arnknecht | Gerlachshausen | Studiosus | 18 |
| Michael Pharo | Bamberg | Mercenarius | 15 |
| And. Baumgaertner | Würzburg | Mercenarius | 44 |
| Antonius Breunes | Schallbach | Scriniarius | 21 |
| Jof. Rosenhoefer | Sternberg | Figulus | 18 |
| Georgius Wiefs | Stalldorf | Laminarius | 18 |
| Georgius Lederer | Würzburg | Scriniarius | 14 |
| Jofephus Pfaff | Zell | Famulus rufticus | 27 |
| Petrus Hildebrand | Rottau | Piftor | 19 |
| Andreas Zipfel | Heffelbach | Cerevifiarius | 28 |
| Nicolaus Karl | Unterdürrbach | Scriniarius | 29 |
| Johannes Hornung | Droffenfort | Coriarius | 20 |
| Nicolaus Schmitt | Ottweiler | Sartor | 44 |
| Michael Arnold | Gamburg | Scriniarius | 17 |
| Johannes Bauer | Koenigsberg | Faber lignarius | 36 |
| Petrus Schulz | Würzburg | Mercenarius | 58 |
| Georgius Senger | Bergtheim | Sartor | 19 |
| Michael Pfeiffer | Veitshöchheim | Hortulanus | 19 |

APRILI.

| Morbus | Dies Receptionis | [Dies] Dimissionis | Eventus curae |
|---|---|---|---|
| Scrofulae | 5 Aprilis | | |
| Peripneumonia fervofa | 8 Aprilis | 21 Aprilis | Sanatus |
| Catarrhus inflammator. | 9 Aprilis | 26 Aprilis | Sanatus |
| Rheumatifmus | 10 Aprilis | 15 Aprilis | Sanatus |
| Peripneumonia nervofa | 10 Aprilis | 30 Aprilis | Sanatus |
| Rheumatifmus | 12 Aprilis | 17 Aprilis | Sanatus |
| Rheumatifmus | 12 Aprilis | 21 Aprilis | Sanatus |
| Anafarca cum Afcite | 12 Aprilis | | |
| Scabies | 12 Aprilis | | |
| Febris tertian. duplicata | 14 Aprilis | | |
| Ophthalmia | 14 Aprilis | | |
| Scabies | 15 Aprilis | 30 Aprilis | Sanatus |
| Afthma pituitofum | 15 Aprilis | 30 Aprilis | Palliative fanatus |
| Scabies | 17 Aprilis | | |
| Peripneumonia | 19 Aprilis | 28 Aprilis | Sanatus |
| Phthifis pituitofa | 21 Aprilis | | |
| Ophthalmia afthenica | 23 Aprilis | | |
| Febris tertiana | 23 Aprilis | | |

4   Table from J. N. Thomann, *Annales Instituti Medico-Clinici Wirceburgensis* (1799).

applied during pregnancy, were very much criticized by the new movement: 'I reject totally the so-called preventative cures with blood-letting, purgatives and such, because I am convinced that very often they do more harm than good and that they disturb natural activity during the delivery.'[11] The method was also applied for cleansing the newborn children of their impurities: 'Such a method was applied for several days and I saw more than one child grow lean and die in convulsions, while everybody assured me that the child had been properly cleaned.'[12]

In the history of medicine progress is very often not so much the introduction of a new, successful treatment, but rather the limitation of the exaggerations of the old. The fundamental difference of the new medicine was the suggestion that therapy does not have a direct influence on the organism but rather an indirect one: a support of the defensive and productive powers. 'When this ability is destroyed and the organism has no efficiency in its struggle against the disease, then the result will be death.'[13]

Thus therapy should consider the whole living being. On the whole, the treatment must support the organism; thus the therapy in most diseases must be stimulating. The opposite treatment was appropriate for only a few cases: when the reaction of the organism is very strong, and disproportionate to the stimulus, for example, in the case of local inflammation, like encephalitis. In these cases what is indicated is calming the organism with the antisthenic method. The calming method can be considered similar to the 'cleansing method' but only coincidentally, because the purpose is quite different; the calming (antisthenic) method was antiphlogistic rather than cleansing.

The stimulating method consisted of drugs, strengthening diets and psychical stimulation. As a stimulating, strengthening diet for acute asthenic diseases, egg-yolk, wine, beer-soup, tea, coffee and spices were used. Ernst Horn, second physician at the Charité hospital under Hufeland, who made a great career as a professor at the University of Berlin and as a public health officer, wrote in his book on medical practice in 1807:

We are happy that the treatment of the severe forms of typhus [its variations as asthenic fevers] has been more successful since we have known the new method and the right application . . . The best doctors today agree that the treatment of these diseases [asthenic fevers] must begin with penetrating stimulants.[14] . . . One of the best stimulants for such cases is opium . . . we have been much happier in the use of opium since we don't use it as a narcotic[15] . . . Many doctors considered meat nourishment dangerous in fever-diseases . . . believing that meat supports the tendency to putridity. The introduction of meat-broth in these cases is the advantage of the new clinic . . . In the hospital of Vienna I have seen the best results from the generous use of Hungarian wines. As to the efficacy of the pure Stein and Leisten wines in similar cases, this was demonstrated by Marcus, Thomann and von Hoven in Bamberg and Würzburg.[16]

Psychical treatment was thought to be a complement to the stimulating treatment.

Psychical treatment is excellent for the ill, especially in severe cases of asthenic fever . . . The more we succeed in animating his (her) spirit and hope, the better we can influence positively his (her) biological suffering. A happy feeling acts like a strong stimulus on the whole organism.[17]

Elias von Siebold, obstetrician in Würzburg, wrote in 1803:

In puerperal fever I applied the stimulating method. I used camphor, opium, musk and vitriol–ether as the best stimulants . . . The diet was also according to the stimulating method of treatment, broth and soups with a good wine and the yolk of an egg.[18]

In his book on pregnancy the physician Christian August Struve, practitioner in Görlitz, criticized traditional therapy:

A great abuse of purgatives during pregnancy has murdered many a pregnant woman . . . The proper way to support the vigour of a pregnant woman is a strengthening diet, consisting of broth, and beer-soup with some wine and the yolk of an egg . . . The most important thing is to protect her from everything which could trouble her soul when the time of delivery is approaching . . . Continuous trouble and sadness weaken the body . . . The most stimulating drugs remain ineffective and are rather harmful when the ill person is in a situation where necessity and trouble suppress the vitality or when frequent occasions for anxiety prevent nature from restoring lost balance.[19]

J. C. Reil was an established and well-known professor at the time the new medicine was introduced. His main work regarding fevers was totally based on traditional medicine, but afterwards he was fundamentally influenced by the *Erregbarkeitstheorie* (theory of excitability), although he did not want to identify himself with the protagonists of the reform. His *Therapy* and *Pathology*, two works quite in the spirit of the new movement, were never published during his life. They were first published in 1816 after his death in 1813. But in his influential book about mental illness Reil shows the characteristic attitude of the new movement and expresses his departure from the traditional treatment. According to his opinion not only the antigastric method but drugs in general are useless in psychical diseases. Reil criticizes doctors who follow traditional medicine for using opium as a sedative to calm patients. 'One can certainly make a raging patient quiet, by giving him opium. But such a treatment is no therapy at all; the only result is to change one variation of mental disorder into another.'[20] Although Reil believed that mental disorders were due to troubles of the vital power of the brain, he insists, nevertheless on the psychical treatment as the only one suitable for mental and psychical illness: 'Feelings, ideas, psychical motions are the means suitable for restoring the disturbed vitality of the brain.' Not only psychical difficulties but also social and political insecurity and uncertainty are frequently responsible for mental illness: 'That's why there have been so many lunatics in France during the anarchy of recent years.'[21] Sadness, need and frustration disturb the soul in the opposite direction, diminishing the vitality of the brain which thus is not able to stimulate the vegetative system efficiently. The suitable treatment of these cases is a strong psychical stimulation in order 'to excite the psychical situation, to make it stormy . . . Try to involve the patient in new situations, to change his mode of living, send him on a journey, prevail on him to marry, show him some dangers.'[22]

The partisans of the new medicine considered hypochondria a typical psychosomatic illness. The hypochondriac has many somatic, organic symptoms but these are not the real disease. Such symptoms are produced by the hypochondriac because he has the wrong idea about his own body:

My experience has shown me that all forms of this disease disappear as soon as the patient is conscious of the deception of his own feelings; when he becomes aware of the fact that his own ideas and judgements about his body are erroneous . . . [Hypochondria] can never be healed with opium, as many think. I myself see a cure in the possibility that the patient be master of his own ideas . . .

I hear patiently the iliad of their suffering, trying to show them that I take their disease as seriously as they describe it. I promise to give them all the help which they, in their anxiety, demand. Only slowly, step by step, do I try very carefully to make them see the possibility that perhaps their own feelings and ideas make the disease seem greater than it really is.[23]

We see clearly that medical practice (as special treatment but also as a general attitude towards the patient) changed fundamentally after 1798.

Bernhard Hirschel, who in the 1840s wrote the most important study about the medicine of the Romantic period, is quite aware of the fact that the Romantic movement, the so-called *Erregbarkeitstheorie*, was a real revolution of medical thought and practice. Hirschel underlines the most important changes in medical practice: the limitation of the antigastric method (cleansing method) and of blood-letting, the demonstration of the possibility of dynamical treatment of suffering due to material changes, a better understanding of the action of remedies, for example opium, a better and more natural treatment of fevers, inflammation, rash and bleedings, and finally the introduction of the psychical method of therapy.[24]

## THE RESTORATION

> The trilling wire in the blood
> Sings below inveterate scars
> Appeasing long forgotten wars.
>
> T. S. Eliot, *Four Quartets*

The period after 1815 is generally considered to be a time of restoration because of the conservative politics of the German governments. The reforms began in the spirit of the French Revolution, which was the source of inspiration for many German intellectuals. But the enthusiasm for France and the Revolution did not last very long. The consequences of French occupation began to oppress the people. The secularization of all religious principles destroyed some of the best centres of liberalism and stopped intellectual activities. So the people fought for its liberation in the year 1813 in the new spirit of general resistance against the power of Napoleon, and the only value attainable at the moment, for which people were willing to die, was patriotism. What people could not know at that time was the fact that their victory over Napoleon and their sacrifices would help the old holders of power in their own country to find a new legitimation again, and that would mean new suppression of their own rights.

All the partisans of the new medical movement came from western and southern Germany, whereas the reaction and polemical attitude against it came from scholars of eastern Germany, especially from Prussia. The main centres of the movement were those supported by the catholic ecclesiastic princes of Bamberg, Würzburg, Mainz. The main partisans were Andreas Röschlaub and Adalbert Marcus in Bamberg, Melchior Adam Weigard in Mainz and Fulda, Friedrich Wilhelm von Hoven in Erlangen and Würzburg,

Joseph Nikolas Thomann and Elias Siebold also in Würzburg, Franz von Walther in Bamberg and Landshut, Ignaz Döllinger in Bamberg.

Foremost among the opponent of the new medical movement were some of the old established authorities who had really been convinced by the old method, for example Professor A. G. Richter in Göttingen. In opposing the new medicine he insisted that there were *gastric* diseases which had to be treated with the *antigastric* method. He suggested that even in asthenic situations, like those after a surgical operation, the patient could be success-fully treated with blood-letting and other cleansing techniques.[25]

The most famous forum against the new efforts of the reform was Hufeland's *Journal der praktischen Heilkunde*. In 1795 a new book was published by Hufeland on the nature and therapy of scrofula which was in the next year the prize-winning book of the Leopoldine Academy of Science. In this work Hufeland considered scrofula to be identical with tuberculosis, a disease due to the corruption of the lymph. He suggested the cleansing therapeutics, mainly emetics and purgatives: 'They are among the most excellent means of combating the disease. Their application has a double purpose: to clean the first paths and to stimulate the organs of the lymphatic systems.'[26]

In the same year, 1795, another medical book, about the nature of fever, was published, which did not win any prize but which did not remain unnoticed. It was the dissertation of a young man, Andreas Röschlaub, which had already caused excitement among the members of the medical faculty in Bamberg. This young man would soon be so famous that 'even the authority of Hufeland faded in view of his activities', as Rudolf Virchow would later judge. Of course he very quickly provoked the reaction of Hufeland, who published as early as 1797 his first attack against those who were attempting a revolution of medicine: 'We assure these gentlemen that not in the least do we need a revolution and we beg them to let us alone to continue this way in peace.'[27]

It was the beginning of a passionate dialogue between the two famous doctors which continued until 1811, the year of a formal reconciliation.

Christoph Wilhelm Hufeland was the son of a rich and respectable doctor from Weimar; he became court physician very soon through recommendation by Goethe and later became professor in Jena. In 1800 he came to Berlin, where he became director at the Charité and court physician and personal friend to the king and the queen, whom he accompanied in exile to Königsberg (after Napoleon's victory over Prussia in 1806). Hufeland was the most prominent and most active representative of conservatism and traditionalism and the dominant personality during the restoration period, the 'Koryphäos und Heros der Eklektiker', as the historian Burkhardt Eble phrased it.

Andreas Röschlaub was the son of a very poor family from Lichtenfels,

5   Christoph Wilhelm Hufeland.

near Bamberg. He studied in Bamberg under very difficult conditions. After
finishing his studies, he was employed as community physician for the poor of
the town in Bamberg; two years later he became professor at the University of
Bamberg; in 1802 he changed to the University in Landshut and finally, in
1824, to the University of Munich. Röschlaub was the leading figure of the
revolutionary medical movement.

During the years of the triumphant success of the movement Hufeland
changed, behaving more diplomatically than before, according to his liberal
principles. In the year 1799 he visited Marcus and Röschlaub at the famous
hospital of Bamberg to inform himself about the practical methods of the new
medicine. In the next year, 1800, he published his new book *A System of
Medical Practice*, where he declared: 'We live in the century of innovations
and reforms, and medicine has also experienced this dominating and vivid
spirit.'[28] Paying homage to this new spirit, Hufeland adopted the main
principles of the new medicine: 'Medicine is based on the principle, according
to which every influence on the living being affects his (her) excitability
[*Erregbarkeit*] as well as his (her) material properties.'[29] He abandoned as well
the antigastric method, suggesting that the main methods of therapy are the
exciting method, which can be direct or indirect strengthening, and the direct
or indirect weakening one. One might assume therefore that, for several years
at least, Hufeland had become a partisan of the new movement, as some of his

6   Andreas Röschlaub.

contemporaries thought. But this was not the case! Hufeland had only changed his vocabulary to conform to the new successful theory of excitability: 'I must confess that I am now using the term *Erregbarkeit* instead of the former *Lebenskraft*, not because I am now of another opinion but because this term *Erregbarkeit* is now more usual . . .'[30] His main purpose was to show that the new medicine was quite similar and compatible with the old therapeutics, the antigastric method which he now called 'antiphlogistic':

In this way we can solve the puzzle of why the partisans of this new method are able to heal, with the stimulating method, all those diseases which we formerly treated so successfully with the antiphlogistic method! The miracle vanishes if we consider the fact that a method of overstimulation is equal to a method of weakening; so the new therapy is the same as the old one, the difference being that the 'weakening' is attained via two different methods![31]

But in fact Hufeland did change his therapeutics in the name of the new movement though in a direction contrary to the movement. He introduced a 'combined method', adding to the old cleansing techniques – which he did not abandon at all – some of the stimulating remedies. Although Röschlaub and

other partisans of the new medicine were struggling against such an interpretation of their doctrines, Hufeland was influential, especially among the lower
practitioners.

These practitioners (the *Wundarzt*, the *Bademeister*, the barber) were
generally against any reform of the medical system. They themselves were a
part of the feudal structure, and its abolition, according to the aims of the
reform, would destroy their own existence. Besides, it was very difficult for
them to change their practice even if they wanted to do so. A document from
Bavaria, where the movement originated, dated 1802, the time of the greatest
popularity of the movement, shows the old antigastric treatment still unchanged. The document is a bill of a *chirurgus* for the treatment of a patient
over a period of twenty weeks. The disease of the patient is described as
'spasms of the limbs and hemorrhoids'. The treatment was the following:
'Ten times laxatives, one blood-letting, four times emetics, for six weeks
blood-cleansing decoctions, one vesicatory, five times clysters and eight times
cupping glasses'.[32]

As we see, there is no deviation from 'antigastric' therapeutics. But the
situation changed when doctors like Hufeland propagated the mixed or
combined method, according to which the new therapy was the same as the
old one, only enriched with stimulating remedies. Hufeland demonstrated
this method in several of his works and made it popular. For example, the
puerperal fever (childbed fever) was said to be due to a *plethora lymphatica* of
the abdominal system, connected with 'gastric impurities'. For the treatment
he began, before the delivery, with a blood-letting, as a strengthener. As soon
as the first symptoms of the disease appeared the purgatives, the clysters and
leeches on the genitals followed. Then, in case of 'gastric turgescence', an
emetic followed. If the disease got worse, a blood-letting was performed
again. Finally (and if the patient was still alive) a dose of opium followed.[33]

Hufeland and Röschlaub are characteristic personalities of their time.
Their relation to one another, their quarrels and even their reconciliation
provide us with an excellent mirror of the medical but also of the
sociopolitical situation. The lower-middle class, awakened from its socio-
political lethargy and apathy by the French Revolution, was willing to
struggle for equality and liberty with the weapons of intelligence, higher
education, verbal disputes and pressure on the state administration.
Röschlaub, who belonged to this class, was himself a genius, conscious of the
chances given at that moment, able to grasp all new ideas with their social
consequences. His success between the years 1798 and 1805 was miraculous.
His new theories spread with an unusual speed, penetrating all medical and
intellectual life. At the same time Röschlaub was also a representative for
those attitudes of his class, which were the legacy of a long tradition of
political apathy and ignorance of social responsibility. He had no idea how

deep the roots of the established power were; he could not be flexible and diplomatic.

In contrast to Röschlaub, these virtues were in abundance in his opponent Hufeland. Coming from the upper class of the bourgeoisie which was traditionally used to social struggles for power, prestige and high positions in court, administration and army, Hufeland was aware of what he could lose in a radical reform (even if he did not belong to the nobility); he was flexible, diplomatic, ingenious in accepting the 'unacceptable' in moments of danger, abandoning it again when the danger was over, and thus changing the situation in his favour. The latter was a common virtue of the German conservatives, never challenging in moments of danger but withdrawing and waiting. It was one of the strongest weapons against the first radical reform at the beginning of the century.

## NOTES

1  Christoph Wilhelm Hufeland, *Bemerkungen über die natürlichen Blattern in Weimar im Jahr 1788* (Leipzig, 1789), p. 113. (All translations from the German in this chapter are mine.)
2  D. Bucholtz, 'Krankengeschichte und Heilung eines Wahnsinnigen', *Journal der praktischen Heilkunde* (edited by C. W. Hufeland and known as *Hufelands Journal*), 2 (1796), 142–50.
3  G. F. Hildebrandt *Hufelands Journal*, 2 (1796), 577–89.
4  Georg Friedrich Hildebrandt, 'Über die Hypochondrie', *Hufelands Journal*, 1 (1795), 56–61.
5  J. N. Rust, *Die Medizinal-Verfassung Preussens* (Berlin, 1838), p. 85. See also Johanna Geyer-Kordesch, 'Court Physicians and State Regulation in Eighteenth-Century Prussia: The Emergence of Medical Science and the Demystification of the Body', in Vivian Nutton (ed.), *Medicine at the Royal Court 1500–1850* (forthcoming).
6  Andreas Röschlaub, *Über Medizin, ihr Verhältnis zur Chirurgie nebst Materialien zu einem Entwurfe der Polizei der Medizin* (Frankfurt-on-Main, 1802), pp. 171–2. See also Nelly Tsouyopoulos, *Andreas Röschlaub und die Romantische Medizin* (Stuttgart and New York, 1982) p. 83.
7  Guenther B. Risse, 'The Brownian System of Medicine: Its Theoretical and Practical Implications', *Clio Medica*, 5 (1970), 45–51; Hans Joachim Schwanitz, *Die Entwicklung des Brownianismus und der Homöopathie von 1795–1845* (Stuttgart and New York, 1983); Nelly Tsouyopoulos, 'The Influence of John Brown's Ideas in Germany', in W. F. Bynum and Roy Porter (eds.), *Brunonianism, Medical History*, Suppl. 8 (London, 1988), 63–74.
8  Adalbert Friedrich Marcus (ed.), *Ephemeriden der Heilkunde* (Bamberg and Würzburg, 1811). J. N. Thomann (ed.), *Annales Instituti Medico-clinici Wirceburgensis* (Cologne, 1799), see fig. 4, above: the several diseases under 'morbus'.

9    Adalbert Friedrich Marcus (ed.), *Ephemeriden der Heilkunde* (Bamberg and Würzburg, 1811), II, pp. 110–11.
10   Georg Adelmann, *Über die Krankheiten der Künstler und Handwerker. Nach den Tabellen des Instituts für kranke Gesellen der Künstler und Handwerker in Würzburg von den Jahren 1786 bis 1802* (Würzburg, 1803), pp. 108ff.
11   Elias von Siebold, *Über praktischen Unterricht in der Entbindungskunst* (Nuremberg, 1803), p. 149.
12   *Ibid.*, p. 177.
13   Andreas Röschlaub, *Magazin zur Vervollkommnung der theoretischen und praktischen Heilkunde*, 7 (1802), 74.
14   Ernst Horn, *Anfangsgründe der medizinischen Klinik*, 2 vols. (Erfurt, 1807), I, pp. 471–2.
15   *Ibid.*, p. 475.
16   *Ibid.*, pp. 500, 506.
17   *Ibid.*, p. 279.
18   Elias von Siebold, *Über praktischen Unterricht*, p. 167.
19   Christian August Struve, *Wie können Schwangere sich gesund erhalten und eine frohe Niederkunft erwarten?* (Hanover, 1807; first edn 1799), pp. 288, 210, 246.
20   Johann Christian Reil, *Rhapsodien über die Anwendung der psychischen Curmethode auf Geisteszerrüttungen* (Halle, 1803), p. 47.
21   *Ibid.*, p. 50.
22   *Ibid.*, p. 290.
23   Andreas Röschlaub, 'Einige Bemerkungen über die Hypochondrie', in his *Magazin*, 9 (1806), pp. 359–97.
24   Bernhard Hirschel, *Geschichte des Brownschen Systems und der Erregungstheorie* (Dresden and Leipzig, 1846), pp. 267–70.
25   *Röschlaubs Magazin*, 6 (1801), p. 188.
26   *Über die Natur, Erkenntnis und Heilart der Skrophelkrankheit* (Weimar, 1795), p. 156.
27   'Bermerkungen über die Brownsche Praxis', in *Hufelands Journal*, 4 (1797), p. 136.
28   *System der praktischen Heilkunde* (Leipzig, 1800), p. v.
29   *Ibid.*, p. xii.
30   *Ibid.*, p. xiv.
31   *Ibid.*, p. 78.
32   'Getreue und mit Belegen versehene Schilderung der noch immer grassierenden Pfuscherei', in G. Oeggl and A. Röschlaub (eds.), *Hygiea* (Frankfurt-on-Main, 1803), pp. 61–5. See also N. Tsouyopoulos, *Röschlaub*, pp. 81–2.
33   *Enchiridion Medicum* (3rd edn, Berlin, 1837), p. 708.

# 8

# *Morphotypes and the historical-genetic method in Romantic biology*

## TIMOTHY LENOIR

In 1802 Gottfried Reinhold Treviranus announced the birth of a new scientific discipline. He called it '*Biologie*', the science whose aim was to determine the conditions and laws under which the different forms of life exist, and their causes.[1] The significance of his declaration was not in denying that biological phenomena had been investigated previously by men such as Harvey, Malpighi, Hales and Haller. Rather Treviranus sought to affirm a set of methods which characterized biology as a discipline in its own right. Although biology was to be aided by physics and chemistry, these were to be ancillary tools subordinate to the guidance of a methodology peculiar to the life sciences: the historical-genetic method. The new discipline would require a new institutional setting; academic medicine should be restructured and linked via the anatomical laboratory to the clinic in a mutually stimulating embrace of theory and practice. The radical methodological core of the new biology was to serve as the basis for transforming medicine, providing it with theoretical underpinnings capable of achieving the failed progress in therapeutics sought by the Hippocratically-inspired medical practitioners of the late Enlightenment. But the role of the new discipline in society was to be no less expansive than its intellectual ambitions. A new scientifically trained, activist physician would be one of the chief agents in spreading knowledge useful for economic and social improvement, thereby preparing the ground for the gradual emergence of a just society.[2] In the present discussion I focus upon the elements and content of this methodological perspective, which served as a principle guiding source for medical reform of the Romantic period in Germany, and which led its practitioners to proclaim the emergence of a new science.

## LIFE AND MECHANICS: THE PROBLEM OF BIOCAUSALITY

Toward the end of the eighteenth century a number of physiologists attempted to provide methodological foundations for their science modelled upon Newtonian mechanics. The most notable of these attempts was that of Albrecht von Haller. Haller sought to explain organ function through forces such as 'irritability', and 'sensibility', and the force of secretion, which he took to be rooted in the material constitution of muscle, nervous and mucous tissue but which were incapable of a further mechanistic reduction.[3] This approach seemed consistent with sound Newtonian practice in the inorganic sciences; for just as Newton had refused to speculate on a mechanical cause for the forces of gravity and electricity and had confined himself merely to investigating the laws governing their effects, so Newtonian physiology was limited to investigating the regulating action of vital forces.

While promising, Haller's program encountered opposition, on the one hand from his arch-opponent, Caspar Friedrich Wolff, and on the other from a man dedicated to further developing (but at the same time modifying) Haller's ideas, Johann Friedrich Blumenbach. These men called attention to certain features of organic bodies which seemed incompatible with Haller's Newtonian force imagery. Central to their concerns was the problem of how causality should be conceived in the organic realm. The issue was to avoid a strongly reductionist approach to biology. Both Wolff and Blumenbach emphasized that the central phenomena of biology, particularly ontogenesis, growth and reproduction, could not be reduced purely and simply to physico-mechanical forces.[4] Organisms, they stressed, are characterized above all by goal-oriented processes. Yet, in spite of the difficulties surrounding a reductionist approach to organic phenomena, Blumenbach and his school were convinced that all talk of souls and non-materialistic causation should be strictly avoided in the organic realm. One hoped to chart a course between the Scylla of reductionistic mechanism and the Charybdis of vitalism.[5] The solution proposed to this problem was to unify mechanistic and teleological principles of explanation in a manner which I have characterized as 'vital materialism'.[6]

The methodological prescriptions for achieving the delicate mixture of teleological and mechanistic explanatory frameworks were set forth by Immanuel Kant. Kant had been following the work of Buffon, Haller, Blumenbach, Wolff and others for several years – Kant himself had published on the question of races, varieties and species – and in 1790 in his *Kritik der Urteilskraft* (*Critique of Judgement*) he gave a definitive analysis and attempted resolution of the problems. Basically Kant concluded that while the goal of science must always be to press as far as possible in providing a mechanical explanation, mechanical explanations in biology must always stand under the higher guidance of a teleological framework.[7] The essential difficulty, he argued, is that mechanical modes of explanation are inadequate to deal with many processes of the organic realm, where the relationship of cause to effect is completely different from that encountered in the inorganic realm. Although even in the inorganic realm there are reciprocal effects due to the dynamic interaction of matter, such phenomena are nonetheless capable of being analysed in some fashion as a linear combination of causes and effects, A→B→C. This is not the case in the organic realm, however. Here cause and effect are so mutually interdependent that it is impossible to think of one without the other, so that, instead of a linear series, it is much more appropriate to think of a sort of reflexive series A→B→C→A. This is a teleological mode of explanation, since it involves the notion of a 'final cause', for, in contrast to the mechanical mode where A can exist and have its effect independently of C, in the teleological mode A causes C but is not also capable of existing independently of C. The final cause is, logically speaking, the first cause. Because its form is similar to human intentionality or purpose, Kant called this form of causal explanation *Zweckmässig* (purposive), and the objects that exhibit such patterns, namely organic bodies, he called *Naturzwecke* (natural purposes):

The first principle required for the notion of an object conceived as a natural purpose is that the parts, with respect to both form and being, are only possible through their relationship to the whole . . . Secondly, it is required that the parts bind themselves into the unity of a whole in such a way that they are mutually cause and effect of one another.[8]

Clearly, Kant went on to argue, organisms qualify as *Naturzwecke*. The laws whereby organic forms grow and develop, he observed, are completely different from the mechanical laws of the inorganic realm. The matter absorbed by the growing organism is transformed into a basic organic matter by a process incapable of duplication by an artificial process not involving organic substances. This organic matter is then shaped into organs in such a way that each generated part is dependent on every other part for its continued preservation: the whole organism is both cause and effect of its parts. 'To be exact, therefore', he emphasized, 'organic matter is in no way analogous to any sort of causality that we know . . . and is incapable of being explicated in terms analogous to the sort of physical capacities at our disposal.'[9]

To be sure, there is, according to Kant, a certain analogy between the products of technology and the products of nature. But there is an essential difference. Organisms can in a certain sense be viewed as similar to clockworks. Thus Kant was willing to argue that the functional organization of birds, for example the air pockets in their bones, the shape and position of the wings and tail, etc., can all be understood in terms of mechanical principles, just as an *a priori* functional explanation of a clock can be given from the physical characteristics of its parts.[10] But while in a clock each part is arranged with a view to its relationship to the whole, and this satisfies the first condition to be fulfilled in an organic explanation as stated above, it is not the case – as it is in the organic realm – that each part is the *generative cause* of the

other as required by the second condition to be fulfilled by such an explanation. The principles of mechanics are indeed applicable to the analysis of
functional relations, but the teleological explanations demanded in the organic realm require an active, productive principle such as the *Bildungstrieb*
(formative drive), postulated by Blumenbach and others, which transcends
any form of natural–physical explanation available to human reason. From
the insight that the forces in the organic realm could not in principle be
constructed from physico-chemical forces, Blumenbach had concluded, correctly in Kant's view, that one is compelled to assume the existence of certain
further irreducible 'ground states' of organization.[11] While it was impossible
to give an account in terms of physico-chemical forces of these ground states
of organization, Kant thought it nonetheless possible to go quite far in
understanding the operation of the forces constitutive of the organic world
with the aid of physics and chemistry.[12] Behind his reasoning stood the
conviction that if a science of the organic were to be possible it was because the
organic realm no less than the inorganic was guided by a fundamental unified
framework of law. These bionomic laws were to be discovered through
empirical research guided by reasonable hypotheses.

The practical implications of this analysis were illustrated by its application
to animal systematics. Kant advocated the construction of morphotypes or
organizational plans to be arrived at through comparative anatomy and
physiology. He argued that the agreement of so many species, not only in their
skeletal structure but in the organization of other parts as well, suggests that
they might all be united by a fundamental ground plan. Through the lengthen-
ing of one part or the suppressed development of another, a multiplicity of
species might be brought forth.[13] The correctness of such speculation, he
urged, could be checked through exact archaeological researches.[14]

## GOETHE, KIELMEYER AND THE HISTORICAL-
## GENETIC METHOD

Goethe was already deeply involved in his own work on the development of
the notion of the morphotype when he read Kant's *Kritik der Urteilskraft*. In
1786 he had circulated his work on the intermaxillary bone in which the
notion of a vertebrate skull morphotype is implicit; and in 1790, just prior to
the appearance of Kant's *Kritik*, Goethe published his work on the metamorphosis of plants in which the continuous transformation of an idealized
primitive organ, the embryonic leaf, is used to establish homologies between
the various structures of plants in different stages of development.[15]

In a short unpublished essay of 1792, entitled 'Der Versuch als Vermittler
zwischen Objekt und Subjekt' ('Experiment as a Mediator between Subject
and Object'), Goethe proposed an experimental methodology especially

suited to the needs of the life sciences, although it might be applied equally well to optics in his opinion. It was a two-tiered methodology. As a general guideline he recommended the construction of experiments which are repeatable whenever the same initial conditions are established. The value of such experiments, however, was only in their connection with others. The problem was how to link isolated experiments through a nexus of necessary conditions so that a natural correspondence obtained between the picture of the subject in question and the object under investigation. In order to insure that the individual experiments would not emerge as an artificially constructed 'aggregate' but rather united in terms of an immanent internal bond, Goethe counselled the construction of a 'higher sort' of experience. It was supposed to emerge from a series of tightly interlinked individual experiments so closely related that each one borders upon and successively flows into the next. Arranged temporally, this series of previously isolated and independent events would then emerge as necessary aspects or moments of a unified, more fundamental experience. 'If one carefully grasps and attends to each of these elements, they constitute only a single experiment, one experience presented under the most manifold aspects.'[16] Goethe saw this method as analogous in the experimental domain to the method whereby a mathematician, through an act of intellectual intuition, seizes upon the formula that links the various statements of a proof into a necessary unity, the *logos* that enables him to see the individual components of the proof as aspects of the whole.

In the inorganic sciences, where, as Kant had argued, strictly mechanical causality prevails, the 'secret, internal links' between the phenomena can be discovered through the theoretical deductive capacity of reason. As Goethe noted in a fragment on 'Analysis und Synthesis', however, the case is different in the organic realm. In the organic realm nature herself solves the problem for the inquirer. The much-prized necessity upon which a scientific knowledge of organic form could be based was built directly into the organism itself. For organisms are the prototype of the secret syntheses described by Goethe in which each part is necessarily related to others and to the whole. No tighter bonds of necessity could exist than the requirements of functional organization in the organic realm.[17] Since inorganic phenomena were essentially passive and only revealed their interconnections through the activity of the experimenter by means of carefully designed technical apparatus and procedure, there is a danger that the logical coherence of the phenomena is one imposed by the activity of the experimenter. Organisms, on the other hand, are their own sources of activity, expressed as an active striving, a formative impulse, a *Bildungstrieb*. Accordingly, as Carl Friedrich Kielmeyer explained to students in his lectures on what he termed the construction of a 'Physik des Thierreiches' ('Physics of the Animal Kingdom') in 1790–3, in the organic realm nature is her own 'analyst'. For she actively unfolds the logical synthetic

unity of organization in both space and time as interrelated groups of animals
and organ functions. Organic nature, Kielmeyer explained, is its own experimental laboratory breaking up the 'integral of life', dissecting the secret
synthesis and its internal organizational logic into spatio-temporal series of
partial organic functions, the flora and fauna of the present as well as past
geological epochs.[18] In order to understand what nature herself constantly
reveals in her experimental laboratory the investigator must come with a
careful grounding in comparative anatomy and physiology. He will then be
prepared to witness the Urphänomen which is the observational core of his
entire science, namely embryogenesis.[19]

Embryogenesis was, in short, the phenomenon that promised to provide the
sort of observational basis needed in order to overcome the difficulties Kant
had disclosed in his Kritik der Urteilskraft.[20] For here the internal principle
linking the various parts into a whole actively manifested itself temporally
through the successive generation of interrelated structures. The formative
principles of biological organization could be revealed observationally as a
nexus of interdependent structures unfolding in space and time, appearing on
the one hand in the ontogenetic development of the individual organisms and
on the other hand in the groups of systematically related organisms differentiated from one another in terms of a more specialized development of one or
more of a common fund of organ systems.

## FORM, FUNCTION AND THE EMBRYOLOGICAL METHOD

By the mid-1790s most researchers agreed that a special method was required
for investigating the phenomena of life and that the genetic method, particularly the careful observation of developing embryos, offered the surest means
for constituting a science of the organic. One of the earliest attempts to apply
the method to both theoretical zoology and medicine was made by Johann
Friedrich Meckel. Meckel had studied with both Blumenbach and Johann
Christian Reil, the leading architects of vital materialism in Germany, and
under Reil's direction he began a dissertation on cardiac malfunctions based
on a study of foetal development and malformations of the heart. After
completing his degree in 1803, Meckel studied in Paris with Cuvier and
Alexander von Humboldt.

The period of study in Paris with Cuvier resulted in the publication of
Meckel's first major work, Abhandlungen aus der menschlichen und
vergleichenden Anatomie (Investigations on Human and Comparative
Anatomy) (Halle, 1806). Meckel argued that further advances in both theoretical and practical medicine could only be expected from the construction of
a general theory of animal organization and the recent developments in this
area could already demonstrate advances about to be made in the practical

knowledge of organ function. The theory of organic form, he argued, must rest ultimately on four areas of investigation: chemical analysis; comparative anatomy and physiology of both vertebrates and invertebrates; pathological anatomy; and comparative embryology. Meckel illustrated the practical advantages to be gained by the comparative embryological method for illuminating the function of the thymus, the thyroid and the adrenal glands.

Methods of chemical analysis, he noted, had not advanced sufficiently to illuminate the functioning of these glands. Moreover, the use of vivisection was not possible in studying these organs, for surgical technique was incapable of exposing them without causing extensive damage to other tissues and organs.[21] The only means of discovering the function of those organs was a comparative study of their presence or absence, their position and relative size with respect to other organs and their development in different classes, genera and species of animals.[22] By coupling the constant interrelationships of these organs with other organs in the animal economy, together with a knowledge of the manner of life and habitat of the organisms in which they appear, inferences could be drawn concerning the function of the organs in question.

Applying the comparative developmental method and utilizing materials supplied him by Cuvier, Meckel argued that the thymus gland must be active primarily during the period of foetal development in man, and shortly after birth, for it is very large in the embryo and is very small in adults.[23] Similar reasoning led him to assert that the suprarenal glands must be connected to the organs of generation. He drew this inference not only from the apparently close material connection of the suprarenal glands and the first signs of the ovaries and testes, but also from a comparative study of malformations during foetal development. His subjects were primarily female. While adjacent organs in the lower half of the body such as the kidneys might be normally developed, he found that if the suprarenal glands were malformed or deficient, so were the sexual parts.[24] He noted exceptions to this pattern, but he thought the inference to be supported by the evidence of numerous anatomies. Anatomies of guinea pigs done by himself and Daubenton appeared to strengthen this correlation; for in the guinea pig's embryonic development the adrenal glands and sexual organs appear simultaneously and retain their relative proportions to one another throughout the development. In other developing rodents, Meckel found that if the sexual parts were relatively large, so were the adrenal glands.[25]

## CONCLUSION: MORPHOTYPES, THE GENETIC METHOD AND MEDICAL DISCIPLINE

Many of the conclusions Meckel drew from his embryological studies were criticized, and some, such as his proposed biogenetic law relating ontogeny and phylogeny, were unceremoniously rejected.[26] Nevertheless Meckel's contemporaries were resolute in their commitment to morphotypes and the genetic method, beautifully illustrated in his work, as providing the disciplinary core for a theoretically invigorated medicine. In concluding this brief survey, I will examine one of the earliest attempts to institutionalize the new disciplinary program, which took place in Kant's home town, Königsberg. The special requirements of the genetic method and its importance for medical training were offered by Karl Friedrich Burdach as justification for the physical layout and the organization of the curricular structure of the first Anatomical Institute in Königsberg, which was established under his directorship in 1817. Burdach explained the rationale behind the design of the Anatomical Institute at its opening in a lecture entitled, 'Über die Aufgabe der Morphologie' ('On the task of morphology').

Morphology, Burdach told his audience, is that higher science concerned with the generation of organic structure, and it is essential to understanding disease; for disease is the result of some abnormality in the operation of the formative force of the organic body. As a general theory of the formative powers of organic structure, morphology is therefore a major branch of natural philosophy as a whole. It aims at finding not only the laws of the generation of individual parts, but also those binding those parts into a whole organism and the relation of that organism to the rest of nature. The philosophically minded physician is permitted access to these internal structuring forces of organic nature, 'for we are not blind products, rather being is an object for us, and to us the world comes to self-consciousness, to self-intuition [Selbstanschauung]'.[27] It is this special relation of man to the structuring forces of organic nature that enables him to grasp the essence and Grundform of the species and see it reflected in a manifold of different individuals, and it is this special relation of man to nature that ultimately makes science possible. In Burdach's discussion, the generation of form and the necessary relationships between different parts of organic systems become manifest through an act of intellectual intuition. The necessary connections between the phenomena lie already preformed, in embryo as it were, in the structure of universals present in reason.

Burdach did not think it possible to spin out the entire structure of the organic world from those ideas without careful and methodically constructed empirical experience, however. Only by immersing oneself properly in the phenomena of organic life would the logical structure latent in the 'idea' of animal organization become fully articulated and manifest, or 'self-conscious', to use his terminology. By presenting the mind with the various external phenomena of organic life – i.e., generation, comparative study of organs and organ systems in different animals, etc. – would the internal, necessary bond linking them into a unified whole become manifest. As Goethe had described it, only then would the Urphänomen emerge. When this occurred, Burdach imagined that a pure, sensuous intellectual intuition would form: 'When in this way morphology is raised to pure science, form no longer appears as a dead product; rather it appears as a coming into being, which emerges out of the externally generative and formative Organism, the knowledge of which is our goal.'[28] In this intellectual intuition, which is the goal of morphology as Burdach conceived it, the systematic interconnections between all the various organic elements unfold almost as if one were to have before him a time-lapse film of embryological development. Burdach's language echoed Goethe's description of his experience of the *Urpflanze* (primordial plant) during his stay in Italy.

Burdach considered the task of the instructor of young physicians to consist in preparing them to have this intellectual intuition, for it formed the spirit and essence of medicine itself, rendering the physician truly capable of understanding and treating disease. The Anatomical Institute and its four-year course of lectures were intended to realize this educational goal. The Anatomical Institute was a three-storey building with a basement in which corpses were stored and prepared. The main lecture hall was directly above the cellar and a dissection table on an elevator allowed the corpses to be brought directly into the middle of the lecture hall. There were spaces in the laboratory for students to work and a laboratory for the prosector (Karl Ernst von Baer) on the first floor. On the next floor was an anatomical museum housing more than 1,600 preparations. Represented in the collection were the various organs of animal and human anatomy in both normal and diseased states. These were arranged in ascending stages of maturation. The museum was divided into several rooms: one for a skeletal and skull collection of various animals and man in different stages of development; a collection of circulatory and respiratory organs, digestive organs, muscles, sensory organs; and a collection of 114 human embryos arranged in sequence beginning with a three-week-old foetus. At the end of his four-year course of study, as a result of the careful arrangement of materials presented in the proper order, the young physician was to have the experience 'in which the living forms and their necessary interconnections swim before his soul'.[29] The entire object of the curriculum at the Anatomical Institute was, in short, designed to initiate the physician into truly scientific medicine by constructing for him the ingredients of an *Urphänomen*.

## NOTES

1  Gottfried Reinhold Treviranus, *Biologie oder Philosophie der lebenden Natur*, 6 vols. (Göttingen, 1802–22), I, p. 4.
2  This theme is nicely treated in Uta Frevert, *Krankheit als politisches Problem*

1770–1880. *Soziale Unterschichten in Preussen zwischen medizinischer Polizei und staatlicher Sozialversicherung* (Göttingen, 1984). Also see Nelly Tsouyopoulos, 'Die neue Auffassung der klinischen Medizin als Wissenschaft unter dem Einfluss der Philosophie im frühen 19. Jahrhundert', *Berichte zur Wissenschaftsgeschichte*, 1 (1978), 87–100. See also Nelly Tsouyopoulos's essay in the present volume, above, pp. 101–18.

3   See Karl E. Rothschuh, *Physiologie. Der Wandel ihrer Konzepte, Probleme und Methoden vom 16. bis 19. Jahrhundert* (Munich, 1961), pp. 134–51; and Shirley A. Roe, *Matter, Life and Generation. Eighteenth-Century Embryology and the Haller–Wolff Debate* (Cambridge, 1981).

4   See Shirley A. Roe, *Matter, Life and Generation.* Also see Wolfgang Lefevre, *Die Entstehung der biologischen Evolutionstheorie* (Frankfurt-on-Main, 1984), pp. 26–68.

5   Timothy Lenoir, 'The Göttingen School and the Development of Transcendental *Naturphilosophie* in the Romantic Era', *Studies in History of Biology*, 5 (1981), 111–205.

6   Timothy Lenoir, 'Kant, Blumenbach, and Vital Materialism in German Biology', *Isis*, 71 (1980), 77–108.

7   Reinhard Löw, *Philosophie des Lebendigen. Der Begriff des Organischen bei Kant, sein Grund und seine Aktualität* (Frankfurt-on-Main, 1980). Also see Timothy Lenoir, *The Strategy of Life: Teleology and Mechanics in Nineteenth-Century German Biology* (Dordrecht and Boston, 1982).

8   Kant, Immanuel, *Kritik der Urteilskraft* (1790), in *Kants Werke* (ed. Prussian Academy of Sciences), V (Berlin, 1908), p. 373.

9   *Ibid.*, pp. 374–5.

10  *Ibid.*, p. 360.

11  *Ibid.*, p. 424.

12  *Ibid.*, p. 418.

13  *Ibid.*

14  *Ibid.*, p. 419.

15  See H. Bräuning-Oktavio, 'Vom Zwischenkieferknochen zur Idee des Typus. Goethe als Naturforscher in den Jahren 1780–1786', *Nova Acta Leopoldina*, new series, 126: 18. Also see Georg Uschmann, *Der morphobiologische Vervollkommnungsbegriff bei Goethe und seine problemgeschichtliche Zusammenhänge* (Jena, 1939).

16  Johann Wolfgang Goethe, 'Der Versuch als Vermittler von Objekt und Subjekt', *Goethes Werke*, ed. Dorothea Kuhn (Hamburg, 1955), XIII, p. 17.

17  Goethe, 'Analysis und Synthesis', *Goethes Werke*, XIII, p. 52.

18  Carl Friedrich Kielmeyer, 'Entwurf zu einer vergleichenden Zoologie', in *Kielmeyers gesammelte Schriften*, ed. F. H. Holler (Berlin, 1938) p. 27.

19  See Kielmeyer's letter to Windischmann in *ibid.*, pp. 203ff. Also see his 'Ideen zu einer allgemeineren Geschichte und Theorie der Entwicklungs-erscheinungen der Organisationen', *Gesammelte Schriften*, p. 107, for Kielmeyer's statement of the biogenetic law. Kielmeyer's use of the embryological criterion is discussed by William Coleman, 'Limits of the Recapitulation Theory: Karl Friedrich Kielmeyer's Critique of the Presumed Parallelism of Earth History, Ontogeny and the Present Order of Organisms', *Isis*, 64 (1973), 341–50.

20  Goethe made this point in his 'Allgemeine Einleitung in die vergleichende Anatomie ausgehend von der Osteologie', (1795), *Goethes Werke*, XIII, p. 181.

21 Johann Friedrich Meckel, *Abhandlungen aus der menschlichen und vergleichenden Anatomie* (Halle, 1806), pp. 1–3.
22 *Ibid.*, pp. 7–8.
23 *Ibid.*, p. 8.
24 *Ibid.*, pp. 174–5.
25 *Ibid.*, pp. 176–7.
26 See Stephen J. Gould, *Ontogeny and Phylogeny* (Cambridge, Mass., 1977), pp. 45–6.
27 Karl Friedrich Burdach, *Über die Aufgabe der Morphologie* (Leipzig, 1817), p. 7.
28 *Ibid.*, p. 21.
29 *Ibid.*, p. 62.

# 9

# 'Metaphorical mystifications': the Romantic gestation of nature in British biology

## EVELLEEN RICHARDS

We know what a masquerade all development is and what peculiar shapes may be disguised in helpless embryos. In fact, the world is full of hopeful analogies and handsome dubious eggs called possibilities.

George Eliot, *Middlemarch*

One of the most significant and distinctive features of the positivist historiographic tradition has been its denial of the positive contribution of Romanticism to science, particularly the life sciences – notoriously susceptible to Romantic contagion. This reached its nineteenth-century apotheosis in the writings of Thomas Henry Huxley, who was probably the single most influential and destructive English-speaking critic of the 'metaphorical mystifications' of *Naturphilosophie*. Huxley's denigration of Romantic science was not only designed to ridicule and undermine his professional rival, Richard Owen (whose morphology was well known to be tarred with the brush of a suspect *Naturphilosophie*), but also to promote the much-vaunted and methodologically guaranteed 'objectivity' and purity of the Darwinian programme. In the Huxley-led drive for Darwinian dominance of nineteenth-century science, he and the Darwinians capitalized on their collective image as 'plain, prosaic inquirer[s] into objective truth', by contrasting this ideologically neutral and sober representation of themselves with the wild-eyed speculations, 'oracular utterances' and general verbal gymnastics of the unruly Romantics – the adversaries of everything that Huxley's reliable and socially efficacious 'natural knowledge' stood for.[1] It was a stratagem that proved very effective in furthering the interrelated social and professional interests of the 'young guard' Darwinians. It also left a deep and enduring impression on evolutionary history, where several generations of historians continued to toe the partisan line established by the dominant Darwinians. It is only comparatively recently that this Huxleyean historiographic heritage has been challenged, and a rich and multilayered history of the deployment of Romantic concepts in nineteenth-century biology is now emerging.

This essay is focused on the Romantic conception of the history of nature, construed as one long gestation analogous to a normal human pregnancy, and its central role in nineteenth-century evolutionary theorizing. It confronts Huxley's assessment of Romanticism head-on by arguing that it was through such 'metaphorical mystification' that Romanticism made its most powerful impact on evolutionary biology. Further, I shall try to show that it was its very ambiguity, its 'mystification', so much cultivated by the Romantics and so much deplored by Huxley, that gave the metaphor of gestation its explanatory power and led to its wide deployment in nineteenth-century biology.

The Romantic preference for analogy and metaphor as the means of conveying, or rather suggesting, ultimate truths is well known. The most important production of *Naturphilosophie* in the life sciences, Lorenz Oken's *Lehrbuch der Naturphilosophie* of 1809–11 (which went through three German editions in his lifetime and was translated into English in 1847), is also an exemplar of this Romantic fetishization of symbolism. To the perplexity of its English readers (and the near apoplexy of some), it was found to consist of 3,652 consecutive aphorisms or *Fragmente*, each of which could be read in isolation as a whole in itself, or as part of an extended argument. Symbols, with all their attendant ambiguities and imprecisions, thus tended to become explanations in Romantic science. And this, together with the reluctance of the Romantics to analyse the meaning of the images they evoked, their deliberate cultivation of mystery in even the simplest things, made much of *Naturphilosophie* almost incomprehensible to the uninitiated. But at the same time this very ambiguity rendered its concepts malleable to the purposes of those who appropriated and employed them in nineteenth-century biology. Ambiguity is manipulable, and it was this aspect of *Naturphilosophie* which made it such a rich and fertile source of ideas and concepts, especially to its British followers who were not bound by the constraints of the formal and systematic philosophy of nature to which Oken aspired, and who were free to combine its elements eclectically with those from a very different cultural tradition.

As it came to fruition within German Romantic philosophy, the metaphor of the gestation of nature encapsulated the organicism, the uncompromising developmentalism, anthropocentrism and insistence on the fundamental unity of all nature, of the Romantics. The Romantic universe being metaphorically an immense animal, it was animal-like in its functions and constitution, and even in its method of procreation. Romantic literature abounds with references to this 'universal gestation of nature' – to the 'impregnation of the terrestrial womb', the 'pregnancy' of the world, the 'generation', 'gestation', 'growth' or 'development' of nature. Above all, through the metaphor of gestation, biological and historical thought could be united so that the rules and concepts of the one could be applied to the other, and this is epitomized in

Oken's very definition of *Naturphilosophie* as 'the generative history of the world'.

It cannot be overemphasized that the task of *Naturphilosophie* was primarily historical, and Oken's definition makes this patent. *Naturphilosophie* had to demonstrate how the universe originated, and to reconstruct its development or *Entwicklung* from the original Idea thought by God to its highest manifestation as man. The task of reconstruction was to be aided by the essential parallel between man's individual history, or gestation, and the universal history. 'It is certain', affirmed Schelling in 1811, 'that whoever could write the history of his own life from its very ground, would have thereby grasped in a brief conspectus the history of the universe.'[2] This certainty was based on the fundamental Romantic tenet that man is the prototype and model of all existence – the microcosm. There is only one developmental tendency, that of producing man, who is, as Oken put it, the 'summit, the crown of nature's development, and must comprehend everything that has preceded him, even as the fruit includes within itself all the earlier developed parts of the plant'.[3] Thus man's individual development or ontogeny necessarily replicates the development of life on earth, the universal development reflected in that abstraction the Romantics called *Entwicklung*. Man and nature share a common *Entwicklungsgeschichte* – a history of development.

Oken, who combined the functions of Romantic ideologue and political and professional activist (he campaigned vigorously for the unification of the German states and of German science) with those of a working (and respected) embryologist, offered a detailed anatomical account of the ideal 'perfect parallelism' he advocated between the forms assumed by the developing human embryo and the ascending sequence of mature forms constituting the animal series. And, typically, he presented the results of his embryological investigations in aphoristic form, invoking the metaphor of gestation: 'Animals are only the persistent foetal stages or conditions of man . . . A human foetus is a whole animal kingdom.'[4]

To the modern eye, the metaphor has obvious and almost inevitable evolutionary or, more precisely, transmutationist implications. If animals are merely embryonic or foetal men, then a lower animal could be transmuted into a higher one through a simple prolongation of development beyond the normal termination of its ontogeny. In this way, metaphor could be realized and the gestation of nature become a literal historical process. According to Karl Ernst von Baer, its leading critic among embryologists, this was the common practice among his contemporaries. 'By degrees', he wrote in 1828, 'it became the custom to look upon the different forms of animals as developed out of one another, and then many appeared to forget that this metamorphosis was after all only a mode of conceiving the facts . . .'[5]

Von Baer's attack on the Romantic 'law of parallelism' and the transmutationism he associated with it was based on his rejection of the unilineal animal series or taxonomy integral to both concepts. Like the eminent French comparative anatomist, Cuvier, and in opposition to the Romantic morphologists, von Baer held that there were four basic types of organization. According to von Baer, the type is manifested in the very early stages of ontogeny, with the result that the simple unilineal parallel insisted upon by Oken and the Romantics no longer applies, and the embryo diverges more and more from other animal forms. It therefore repeats not the adult forms of lower animals, but their embryonic ones. However, von Baer failed to dislodge the deeply entrenched embryological law of parallelism. In spite of the very real differences between it and von Baer's 'law of divergence', they had an underlying continuity and essential similarity (recognized by von Baer himself) which facilitated their confusion and conflation. As well, parallelism as conceived by its *Naturphilosophen* exponents was an idealization, as Oken made clear. It was not, in this sense, amenable to von Baer's empirical investigation and criticism, and the two laws therefore coexisted or, more often, were conflated, until their ultimate conflation in the post-Darwinian theory of recapitulation.[6]

Similarly, for reasons implicit in Romantic philosophy, the animal series was not usually assumed to constitute a genealogical sequence. *Entwicklung* was primarily an archetypal development, not an evolutionary one, and for the majority of the *Naturphilosophen* it remained an abstraction like the Romantic 'archetype', not to be materially realized in nature. Earlier interpretations to the contrary, Oken himself, as more careful scholarship has shown, was no transmutationist. He offered an elaborate alternative to transmutation of species by positing the spontaneous generation of new and higher organisms from a universal organic 'infusorial mucus' or *Urschleim*, through a process of polar conflict which necessitated the interaction of material nature with the spiritual or Absolute. For all the suggestive ambiguity of his expression, Oken rejected the idea that nature could be explained in terms of material causality and therefore rejected the idea of organic evolution as a real physical possibility.[7] The problem of man's origin could not be resolved by making him emerge from an ape as the transmutationists required, and the *Naturphilosophen* generally seem to have been as unwilling as Bishop Wilberforce (in his celebrated confrontation with Huxley at the 'Oxford Debate' of 1860) to countenance a 'miserable ape' as grandparent. Indeed, as Richard Owen was shortly to demonstrate, *Naturphilosophie* was not incompatible with British natural theology. And it was through its thorough assimilation into natural theology via the ideology of progressionism that the metaphor of gestation assumed its prominent role in pre-Darwinian British biology and paleontology, rather than through direct evolutionary speculation.

Nevertheless, as von Baer attests, some of the less idealistic contemporaries

of the *Naturphilosophen* did give the metaphor of gestation a literal interpretation, and they were aided in this by the ambiguity of Romantic symbolism
and the failure of the leading *Naturphilosophen* to clarify the meaning of this
potent metaphor. The French transcendentalist, Etienne Geoffroy Saint-
Hilaire, for one, found its evolutionary and materialistic implications
irresistible. Geoffroy founded his morphology (which emphasized serial
development, parallelism, transmutation and unity of composition), on the
sovereignty of material laws, and directed it against Cuvier and towards the
young medical reformers and republicans of Paris. For Geoffroy, to view the
stages of embryonic development was to see in summary form the 'spectacle of
the evolution of the terrestrial globe', to 'catch nature in the act'. He was
convinced that in provoking the birth of monstrosities from hens' eggs by
artificially varying the conditions of incubation, he had experimentally illustrated the way new species arose in nature, and he offered a materialistic
account of this process. According to Geoffroy, mechanical and chemical
changes in the environment (especially in the respiratory milieu), induced
changes in the organism during the embryonic stage which were akin to
monstrous development. Through their propagation by inheritance, these
embryonic changes brought about the transmutation of species.[8]

Geoffroy's evolutionary speculations were taken up and adapted in various
ways for different social and institutional ends by the three most prominent
and influential British transcendental anatomists, Robert Edmond Grant,
Robert Knox and Richard Owen. Each of these patterned his version of
organic descent on variations of embryological development, and in ways
consistent with his politico-institutional position. Grant and Knox, who were
both attracted by the contingent radical political ties of Geoffroy's morphology, tried to produce self-consistent materialistic theories of life: Grant, the
radical democrat of the University of London, modelled his progressivist
transmutation on Geoffroy's traditional unilineal version of embryogenesis
and made it do service to reformist institutional and social interests,[9] while
Knox, institutional and social outcast, developed a theory of 'generic descent',
emphasizing the abrupt non-linear embryogenesis of new species as a constitutive part of his anomalous ideology of radical racism.[10] Owen, on the other
hand, doyen of the Royal College of Surgeons (at that stage under heavy
attack from the medical reformers) and aspirant to the coveted title of the
'British Cuvier', harnessed Geoffroy's transcendental anatomy to conservative institutional and social needs by reconciling it with Cuvierian functional-
ist teleology. At the same time, in the 1840s, he also attempted to adapt
Geoffroy's teratological speculations to these same conservative concerns, by
devising a non-materialist theory of gross embryonic evolutionary change,
analogous to 'anomalous monstrous births', and divinely pre-programmed to
a divergent von Baerian embryological model.[11]

However, in 1849 Owen's evolutionism, conservative as it was, was brought up short against the larger social forces of the day. The cautious evolutionary hints of his *On the Nature of Limbs* of the same year (cast in the form of the traditional Romantic ambiguity of an archetypal development) were lumped together with the recent translation (well known to have been instigated by Owen) of Oken's *Lehrbuch der Naturphilosophie*, and the popular evolutionary work, *Vestiges of the Natural History of Creation*. All were publicly castigated in the *Manchester Spectator* for their promotion of a 'desolating Pantheism' in the form of the 'THEORY OF DEVELOPMENT' which was undermining religious belief and contributing to the contemporary political and social unrest. In the socially troubled forties evolution was virtually synonymous with revolution (as indeed the theories of Grant and Knox were constitutive of their radical politics). The ambitious Owen reacted to these 'hard epithets' by muting his evolutionism until the more liberal climate of the prosperous and secular sixties and the appearance of *The Origin of Species* made possible the voicing of his (by then socially and intellectually outmoded) views.[12]

Owen's major difficulty lay in the closeness of his views to those of the heretical and much maligned, but widely read, *Vestiges*. It was the *Vestiges*, first published anonymously in 1844 and reissued in at least twelve subsequent English editions, which brought the transmutationist implications of the 'universal gestation of nature' before the middle-class British public to an unprecedented degree. Its author, Robert Chambers, the Edinburgh publisher and essayist, drew explicit inspiration for his 'development hypothesis' and its mechanism from the idea that the 'ordinary phenomenon of reproduction was the key to the genesis of species'. In effect, Chambers demystified and 'domesticated' the Romantic metaphor of the gestation of nature. The production of new species, he reassured his readers,

has never been anything more than a new stage of progress in gestation, an event as simply natural, and attended as little by any circumstances of a wonderful or startling kind, as the silent advance of an ordinary mother from one week to another of her pregnancy.[13]

This naturalistic idea, of course, had not come 'unpromptedly' into Chambers's mind as he claimed, but through his exposure to transcendentalist conceptions. By the time of the writing of the *Vestiges* these were common currency, not only in British biology (and especially so in Edinburgh, the major centre for the dissemination of French transcendental anatomy into British biology and medicine), but also in natural theology. For these malleable concepts appealed not only to the nonconformist medical men like Knox and Grant, who naturalized and 'preached' them to their students, but equally to the orthodox followers of William Paley, who absorbed them into a

7  'Cutting into the stony womb of nature' the progressionist Hugh Miller
   found fossil testimony of man's own preordained creation. This
   romanticized image of nature's 'gestation' as revealed by the geological
   hammer adorned the first page of Miller's *Footprints of the Creator* of
   1849.

progressivist natural theology and employed them to demonstrate God's
providential design and continuing stabilizing presence in nature and the
social order. Even before Owen's formal synthesis of Geoffroyan morphology
and Cuvierian teleology, British natural theology had managed to reconcile
*Entwicklung* with Christianity in a progressionist philosophy of nature which
was both discontinuous and historical.

'Progressionism', or 'developmentalism' as its exponents sometimes
termed it, may be described as evolution without physical continuity. The
continuity exists only in the mind of God who has created a succession of new
and higher life forms after the extinction of all their predecessors through a
series of divinely invoked universal catastrophes. Progressionism, that pecu-
liarly British phenomenon, formed a conceptual framework surprisingly
congenial to many of the ideas, if not the Idea, of Romantic *Naturphilosophie*.
This is particularly evident in the writings of Hugh Miller, the Scottish
stonemason, whose popular works outsold the *Vestiges* many times over. In
the tradition of *Naturphilosophie*, Miller construed the whole of nature as a
pre-ordained progress towards man, but he gave this familiar Romantic
theme revelatory significance, and posited a succession of special creations

consistent with the fossil record. For Miller, fossils were 'mute prophesies' pointing towards man's coming, and embryos the 'epitome of geologic history'. He found the Romantic metaphor of gestation an especially beguiling 'evidence' of God's handiwork and revelation to man, and he delighted in the 'wonderful analogies' he constantly invoked between the fossil progress and the 'individual history' of organisms: 'analogies that point through the embryos of the present time to the womb of Nature, big with its multitudinous forms of being'.[14] Such transcendental concepts and expressions even found their way into that compendium of natural theology, the *Bridgewater Treatises*, in one of which the physician Peter Mark Roget also distilled Christian comfort from heterodox German Romanticism. Roget even went so far as to ensure man's spiritual progress via *Entwicklung*, finding in man's 'spiritual constitution' some 'embryo faculties which raise us above this earthly habitation', 'the traces of higher powers, to which those we now possess are but preparatory': a notion which Tennyson, in the true Romantic tradition, was shortly to celebrate in verse.[15]

This progressionist religion of embryology culminated in the paleo-biology of the famous Swiss–American naturalist and devout teleologist Louis Agassiz, who, while a student at Munich from 1827 to 1829, cut his teeth on the concepts of *Naturphilosophie* under Oken's direct tutelage. Agassiz made explicit the identification of ontogenetic stages with the fossil sequence, generalized it into a law of nature and familiarized it by reiterating it again and again in his writings, applying it with a thoroughness that bordered on the obsessive to all the phenomena uncovered by his extensive paleontological and embryological researches. Like Owen, Agassiz adopted von Baer's law of divergent development, but the none-too-muted undertones of the familiar metaphor of gestation impressed a unilinearity on his conception of nature which persistently neutralized von Baer's divergent schema of development. For Agassiz, the mastodon was an embryonic elephant, and fossils generally were the 'embryonic types' of the living organisms of the 'present creation', foetal structures which prophetically announced man's own creation.[16]

It was this dedicated life-long opponent of evolution in all its forms who, with all the force of his powerful reputation, made ontogenetic development the paradigm of all of nature's history and the leading argument for premeditated design amongst his British devotees. But in the process Agassiz also brought progressionism too close for comfort to the 'development hypothesis' of the *Vestiges*. Chambers could gather almost all the 'vestiges' he required from this unimpeachable source, and he could, much to their consternation, cite chapter and verse from some of these most respected and influential natural theologians in support of his hypothesis. Adam Sedgwick, the Cambridge cleric and progressionist, who abhorred the *Vestiges*, this 'foul book' which he brutally satirized as a 'deformed progeny of unnatural

conclusions' and 'a filthy abortion', and held responsible for the imminent
insurrection of the lower orders and the corruption of Victorian womanhood,
belatedly was made aware of the pitfalls of ambiguity and metaphor. As he
warned Agassiz in 1845, a great deal hinged on the troublesome term 'development' and its expression: 'Now I allow (as all geologists must do) a *kind* of
*progressive development*.' But 'Generation and creation are two distinct
ideas, and must be described by two distinct words, unless we wish to
introduce utter confusion of thought and language.'[17]

For Sedgwick, the publication of the translation of Oken's *Lehrbuch der
Naturphilosophie* by the Ray Society in 1847 was a most unwelcome revelation of the proximity of the progressionist faith to such ungodly (and almost
unintelligible) notions. Although they had become so familiar with certain of
its concepts in the guise of the argument from design, this was the first
opportunity most English readers had had for any close scrutiny of the
productions of German *Naturphilosophie*. A number, who deciphered 'symptoms of unsound religious principle' in Oken's hieroglyphs, now identified the
work as the 'polluted source' of the *Vestiges*, and the Ray Society was
censured publicly for its sponsorship of this 'obscene and atheistical work'.
Sedgwick, who denounced its author as 'a profane mystic, and a babbler',
even going so far as to exonerate the author of the odious *Vestiges* from
Oken's supposed excesses of atheism and materialism, expressed the pious
hope that God might forever save the University of Cambridge from this 'base,
degrading, demoralizing, hopeless creed!' But as Sedgwick well knew, it was
too late for divine intervention. The 'Romantic tide' had already reached
Trinity and spilled over into the progressionist faith. Canute-like, Sedgwick
could only join forces with the *Manchester Spectator* in warning his colleagues and protégés like Owen of the pressing political need to clarify the
ambiguity of their expressions, lest impressionable readers should misinter-
pret them and think (as Sedgwick himself had done in Owen's case!) that they
were endorsing 'some theoretical law of generative development from one
type of animal to another along the whole ascending scale of Nature'. Owen
took the hint, while Sedgwick schooled himself in the latest von Baerian
embryology and attempted to dissociate progressionism from the insidious
and newly suspect embryological law of parallelism: 'false at every step . . . an
idle dream of the philosophy of resemblances'.[18]

However, some, like Hugh Miller (who also did a good deal of hasty
backtracking from the precipice of progressive development), refused to
abandon such 'wonderful' analogies to the infidel: '[T]his strange fact of the
progress of the human brain is assuredly a fact none the less worth looking at
from the circumstances that infidelity has looked at it first.' For Miller, such
analogies gave 'fair warning that, in tracing to its first principles the moral and
intellectual nature of man, what is properly his "natural history" should not

be overlooked'.[19] Charles Lyell, the eminent uniformitarian geologist, who earlier had argued against both the progressionist and transmutationist implications of such transcendental embryological conceptions, now began to waver before the force of these same embryological arguments for the 'transmutation hypothesis'. As he thoughtfully concluded while reviewing the 'species question' in his private journal:

There are none who so much aid the transmutationist as they who push the doctrine of progressive development farthest – since their assimilation of the successive appearance of species and genera to embryonic develop[t] in an individual is the very opposite of that arbitrary fiat which at other times they invoke, it is creation working by law, according to a prescribed pattern & by a force analogous to that displayed in an individual from the embryo to the adult.[20]

One committed anti-progressionist who exploited the narrowing of the gap between transmutationists and progressionists to the full was the young and aggressively ambitious Huxley. During the fifties, he impartially lambasted what he called the 'ideal quasi-progressionism' of Agassiz and Owen, the development hypothesis of the *Vestiges* and, above all, Owen's morphology and paleontology, opportunistically conflating them and capitalizing on the current distrust of the embryological law of parallelism and its associations with an ideologically suspect *Naturphilosophie* and a politically dangerous evolutionism. The anticlerical and iconoclastic Huxley found Oken, Owen and the *Vestiges* as obvious a target as had the *Manchester Spectator*, if for rather different reasons. Setting himself in professional opposition to Owen's Oken-inspired 'osteological extravaganzas', Huxley aspired to become the British von Baer, and he determinedly applied von Baer's embryological criteria and methodology to the demolition of the pervasive 'popular notion' of the parallel between ontogeny and the fossil sequence in paleontology – a 'fallacious doctrine' which he foisted onto his arch-rival Owen. However, the budding positivist who refused even to use the term 'archetype', whose 'connotation is so opposed to the spirit of modern science', did not escape baptism in the Romantic tide. In this crucial period, Huxley himself toyed with some very Romantic 'mystifications', including this pertinent one: 'The individual animal is one beat of the pendulum of life, birth and death are the two points of rest . . . The different forms which an animal may assume correspond with the different places of the pendulum.'[21]

For Huxley before 1859 there was no evidence, embryological or paleontological, to support any hypothesis of the 'progressive development of animal life in time'. After 1859 it was, of course, a different story, as Huxley became the chief expounder and champion of a Darwinian 'world view' of 'harmonious order governing eternally continuous progress'. By the end of the sixties, Huxley and the Darwinians were virtually running British science from the epicentre of the influential X-Club and recruiting social support from an increasingly secular and 'progressive' middle class. By this time von Baer had disappointed Huxley's earlier expectation of his support for Darwinism, and Ernst Haeckel had emerged as Darwin's 'German bulldog'. Huxley relaxed his vigilance against the erstwhile 'fallacious doctrine' once Haeckel and others (including Darwin and Herbert Spencer) had given it an acceptable Darwinian gloss and begun to put it to socio-political use. Translated into the theory of recapitulation in the form of Haeckel's triumphal slogan, 'Ontogeny recapitulates Phylogeny', it went on to dominate embryological research until well into the twentieth century. And one of its leading British exponents was that selfsame Huxley, the prominent exorcist of Romantic influence from science.

By the seventies, Huxley was busily deducing ancestral structures from living embryos and using the theory of recapitulation to support the 'natural' inferiority of women and blacks. While he relentlessly maintained his vendetta against Owen's 'Okenism', ridiculing his embryogenetic theory of evolution and mocking the ambiguity of Owen's 'continuous creation', Huxley was praising Haeckel's work for having 'all the force, suggestiveness, and, what I may term the systematizing power, of Oken, without his extravagance'.[22] A Darwinized Oken was permissible. Huxley was nothing if not selective in his Romantic targets. When the journal *Nature* was founded in 1869, it was Huxley who responded to the prestigious invitation to contribute the opening article with a few pet Romantic metaphors – a selection of 'Aphorisms by Goethe', translated by himself – and who took a malicious delight in the ensuing mystification of the 'British Philistines' who thought that he had 'suddenly gone mad!'[23]

Historians increasingly see Darwinism as representing not so much a revolutionary break as a 'subtle accommodation' with natural theology, and the history of the metaphor of gestation offers an exemplary instance of this historical contention. Its long assimilation into British natural theology via progressionism left its indelible impress on the post-Darwinian theory of recapitulation in the conflation of von Baer's law with the earlier and dominant Romantic law of parallelism, and in the progressionist trend of the whole of nature's gestatory history. Let us permit the older and mellower Huxley the last word on these complex interactions of German Romanticism and British evolution and natural theology, as in 1888 he complacently retools the leading article of the by then redundant progressionist faith for Darwinian use:

It is quite certain that a normal fresh-laid egg contains neither cock nor hen; and it is also . . . as certain as any proposition in physics or morals, that if such an egg is kept under proper conditions for three weeks, a cock or hen chicken will be found in it . . . Therefore Evolution, in the strictest sense, is actually going on in this and analogous millions and millions of instances, wherever living creatures exist. Therefore, to borrow an argument from Butler, as that which now happens must be consistent with the attributes of the Deity, if such a Being exists, Evolution must be consistent with those attributes. And, if so, the evolution of the universe, which is neither more nor less explicable than that of a chicken, must also be consistent with them.[24]

## NOTES

1  See T. H. Huxley, 'Owen's Position in the History of Anatomical Science', in Rev. R. Owen (ed.), *The Life of Richard Owen*, 2 vols. (London, 1894), II, pp. 273–332 (p. 315); and A. Desmond, *Archetypes and Ancestors: Palaeontology in Victorian London, 1850–1875* (London, 1982).
2  F. W. J. Schelling, *The Ages of the World*, trans. F. de Wolfe Bolman (New York, 1942), p. 94. For the classic analysis of the relation of Romantic embryology to the rise of historicism, see O. Temkin, 'German Concepts of Ontogeny and History around 1800', *Bulletin of the History of Medicine*, 24 (1950), 227–46.
3  L. Oken, *Elements of Physiophilosophy* (London, 1847), p. 2. Alexander Gode-von Aesch has argued that the embryological law of parallelism was 'part of the intellectual equipment of every good Romantic thinker'; see his *Natural Science in German Romanticism* (New York, 1941, repr. 1966), p. 120.
4  Oken, *Elements of Physiophilosophy*, pp. 491–2.
5  K. E. von Baer, *Über Entwickelungsgeschichte der Thiere: Beobachtung und Reflexion* (Königsberg, 1828), pp. 129–201; English rendering by T. H. Huxley, 'Fragments Relating to Philosophical Zoology, Selected from the Works of K. E. von Baer' (trans. T. H. Huxley), in A. Henfrey and T. H. Huxley (eds.), *Scientific Memoirs* (London, 1853), pp. 176–238, esp. 186–8.
6  See E. Richards, 'The German Romantic Concept of Embryonic Repetition and its Role in Evolutionary Theory in England up to 1859' (unpublished University of New South Wales Ph.D thesis, 1976); E. S. Russell, *Form and Function: A Contribution to the Study of Animal Morphology* (London, 1916); and S. J. Gould, *Ontogeny and Phylogeny* (Cambridge, 1977).
7  See Oken, *Elements of Physiophilosophy*, pp. 185–93.
8  See T.A. Appel, *The Cuvier–Geoffroy Debate: French Biology in the Decades before Darwin* (Oxford, 1987).
9  See A. Desmond, 'Robert E. Grant: The Social Predicament of a Pre-Darwinian Transmutationist', *Journal of the History of Biology*, 17 (1984), 189–223.
10  See E. Richards, 'The "Moral Anatomy" of Robert Knox: The Interplay Between Biological and Social Thought in Victorian Scientific Naturalism', *Journal of the History of Biology*, 22 (1989), 373–436.
11  See E. Richards, 'A Question of Property Rights: Richard Owen's Evolutionism Reassessed', *British Journal for the History of Science*, 20 (1987), 129–71.
12  *Ibid.* The radical political associations of transmutation in this period are brought out in A. Desmond, 'Artisan Resistance and Evolution in Britain, 1819–1848', *Osiris*, 2nd ser., 3 (1987), 77–110.
13  [R. Chambers], *Vestiges of the Natural History of Creation* (London, 1844; repr. Leicester, 1969), p. 223. See also the excellent discussion by J. A. Secord, 'Behind the Veil: Robert Chambers and the Genesis of the *Vestiges of*

*Creation*', in J. R. Moore (ed.), *History, Humanity and Evolution* (Cambridge, 1989); and M. J. S. Hodge, 'The Universal Gestation of Nature: Chambers' *Vestiges* and *Explanations*', *Journal of the History of Biology*, 5 (1972), 127–51.

14  H. Miller, *The Old Red Sandstone* (Edinburgh, 1841), pp. 244–6. For a discussion of 'transcendental progressionism', see P. J. Bowler, *Fossils and Progress: Paleontology and the Idea of Progressive Evolution in the Nineteenth Century* (New York, 1976), pp. 47–62.

15  P. M. Roget, *Animal and Vegetable Physiology Considered with Reference to Natural Theology*, 2 vols. (3rd edn, London, 1840), II, p. 573. On Roget's transcendentalism, see P. F. Rehbock, *The Philosophical Naturalists: Themes in Early Nineteenth-Century British Biology* (Madison, Wisc., 1983), pp. 56–9. Tennyson's reference to the Romantic law of parallelism is to be found in the Prologue to 'In Memoriam'.

16  L. Agassiz, *Twelve Lectures on Comparative Embryology* (Boston, 1849), pp. 26–7, and 'On the Differences between Progressive, Embryonic, and Prophetic Types in the Succession of Organized Beings through the whole Range of Geological Times', *Edinburgh New Philosophical Journal*, 49 (1850), 160–5. On the relation between Agassiz and Oken, see E. Agassiz (ed.), *Louis Agassiz, his Life and Correspondence*, 2 vols. (Cambridge, 1887), I, pp. 34, 52–4, 150–4; and E. Lurie, *Louis Agassiz, a Life in Science* (Chicago, 1960), pp. 20–63.

17  Sedgwick to Agassiz, 10 April 1845, in J. W. Clark and T. M. Hughes (eds.), *The Life and Letters of Adam Sedgwick*, 2 vols. (Cambridge, 1890), II, p. 86.

18  See Richards, 'A Question of Property Rights', pp. 163–5.

19  H. Miller, *Footprints of the Creator: or, The Asterolepsis of Stromness* (London, 1849), pp. 291–3.

20  Lyell, 27 May 1859, in L. G. Wilson (ed.), *Sir Charles Lyell's Scientific Journals on the Species Question* (New Haven and London, 1970), pp. 66–7. For Lyell's earlier views on transcendentalism, see P. Corsi, 'The Importance of French Transformist Ideas for the Second Volume of Lyell's Principles of Geology', *British Journal for the History of Science*, 11 (1978), 221–44.

21  T. H. Huxley, 'Upon Animal Individuality', *Proceedings of the Royal Institution*, 1 (1851–4), 186; and his 'On Certain Zoological Arguments Commonly Adduced in Favour of the Hypothesis of the Progressive Development of Animal Life in Time', *Proceedings of the Royal Institution*, 2 (1854–8), 82–5. See Richards, 'A Question of Property Rights', pp. 167–71; and D. Ospovat, 'The Influence of K. E. von Baer's Embryology, 1828–1859: A Reappraisal in Light of Richard Owen's and William B. Carpenter's Paleontological Application of "von Baer's Law"', *Journal of the History of Biology*, 9 (1976), 1–28.

22  T. H. Huxley, 'The Genealogy of Animals', *The Academy* (1869), reprinted in his *Critiques and Addresses* (London, 1873), p. 301. For Huxley's use of recapitulatory arguments to endorse the inferiority of women and blacks, see E. Richards, 'Huxley and Woman's Place in Science: The "Woman Question" and the Control of Victorian Anthropology', in Moore (ed.), *History, Humanity and Evolution*.

23  T. H. Huxley, 'Aphorisms, by Goethe', *Nature*, 1 (1869), 9–11. See Huxley to Anton Dohrn, 30 Jan. 1870, in L. Huxley, *Life and Letters of Thomas Henry Huxley*, 2 vols. (London, 1900), I, pp. 326–7.

24  T. H. Huxley, 'On the Reception of the "Origin of Species"', in F. Darwin
    (ed.), *Life and Letters of Charles Darwin*, 3 vols. (London, 1888), II, pp. 202–
    3. On the 'subtle accommodation' of Darwinism with natural theology, see
    R. M. Young, *Darwin's Metaphor: Nature's Place in Victorian Culture*
    (Cambridge, 1985).

## FURTHER READING

A. Gode-von Aesch, *Natural Science in German Romanticism* (New York, 1966)

T. A. Appel, *The Cuvier–Geoffroy Debate: French Biology in the Decades before
Darwin* (Oxford, 1987)

A. Desmond, *Archetypes and Ancestors: Palaeontology in Victorian London, 1850–
1875* (London, 1982)
    *The Politics of Evolution: On Comparative Anatomy and the Political Order in the
    1820s and 1830s* (Chicago, 1989)

S. J. Gould, *Ontogeny and Phylogeny* (Cambridge, 1977)

T. Lenoir, *The Strategy of Life: Teleology and Mechanics in Nineteenth-Century
German Biology* (Dordrecht, 1982)

D. Ospovat, *The Development of Darwin's Theory: Natural History, Natural Theology and Natural Selection* (Cambridge, 1981)

P. F. Rehbock, *The Philosophical Naturalists: Themes in Early Nineteenth-Century
British Biology* (Madison, Wisc., 1983)

E. Richards, 'A Question of Property Rights: Richard Owen's Evolutionism Reassessed', *British Journal for the History of Science*, 20 (1987), 129–71.
    'The "Moral Anatomy" of Robert Knox: The Interplay between Biological and
    Social Thought in Victorian Scientific Naturalism', *Journal of the History of
    Biology*, 22 (1989), 373–436

E. S. Russell, *Form and Function: A Contribution to the History of Animal Morphology* (London, 1916)

J. B. Stallo, *General Principles of the Philosophy of Nature* (Boston, 1848)

# IO

## Transcendental anatomy

### PHILIP F. REHBOCK

It seems that nature has enclosed herself within certain limits, and has formed all living beings on only one unique plan, essentially the same in its principle, but which she has varied in a thousand ways in all its accessory parts.

Etienne Geoffroy Saint-Hilaire, 1795[1]

Transcendental Anatomy – The highest department of anatomy; that which, after details have been ascertained, advances to the consideration of the type or plan of structure, the relations between the several parts, and the theoretical problems thus suggested.

*The American Encyclopedic Dictionary*, 1894[2]

The sciences, like other human endeavours, proceed not by lock-step logic but by wonderfully varied, ofttimes inexplicable and sometimes admittedly fruitless pathways, as other chapters in this volume attest. The vogue of 'transcendental anatomy', one of the most curious – and for the historian one of the most elusive – episodes in nineteenth-century biology, provides rich confirmation of the pluralistic nature of natural science.

Straightforward, unambiguous definitions of 'transcendental anatomy' are rarely to be found either in dictionaries or in the literature of those who claimed to practise it. Even the venerable *Oxford English Dictionary* did not define it. The term seems to have been popularized in the 1820s by the French anatomist Etienne Reynaud Augustin Serres (1786—1868), who used it to designate collectively the morphological laws of animal development.[3] The formal definition quoted above appeared only at the end of the nineteenth century, however, and it gives but the barest hint of the controversies that transcendental anatomy evoked, and the careers it enhanced and tarnished, in the first half of that century. Roughly equivalent terms of that period, such as 'higher anatomy', 'philosophical anatomy' and 'transcendental morphology' will be treated as synonymous with 'transcendental anatomy' in this chapter.

The distinguishing characteristics of transcendental anatomy, throughout its history, were (a) the presupposition that a single Ideal Plan or Type (or, at most, a few such plans) lay behind the great multiplicity of visible structures in the animal and plant kingdoms, and that this Plan determined an organism's functional capacities rather than being determined by them; (b) the further presupposition that the Ideal Plan acted as a force for the maintenance of anatomical uniformity, in opposition to the diversity-inducing (some would argue *degenerating*) forces of the physical environment; (c) the belief that this *a priori* Plan, though it had no physical existence in its pure state, was nevertheless discoverable; and (d) the aspiration to discover additional concepts ('laws') which would support and elaborate the Ideal Plan by specifying how apparent anatomical diversities may be seen as uniformities. Despite the variety of other beliefs which the practitioners of transcendental anatomy brought to, or derived from, the enterprise, the above elements seem fundamental and common to their work.

Studies of the past decade have examined the development of transcendental anatomy in the writings not only of the best known representatives and spokesmen of the tradition – Goethe, Lorenz Oken, Geoffroy Saint-Hilaire and Richard Owen – but of a number of lesser-known naturalists of the era, such as Henri de Blainville, Edward Forbes, John Goodsir, Robert Edmond Grant, Joseph Henry Green, Robert Knox and Jeffries Wyman.[4] Surveying these works, several themes emerge. First, the propositions of transcendental anatomy were compatible with a surprisingly wide variety of other biological, metaphysical and socio-political doctrines; there was thus a malleability to transcendental anatomy which no doubt contributed to its popularity and longevity. Second, some advocates regarded transcendental anatomy as the ultimate explanation for biological structures, while others saw it as one of several necessary explanatory devices. One might call these two approaches the 'strong program' (high transcendentalism) and the 'eclectic program' (low transcendentalism). Third, transcendental anatomy, initiated primarily in the German provinces, elsewhere assumed a variety of forms. As it expanded in Paris and reached a culmination in Britain, transcendental anatomy grew more diverse, more complex and more interesting. In his classic exposition of anatomical traditions, *Form and Function* (1916), E. S. Russell said of transcendental anatomy that 'The philosophy seems to have come chiefly from Germany, the science from France.' To that terse assessment we might add that its variety and longevity seem most manifest in Britain. The purpose of this chapter will thus be to elaborate briefly these themes, and to suggest possible avenues for future exploration.

## GERMAN BEGINNINGS

The underlying spirit of transcendental anatomy derived from the same late-eighteenth-century sources as the rest of the Romantic movement: the intellectual alternative to empiricist, Newtonian philosophy presented by Kant, the German idealist philosophers and *Naturphilosophie*, and (to a lesser extent) the political attacks upon established autocratic rule presented by the apostles of nationalism and liberalism. We can best appreciate these initiating forces through an examination of the works, and lives, of Goethe and Oken.

If there must be a father (or mother) for every intellectual movement, then there is ample rationale for regarding Johann Wolfgang von Goethe (1749–1832) as the father of transcendental anatomy. The notion of an Ideal Plan or Type (*Urtyp*) for all organisms appears in the writings of various eighteenth-century naturalists and anatomists, both French and German, including Georges Buffon, L. J.–M. Daubenton, Albrecht von Haller, J. F. Blumenbach, C. F. Wolff and A. J. G. C. Batsch.[5] But the first passionate search for ideal plans has come to be identified with Goethe, and it was he who designated this search 'morphology'.

Goethe's fascination with anatomy began at least as early as the 1780s, when he studied botany with Batsch and comparative anatomy with Just Christian Loder at the University of Jena. By the end of that decade he was fully caught up in attempts to construct the *Urpflanze* and the *Urtier* – ideal archetypes for the plant and animal kingdoms. Moreover, in the case of plants Goethe extended the search for unity from the whole organism down to the level of individual organs, arguing that plant organs, especially floral parts, are all modifications – *metamorphoses* – of a primal leaf form.[6] Although these ideas did not reach print until years later, they were the first on the subject to be widely known and cited, and they had a powerful effect on the younger generation of naturalists in the 1790s and early 1800s.

For Goethe, the archetype may have had only an ideal not an actual existence – it was a tool for understanding phenomena, not a reality of nature; in Kantian terms the archetype had a *regulative*, not a *constitutive* function. But for subsequent transcendental anatomists, including the *Naturphilosophen* in general and Oken in particular, this distinction broke down: the archetype was not just a schema of reason but an objective, historical entity; not regulative but constitutive of nature. Lenoir argues that this deviation, in effect an abandoning of the rules established by Kant, was common to the generation of German naturalists who reached intellectual maturity during the French Revolution, and that their altered approach to transcendental anatomy was shaped, at least in part, by the socio-political aspirations of the new era. The program of the *Naturphilosophen* was thus the product of Kant's transcendental idealism coupled with the un-Kantian belief that transcendental ideas, like the Ideal Plan, have an objective existence in Nature. Put another way, *Naturphilosophie* was German philosophical idealism yielding to French political radicalism. Friedrich von Schelling, G. W. F. Hegel and Friedrich Schlegel were among the principal fomenters of this transformation in the 1790s. For its anatomical ramifications, however, an examination of the work of Lorenz Oken (1779–1851) is especially intriguing.

In 1807 the young Oken was recommended by none other than Goethe for his first post – at the cauldron of transcendentalism, the University of Jena. Thus, if Goethe's morphological notions did not initially reach a wide audience in printed form, he contributed nevertheless to ensuring the longevity of transcendental anatomy by other means. The '*doyen* of *Naturphilosophie*',[7] Oken's twin dreams were the realization of a strong German natural scientific tradition founded upon *Naturphilosophie* and the establishment of a unified German state founded upon enlightened political ideals. His journal *Isis*, begun in 1816, was intended as a vehicle for the simultaneous pursuit of these two goals. Not surprisingly, the intemperance of his political activities brought on the suppression of that side of the enterprise (and censure by Goethe). But his work for German natural science continued unabated. By the end of his career Oken could boast the publication of his thirteen-volume *Allgemeine Naturgeschichte für alle Stände* and the establishment of the annual meetings of the *Gesellschaft Deutscher Naturforscher und Ärzte* (1822, a model for the British and American Associations for the Advancement of Science). A host of later scientists and historians has attested to the brilliance of many of his insights, such as his conception of a prototypical cell theory some thirty years prior to the creation of the widely accepted version of Schleiden and Schwann. Often these insights were couched more in the mystical language of a cosmologist than in the technical details of the anatomist, but they were enormously suggestive, perhaps for that very reason.

To transcendental anatomy Oken contributed the theory of the ideal vertebrate archetype. A single generalized vertebra was the basic unit of animal design, he believed, and a sequence of such ideal units constituted the *Urtyp* or primitive model for vertebrates, for animals the equivalent of Goethe's primal leaf form for the plant kingdom. A component of the archetype doctrine was the vertebrate theory of the skull, the notion propounded by Goethe and others that the skull consists of a number of modified and fused vertebrae. Oken made this doctrine the topic of his inaugural dissertation at Jena.[8] In all, the vertebral archetype proved to be one of the most enduring concepts of transcendental anatomy, receiving its extension toward the invertebrate classes by Geoffroy Saint-Hilaire and its most substantial explication by Richard Owen, as we shall see.

Before leaving the German segment of our story, we should note that one of the most appealing and long-lived doctrines of transcendental anatomy, that of embryological recapitulation, also had its origins in Germany. Eighteenth-century anatomists had noted that many animal species, as they develop embryologically, temporarily manifest structures which closely resemble those exhibited by the *adult* forms of species lower down in the scale of animal organization. Possibly the first to pronounce this phenomenon to be a law and

8    Lorenz Oken: the frontispiece to his *Allgemeine Naturgeschichte für alle Stände* of 1839.

to emphasize it in his teaching (1793) was Carl Friedrich Kielmeyer (1765–1844), instructor at the Karlsschule in Stuttgart (where he tutored Georges Cuvier) and later professor at Tübingen. The detailed explication of the law was carried out by Johann Friedrich Meckel (1781–1833) in Germany, and by Serres in France, and has often been referred to as the Meckel–Serres Law. To distinguish it from Ernst Haeckel's 'biogenetic law' of evolutionary recapitulation, the Meckel–Serres Law is sometimes called the Law of Parallelism.[9]

## FRENCH DEVELOPMENTS

From its origins in Kantian philosophy, Goethian inspiration and *naturphilosophisch* enthusiasm, transcendental anatomy passed from Romantic Germany to Napoleonic France, where it received its strongest advocacy from Etienne Geoffroy Saint-Hilaire (1772–1844) and Henri de Blainville (1777–1850). Geoffroy worked his influence during a forty-eight-year career (1793–1841) as professor of vertebrates at the Paris Muséum d'Histoire Naturelle. He thus preceded *and* outlived his nemesis, Georges Cuvier (1769–1832), the Muséum's professor of comparative anatomy and the panjandrum of French natural science during the first third of the century. Had it not been for Cuvier's domination of both the content and the patronage of natural history during this period, Geoffroy's influence might have been far greater than it was. On the other hand, had it not been for the decade-long controversy between Geoffroy and Cuvier, transcendental anatomy might never have achieved the visibility outside of Germany that it clearly did.

As the champion of transcendental anatomy, Geoffroy was in diametric opposition to Cuvier. The latter's career was early established upon the careful analysis of animal *function*, rather than form; and on this basis he erected a biological determinism in which zoological structure was strictly dictated by functional necessity. Thus, Geoffroy epitomized the morphological point of view, Cuvier the teleological. The controversy was a multidimensional affair, however, reaching from scientific fact and method to religion, politics and popular views of nature. In her recent book, *The Cuvier–Geoffroy Debate*, Toby Appel addresses these many facets admirably. 'Geoffroy was depicted by some', she summarizes, 'as a philosopher dedicated to unraveling the mysteries of nature for the common man, while Cuvier was seen as an elitist fact collector, upholder of Biblical orthodoxy, manipulator of patronage, and suppressor of the ideas of men like Lamarck and Geoffroy.'[10]

Geoffroy, whose philosophical sympathies lay with Enlightenment materialism and deism rather than *naturphilosophisch* idealism and pantheism, was nevertheless France's chief practitioner of transcendental anatomy. As early as 1795, his publications began to refer to the unity of plan in the animal kingdom (recall the opening quotation of this chapter), a morphological orientation he absorbed more likely from his countrymen and mentors Daubenton and especially Buffon than from the German *Naturphilosophen*.[11] Cuvier, for his part, had just then been appointed (with Geoffroy's help) to a position at the Muséum, was beginning to collaborate with Geoffroy on joint publications and had not yet established a biological philosophy of his own. The latter was not slow in coming, however. By 1802, when Geoffroy returned from a four-year scientific expedition with Napoleon in Egypt, their positions had diverged radically.

Geoffroy's most enduring contribution was the concept of *homology*. 'Homologous' parts or organs are anatomical elements of different species which have the same relationship to the ideal plan. They may have differing shapes or perform different functions, but their position in the overall design of the body is identical. Once the multiplicity of homologues throughout the animal world becomes apparent, one is led inductively and inevitably to what Geoffroy called the 'principle of the unity of organic composition for all the vertebrates' – the same parts, the same building blocks, are used in the construction of all vertebrates.[12]

Geoffroy began in earnest the process of establishing homologies among

9    Etienne Geoffroy Saint-Hilaire in the 1820s.

the higher vertebrates in a series of publications on the bones of fishes in 1807. Here emerged what Geoffroy would later call the 'principle of connections': the essence of the homologue is not its function or its shape but its connections with surrounding parts. Geoffroy's ability to identify totally unexpected homologues in very different animals advanced him to a new level of professional prominence among European naturalists. By 1815, associates such as Jules-César Savigny and Henri de Blainville had followed his lead by extending the search for ideal plans into the insect world. But the apex of Geoffroy's career came in 1818 with the appearance of the first volume of his two-volume *magnum opus*, the *Philosophie anatomique*. Here Geoffroy attempted to correlate, on the grand scale, the homologues of all vertebrate animals.

The *Philosophie anatomique* also brought on the beginnings of overt disagreement with Cuvier. As Geoffroy attempted, through the 1820s, to demonstrate a unity of plan for larger and larger portions of the animal kingdom, Cuvier became increasingly critical of the entire endeavour. For example, Geoffroy claimed the exoskeletons of insects and crustacea were homologous with the skeletons of vertebrates. Finally in 1829 two of Geoffroy's students attempted to bring the mollusks into the fold by establishing homologies between cephalopods and vertebrates. This work ignited an open debate with Cuvier before the Académie des Sciences which lasted for

10  Interior of the amphitheatre of comparative anatomy at the Jardin des Plantes in the nineteenth century.

two months. Although Cuvier's easy demolition of the excesses of transcendentalism made him appear an immediate victor, the debate highlighted the need for a compromise position which would acknowledge the value of both formal and functional approaches in anatomical research.

For British followers of the tradition, and for many subsequent historians, Geoffroy was unquestionably transcendental anatomy's leading theorist and spokesman. But as Appel has argued,[13] Henri de Blainville must be regarded as a close competitor for that role. Blainville was professor of zoology and comparative anatomy of the Paris Faculté des Sciences, and upon Cuvier's death in 1832 assumed the latter's chair of comparative anatomy at the Muséum. He thus concluded his career as he had begun it, in a position close to Cuvier, since his first post (1810) had been as Cuvier's disciple and collaborator in comparative anatomical research. But by 1816 a conflict of strong personalities had developed, and from then on Blainville was forced to make his way against the tide of patronage that Cuvier commanded.

The conflict was probably inevitable for ideological as well as for personal reasons, because Blainville's biological philosophy soon diverged from the functional anatomy, *embranchement* taxonomy and catastrophism that were the hallmarks of Cuvier's scientific career. Blainville's thinking was coloured throughout by a belief in the animal series, a modified conception of the eighteenth-century chain of being doctrine which he rescued from the destructive attempts of Daubenton, Geoffroy and especially Cuvier. According to this conception, all animals can be ordered in a single linear hierarchy, intermediate forms (some still living, others only fossil) always linking what appear to be distinct groupings.

To support his belief in the animal series, Blainville adopted elements from several sources, including *Naturphilosophie* and Geoffroy's transcendentalism. As early as the 1810s he subscribed to the transcendental principles of parallelism, serial homology and the vertebral origin of the skull. For example, he looked for parallelism between the cardiac structures of mammalian foetuses and adult fishes and reptiles. And, with Savigny, he extended the principle of serial homologies from the vertebrates to the articulated animals (insects, crustacea, etc.).

Commitment to the animal series forced Blainville to disagree with Geoffroy on some points, especially Geoffroy's revered unity of plan. A hierarchical system of animal forms implied that superior types must differ from inferior ones by the addition of new and distinct organs (most importantly in Blainville's scheme, organs of sensation and locomotion), thus destroying any overall unity. As a result Blainville did not feel compelled to support Geoffroy in the debate with Cuvier. Transcendental anatomy, for him, was not an end in itself nor a tool for realizing the unity of nature; it was a means of elucidating the animal series.

The strength of Blainville's devotion to the animal series derived from religious and political as well as scientific sources. As a committed Roman Catholic and royalist, he believed that a hierarchical structure was intrinsic to both nature and society. This structure had been established, Blainville believed, at the original (and only) Creation; thus he was firmly opposed to transformism. This conventionalism was bound to make Blainville's modified transcendentalism more attractive (or at least palatable) to conservative Tory/Anglican naturalists in Britain, than Geoffroy's strong program with its connotations of deism and liberalism. As we shall see, the transcendentalism of Richard Owen shows many affinities with that of Blainville, and indeed there survives a substantial exchange of letters between them during the period 1833–50.[14]

## BRITISH CONCLUSIONS

Transcendental anatomy began to influence biological thought in Britain in a significant way within about a decade of its début in France. By the late 1820s, as the debate between Cuvier and Geoffroy was heating up, transcendental anatomy was being incorporated into lectures, and to a lesser extent into research, by British anatomists and naturalists. By the 1850s, Britain had become the principal remaining stronghold of the tradition.

The paths by which transcendental anatomy reached Britain at this time were several. Until recently historians of biology regarded Richard Owen (1804–92) as the key to understanding transcendental anatomy in Britain. First as conservator of the Royal College of Surgeons' Hunterian Museum and then as director of the Natural History section of the British Museum, Owen exercised a power over both theory and politics in zoology, comparative anatomy and paleontology, that was unequalled during the middle decades of the nineteenth century. In this respect he occupied a position comparable to that of Cuvier in France during the first third of the century.

Although he expressed a preference for formal, *vice* final, causes in lectures as early as 1837, Owen's most influential contributions to transcendental anatomy came in the 1840s. First in a long analysis presented before the British Association (1846) and then in two books, *On the Archetype and Homologies of the Vertebrate Skeleton* (1848) and *On the Nature of Limbs* (1849), Owen elaborated the skeletal homologies of the vertebrates. Especially eye-opening were his diagrams of the 'ideal typical vertebra' – the goal of Oken finally given pictorial reality (see Fig. 11); and the 'archetypus' of the vertebrates – the skeleton of Goethe's long-sought *Urtier* (see Fig. 12). These feats would have entitled Owen to the appellation 'the British Geoffroy' had it not been for the fact that he had already become known as 'the British Cuvier'. His approach was indeed an eclectic one, employing transcendentalism and teleology as the situation warranted.[15]

To his contemporaries, Owen was easily Britain's most audible apostle of

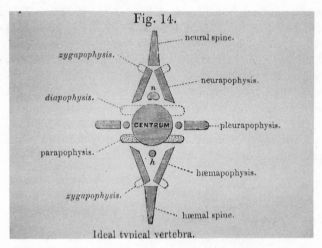

Fig. 14.

Ideal typical vertebra.

11   The ideal typical vertebra, according to Richard Owen. From *On the Archetype and Homologies of the Vertebrate Skeleton* (1848).

transcendental anatomy. In truth, however, he was only one of the officers of a small battalion of devotees of transcendentalism in that country. Well before Owen's stardom in the 1840s, transcendentalism had been imported to both London and Edinburgh by J. H. Green, Robert Knox and R. E. Grant.

Joseph Henry Green (1791–1863) is probably the least well-known of all the participants in this saga.[16] As professor of anatomy at the Royal College of Surgeons, King's College, London, and the Royal Academy successively, Green lectured to a wide audience. His attachment to transcendental anatomy came not through apprenticeship with like-minded anatomical mentors, but from an abiding fascination with German idealism. His early education had included three years' study in Germany, and in 1817 he returned to Berlin to read transcendental philosophy. This avocation was reinforced at the same time by an acquaintance with Coleridge, whose role in spreading *Naturphilosophie* to British friends like Humphry Davy is well known. Green became such a close associate of Coleridge that when the poet died in 1834 Green took on the tasks not only of literary executor but of 'systematizing, developing and establishing the doctrines of the Coleridgian philosophy'.[17]

Green seems to have introduced no novel concepts into the transcendental tradition. His eloquence and enthusiasm in lecturing on the subject, however, were pivotal to its dissemination in London. Owen himself heard Green's lectures while a medical student in the 1820s and later wrote:

For the first time in England the comparative Anatomy of the whole Animal Kingdom was described, and illustrated by such a series of enlarged and coloured diagrams as had never before been seen. The vast array of facts was linked by reference to the underlying Unity, as it had been advocated and illustrated by Oken and [Carl Gustav] Carus.[18]

At the opposite extreme from Green, Robert Knox (1793–1863) has attracted probably the most interest in the recent historiography of transcendental anatomy – and not without reason, for he was a colourful, outspoken individual whose life was packed with radical notions, tragic events and puzzles for the historian. After an Edinburgh medical training, service as an army surgeon (including a South African tour) and a year of study with Geoffroy, Cuvier and Blainville at the Paris Muséum, Knox settled in Edinburgh as an extramural teacher of anatomy. To his enthusiastic students, and in later years to his reading audience, he energetically purveyed the 'philosophy of osseous form'.[19]

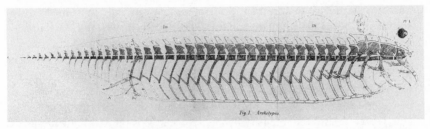

12 Archetype of the vertebrate skeleton, according to Richard Owen. From *On the Archetype and Homologies of the Vertebrate Skeleton* (1848).

Knox's biological philosophy was a unique blend of transcendentalism, materialism and polemic, poorly appreciated by earlier scholars partly because he failed to set forth his ideas systematically in print. They must be gleaned largely from his translation of Blainville's lectures and from miscellaneous articles which were published only in the 1850s. In an excellent study of 'The "Moral Anatomy" of Robert Knox', however, Evelleen Richards substantially advances our understanding of Knox's thought.[20] Arguing that his ethnological notions and political radicalism formed a consistent system with his biological ideas, Richards has shown that many of the anomalies in Knox's transcendentalism can be resolved by careful attention to his writings on race. Unlike most transcendental anatomists, but in common with Geoffroy's later thought, Knox subscribed to a limited evolutionism. He opposed both the direct environmental determinism of Lamarck and the randomness of Darwin's natural selection, along with the implications of progress in both hypotheses. But he insisted on the genetic continuity of all life, and theorized that evolution resulted not from transmutation but through a process of saltatory descent. Every plant and animal *genus*, he claimed, carried all the traits of its past, present and future species; thus it had the capacity to replace one species with another simply by arresting the development of some traits

and provoking the development of others in the course of embryonic development. This theory did not of course solve the question of the origin of *genera*,
but Knox was sure that the answer lay in material, not divine, causes. More
important to Knox, however, was that the theory, at least in his view,
supported both the unity of origin of the races of man and their distinction as
separate species. He was especially vehement on the latter doctrine, insisting
that racial differences were the key to understanding not only Victorian
colonial difficulties, but all of human history. The intersection of transcendental anatomy and radical politics in Knox's thinking produced a 'moral
anatomy', which would colour anthropological and Darwinian debates well
after his death, though often in ways he would not have endorsed.

The admixture of transcendentalism and materialism in Knox demands
further explanation. To this point we have associated transcendental
anatomy with various forms of idealism, which generally presumed a transcendent God as the ultimate cause of sensible events. Here L. S. Jacyna's
analysis of transcendence and immanence in early-nineteenth-century British
physiology provides a possible key to Knox's thought.[21] The issue of whether
life was inherent in organized matter, as immanentists claimed, or superadded
to matter by divine power, as transcendentalists believed, was an issue of
considerable gravity among British medical thinkers during the Napoleonic
era. Jacyna relates the intensity of the debate at this time to the widespread
apprehensions of religious and political conservatives who saw their Old
Order threatened by the same radicalisms that had brought down the *ancien
régime*. Immanentism in their eyes was inimical to morality and thus tantamount to Jacobinism. For their part, the immanentists saw transcendentalism
as sanctioning the authoritarianism and corruption of the traditional structures of power and belief.

Prominent among the transcendentalists in this debate was John Barclay,
Knox's senior partner in the Edinburgh extramural school of medicine. But by
the time Knox entered into teaching with Barclay his own radical democratic
sympathies were already well established. To subscribe to the traditional,
idealist form of transcendentalism would have carried with it a commitment
to Tory political and religious doctrines that Knox found anathema. Thus, on
the question of the source of life, he could only have felt comfortable with a
materialist, immanentist metaphysics.

The reputations of Owen, Green and Knox benefited substantially from
their having had sympathetic Victorian biographers. Such was not the case
with Robert Edmond Grant (1793–1874). Adrian Desmond has pieced
together an excellent picture of Grant, however, in spite of the poverty of
original source materials.[22] Like Knox, Grant grew up in Edinburgh, began his
career there, then moved to London. His early work showed considerable
promise and in 1827 he was appointed the first professor of comparative

anatomy and zoology at the University of London. But financial exigency and other difficulties kept him at lecturing and prevented further research and publication. Like those of Knox, his fortunes began to decline in the 1830s. His Fellowship in the Zoological Society, easily achieved in 1833, was just as quickly terminated in 1835, due in large measure to growing opposition from Owen. Though living in relative poverty by the 1840s, Grant did not abandon the radical ideals which were responsible, at least in part, for his misfortunes.

Again like Knox, Grant was an enthusiastic supporter of the French transcendentalists Geoffroy and Blainville. From 1815 until the 1830s Grant made regular trips to Paris (living and studying with Owen on at least one occasion). In England he was lecturing on transcendental anatomy at the University of London by the 1830s. Long after Geoffroy's defeat in the 1830 debate with Cuvier, Grant persisted in the search for evidence of the unity of plan in the mollusks and vertebrates. At the same time he subscribed to Blainville's conception of the animal series. But Grant's transcendentalism was unique in that, for him, the unity of plan was not so much testimony of a divine plan or an ideal pattern as it was evidence for transformism. For, along with the political and religious heresies he shared with Knox, Grant added a commitment to Lamarckianism. The unity of plan throughout the animal kingdom, evident to Grant in his early studies of invertebrates from the Firth of Forth, could only be the result of a unity of descent. All species were the evolutionary end-products of the first living forms which had been spontaneously generated and then altered, he thought, as a result of the earth's declining central heat. But throughout this process the unity of plan of the animal series had been preserved.

Thus, ironically, whereas Knox was delivered unto a belief in descent (at least at the species level) by his commitment to materialism, Grant arrived at a full-blown Lamarckian transformism via a dedication to the idealist doctrine of the unity of structure – varying combinations of unorthodoxies having transcendental anatomy and an opposition to Paleyite teleology as their common elements.

If transcendentalism and especially transformism were heterodox notions in Britain in the 1820s and 1830s, where did controversial advocates like Knox and Grant find a sympathetic forum for voicing their ideas? (It should be no surprise that their names are rarely found in the *Reports* of the British Association.) In Edinburgh in the 1820s the Plinian Natural History Society and Robert Jameson's *Edinburgh Philosophical Journal* were available and used by both Knox and Grant. In their later years in London they found *The Lancet* a ready outlet, while Knox was also able to expound his views in *The Zoologist*. The other source of support to these 'professors' of transcendental anatomy was of course their students. Before abandoning Edinburgh, Knox passed on the fervour for 'higher anatomy' to many of his hearers, most notably to the naturalist Edward Forbes, to the anatomist John Goodsir and to his biographer Henry Lonsdale. For his part, Grant could claim as converts physiologists P. M. Roget and W. B. Carpenter, entomologist George Newport and even Richard Owen (who of course in later years would have eschewed any serious indebtedness to Grant).

# AFTERWORD

In common with most other scientific creations of the Romantic era, transcendental anatomy would be superseded by other less empirically elusive programs. Ideal plans would be replaced by branching phylogenetic trees, archetypes by ancestors and speculative 'excesses' by 'sober' observation. Newtonian natural science did not survive unscathed the challenges of Kant and the *Naturphilosophen*, however; both physics and biology had acquired a host of new concepts by the end of the nineteenth century, concepts that would have been intractable for Enlightenment men of science. Those concepts were in large measure a tribute to the Romantic ideals of a science that operated intuitively, synthetically and anti-mathematically and viewed nature holistically, organismically and as the product not of atoms but of forces.

In common with the Romantic movement generally, transcendental anatomy owed much of its appeal to the grand goal of finding unity in nature, and to the expression of its results in aesthetically pleasing, almost poetic form. Thus, its demise must be attributed not only to the discovery of newer facts and theories, but also to the spread of a more cautious and parsimonious style of prose (and thought) throughout science as the nineteenth century advanced, a style we now identify automatically as that of the professional scientist. The style of a Buffon or a Schelling was simply inappropriate once science had become the domain of the specialist.

But the alternative, Romantic view of natural science has retained its appeal, occasionally among scientists and more frequently among laymen. Much of the original spirit of transcendental anatomy recurs from time to time in works of pure morphology: D'Arcy Thompson's *On Growth and Form* (1917) is twentieth-century zoology's version of *Naturphilosophie*, while Goethe's transcendental botany finds its modern exposition in Agnes Arber's *The Natural Philosophy of Plant Form* (1950).[23] But such works are exceptional. Far more common are the censures of the values and methodology of established science that come from outside the scientific community, and it is fascinating to note how frequently they resonate with the criticisms posed by the original Romantics. Kant's critique endures.

## NOTES

1 Quoted in Toby A. Appel, *The Cuvier–Geoffroy Debate: French Biology in the Decades before Darwin* (Oxford, 1987), p. 28.
2 Robert Hunter (ed.), *The American Encyclopedic Dictionary*, 4 vols. (Chicago, 1894), IV, p. 4151.
3 Appel, *The Cuvier–Geoffroy Debate*, p. 122.
4 Blainville, Grant, Green and Knox are discussed in this chapter. On Forbes see Philip F. Rehbock, *The Philosophical Naturalists: Themes in Early Nineteenth-Century British Biology* (Madison, Wisc., 1983); and Eric L. Mills, 'A View of Edward Forbes, Naturalist', *Archives of Natural History*, 11 (1984) 365–93. On Goodsir, see Peter D. James, 'John Goodsir (1814–1867) and the Edinburgh School of Anatomy', (unpublished University of Cambridge M. Phil. thesis, 1984). And on Wyman, see Toby A. Appel, 'Jeffries Wyman, Philosophical Anatomy, and the Scientific Reception of Darwin in America', *Journal of the History of Biology*, 21 (1988), 69–94.
5 Timothy Lenoir, 'Generational Factors in the Origin of *Romantische Naturphilosophie*', in *Journal of the History of Biology*, 11 (1978), 57–100. See also Lenoir's 'The Göttingen School and the Development of Transcendental *Naturphilosophie* in the Romantic Era', *Studies in History of Biology*, 5 (1981), 111–205, and *The Strategy of Life: Teleology and Mechanics in Nineteenth-Century German Biology* (Dordrecht, 1983). Lenoir's writings are among the most useful and stimulating explications of the German transcendental anatomists, and I have depended heavily upon them in this section.
6 On Goethe's natural history, see Agnes Arber, trans., 'Goethe's Botany', *Chronica Botanica*, 10 (1946), 63–126 (a translation of Goethe's *Versuch, die Metamorphose der Pflanzen zu erklären*); Robert Bloch, 'Goethe, Idealistic Morphology and Science', *American Scientist*, 15 (1952), 317–22; R. H. Eyde, 'The Foliar Theory of the Flower', *American Scientist*, 63 (1975), 430–7; James L. Larson, 'Goethe and Linnaeus', *Journal of the History of Ideas*, 28 (1967), 590–6; R. Magnus, *Goethe as a Scientist* (New York, 1949); H. B. Nisbet, *Goethe and the Scientific Tradition* (London, 1972); George A. Wells, *Goethe and the Development of Science, 1750–1900* (Alphen aan den Rijn, Netherlands, 1978).
7 Pierce C. Mullen, 'The Romantic as Scientist: Lorenz Oken', *Studies in Romanticism*, 16 (1977), 381–99, p. 382. See also Alexander Ecker, *Lorenz Oken: A Biographical Sketch* (London, 1883), and Richard Owen's entry, 'Oken, Lorenz', *Encyclopaedia Britannica*, 11th edn.
8 E. S. Russell, *Form and Function: A Contribution to the History of Animal Morphology* (London, 1916), ch. 7.
9 *Ibid.*
10 Appel, *The Cuvier–Geoffroy Debate*, pp. 22–7.
11 *Ibid.*, p. 12.
12 *Ibid.*, p. 71.
13 Toby A. Appel, 'Henri de Blainville and the Animal Series: A Nineteenth-Century Chain of Being', *Journal of the History of Biology*, 13 (1980), 291–319.
14 *Ibid.*, p. 318, note 95.

15  On Owen see Adrian Desmond, *Archetypes and Ancestors: Palaeontology in Victorian London, 1850–1875* (London, 1982); Roy M. MacLeod, 'Evolutionism and Richard Owen, 1830–1868: An Episode in Darwin's Century', *Isis*, 56 (1965), 259–80; Rev. Richard Owen, *The Life of Richard Owen*, 2 vols. (London, 1894); Rehbock, *The Philosophical Naturalists*, ch. 3; Russell, *Form and Function*, ch. 8; and Nicholas Rupke, 'Richard Owen's Hunterian Lectures on Comparative Physiology, 1837–55', *Medical History*, 29 (1985), 237–58.

16  On Green see Joseph Henry Green, *Spiritual Philosophy: Founded on the Teaching of the Late Samuel Taylor Coleridge*, edited, with a memorial of the author's life, by John Simon, 2 vols. (London and Cambridge, 1865); Phillip R. Sloan, 'Darwin, Vital Matter, and the Transformism of Species', *Journal of the History of Biology*, 19 (1986), 369–445. I am indebted to Professor Sloan for bringing Green to my attention some years ago.

17  Green, *Spiritual Philosophy*, I, pp. xxxiv–xxxv.

18  *Ibid.*, I, p. xiv.

19  Henry Lonsdale, *A Sketch of the Life and Writings of Robert Knox the Anatomist* (London, 1870), p. 144. On Knox, see also Isobel Rae, *Knox, The Anatomist* (Springfield, Ill., 1964); Rehbock, *The Philosophical Naturalists*, ch. 2.

20  Evelleen Richards, 'The "Moral Anatomy" of Robert Knox: The Interplay between Biological and Social Thought in Victorian Scientific Naturalism', *Journal of the History of Biology*, 22 (1989), 373–436.

21  L. S. Jacyna, 'Immanence or Transcendance: Theories of Life and Organization in Britain, 1790–1835', *Isis*, 74 (1983), 311–29.

22  Adrian Desmond, 'Robert E. Grant: The Social Predicament of a Pre-Darwinian Evolutionist', *Journal of the History of Biology*, 17 (1984), 189–223; see also his 'Robert E. Grant's Later Views on Organic Evolution: the Swiney Lectures on "Palaeozoology", 1853–1857', *Archives of Natural History*, 11 (1984), 395–413, and *Archetypes and Ancestors*, ch. 4.

23  D'Arcy Wentworth Thompson, *On Growth and Form* (Cambridge, 1917); Agnes Arber, *The Natural Philosophy of Plant Form* (Cambridge, 1950).

## FURTHER READING

Appel, Toby A., *The Cuvier–Geoffroy Debate: French Biology in the Decades before Darwin* (New York and Oxford, 1987)

Arber, Agnes, *The Natural Philosophy of Plant Form* (Cambridge, 1950)

Coleman, William, *Biology in the Nineteenth Century: Problems of Form, Function, and Transformation* (New York, 1971)

Desmond, Adrian, *Archetypes and Ancestors: Palaeontology in Victorian London, 1850–1875* (London, 1982)

Lenoir, Timothy, *The Strategy of Life: Vital Materialism in Nineteenth-Century German Physiology* (Dordrecht: 1983)

Rehbock, Philip F., *The Philosophical Naturalists: Themes in Early Nineteenth-Century British Biology* (Madison, Wisc., 1983)

Russell, E.S., *Form and Function: A Contribution to the History of Animal Morphology* (London, 1916)

# II

# Romantic thought and the origins of cell theory

## L. S. JACYNA

### INTRODUCTION

Every science possesses an 'official' history: the terse account generally found in the early pages of textbooks or in introductory lectures. Sometimes versions also appear in works that purport to be serious essays in the history of science. In the case of histology the story goes something like this. 'The history of histology', we are told, 'begins for all practical purposes with Bichat's work'[1] – that is, the work of the French anatomist François Xavier Bichat (1771–1802) at the turn of the nineteenth century. Bichat's contribution was to divert the attention of anatomists away from organs *en masse* and to direct it instead to the constituent *tissues* of the animal body. These tissues were the true seats of vital properties and the loci of pathological change.

Bichat was a martyr to his science: he died from an infected wound sustained in the dissection-room. After his death a long hiatus followed; 'nothing of much consequence for animal tissues followed [Bichat's researches] for many years'.[2] After 1830, however, a new generation of researchers took up the torch. They advanced upon Bichat's achievement because, while he had contented himself with investigating the constituent parts of animals by naked-eye examination, they undertook histological researches with the aid of the microscope. These inquiries culminated in the publication in 1839 of the *Mikroskopische Untersuchungen (Microscopical Researches)* of Theodor Schwann (1810–82) in which the cell theory received its first full exposition. Bichat's reluctance to use the microscope is explained by the optical deficiencies of the instruments available in his day; these had to be overcome before histology could advance from the gross to the minute level.

A number of characteristics of this account of the history of histology should be noted. First, it assumes an essential *continuity* between the work of Bichat and of Schwann: both were engaged in the same enterprise – the creation of modern histology. Bichat laid the necessary foundations for this science; but his work had to be completed by Schwann and other later workers because of the limited means of investigation available around 1800.

There is little room in this account for Romanticism – or indeed any intellectual or cultural movement – to have played a role in the inception and development of tissue doctrines in the nineteenth century. The only explanation proferred for the manifest differences between what Bichat and Schwann had to say about animal tissues is a technological one: the former could not make all the observations available to the latter because he did not use the same instruments.

A more sophisticated analysis belies many of these assumptions. Historians who have closely studied the texts of Schwann and of Bichat have challenged the continuity supposed to exist between their researches. They have pointed out fundamental differences in outlook between the two, which cannot be accounted for merely by saying that one pursued histology at the macroscopic and the other at the microscopic level. These divergent approaches to the animal body can only be explained by an attention to the theoretical structures with which Bichat and Schwann worked; and these structures must, in turn, be seen as aspects of a wider cultural milieu.

## BICHAT'S GENERAL ANATOMY

Bichat's concern was to offer a topographical or natural-historical account of tissues to which was subjoined an analysis of their vital properties.[3] His plan was 'to consider in isolation and to present with all their attributes each of the simple systems, which by their diverse combinations, form our organs'.[4] These 'simple systems' comprised the cellular, arterial, venous, exhalant and absorbent tissues, which provided 'a common and uniform basis for every organised part'.[5]

But, although Bichat called these systems 'simple', this was true only in a relative sense. In fact, they were all *compound*, each consisting of a combination of vessels and fibres: there was a 'general interlacing wherein each [tissue] both gives and receives'.[6] There was, in short, no basal element in Bichat's general anatomy – 'no closure of the regression: no point at which anatomy and physiology met in some basic structure or stuff which was accorded all the basic properties of life'.[7]

This failure to identify any fundamental unit or material of life can be related to another feature of Bichat's system. His was essentially a *spatial* arrangement of animal tissues, which showed little regard for how these structures emerged through time. Bichat did devote some attention in the general anatomy to questions of embryogeny: he maintained that 'shortly after conception, the foetus is merely a mass of apparently homogeneous mucous matter, in which the cellular tissue appears to predominate almost exclusively'.[8] Such excursions into embryology were, however, incidental to the main structure and goals of the work. Moreover, it is clear that in Bichat's view this mucous substance could only begin to perform the role of developmental parenchyma if an 'entrelacement' of vessels and nerves was present within it. This, however, implied that the 'apparently homogeneous' stuff of the early embryo itself contained – albeit invisibly – the nutritional parenchyma for this later network.[9] Embryology therefore supplied no escape from the infinite regress of similar parts.

It is beyond the scope of this chapter to offer any explanation of these distinctive characteristics of Bichat's tissue doctrine.[10] What matters for the present discussion is the distance between these assumptions and those that informed the work of the authors of the cell theory. The latter sought the basic element of life and organization that was so strikingly absent from Bichat's scheme; their researches were, moreover, conducted within an explicitly developmental framework. It is in seeking to account for these differences that the importance of the Romantic outlook to nineteenth-century life science becomes apparent.

## THE 'GENETIC METHOD'

Some historians have recognized that while the technological improvement in microscope construction may have been necessary to the upsurge of research into the minute structure of tissues after 1830, it is not sufficient to explain the orientation or outcome of those inquiries. The focus of this work lay in Germany, and it has been argued that the intellectual climate prevalent there made a crucial contribution to the investigations that led to the formulation of the cell theory.[11]

In the German-speaking states around 1800 there was a movement to create a new synthetic science of life, which would uncover the fundamental laws underlying organic phenomena. This project rested upon the assumption that a fundamental identity underlay the diversity of the biological realm, an assumption derived from the doctrine of the unity of nature expounded by Romantic nature philosophy.[12] In 1802 Gottfried Reinhold Treviranus stated that in this new science of life, in contrast to the disjointed products of the old 'natural-history', 'the observations of the vital phenomena [*Lebensweise*] of animals and plants . . . receive their proper place and unite themselves into a whole, wherein the spirit perceives unity and harmony'.[13] Moreover, Treviranus made clear the *a priori* character of the assumptions underlying this endeavour towards a synthetic view of organic nature; in his view, 'the spirit strives towards unity in multiplicity, and attains this through suppositions where experience cannot supply it'.[14]

The unity of the organic realm was supposed to manifest itself in certain 'typical' or fundamental elements of structure which, although endlessly modified, retained an essential identity. There was considerable vagueness about the nature of such types. For an author like Goethe, they were no more than ideal concepts with no real representative in nature.[15] For others, however, the type corresponded to determinate elements which combined to constitute living forms. A further distinction can be made between those who interpreted type in a purely morphological fashion – for example, the French naturalist Etienne Geoffroy Saint-Hilaire (1772–1844) – and those who sought a unit of *function* as well as of structure.

The second characteristic of the Romantic approach to the living world was its preoccupation with the *temporal* aspect of organisms. For Romantic nature philosophy the living world was in a state of continuous flux; this was true both of the individual and of the totality of organisms. This process of constant change was not, however, random: there was a fixed tendency for organisms to *develop* from the less to the more perfect. Just as each individual commenced 'with the simplest form and during its metamorphosis becomes more developed and elaborated, so the collective animal-organism seems to have begun its development with the simplest form or the lowest class of animal'.[16] The study of these processes of development at all the levels at which they were manifested supplied *the* path to an understanding of life.[17] On this view, biological *explanation* was equivalent to a *history* of a given species, individual or part.[18]

When the quest for types was conjoined to the emphasis upon development the result was an attempt to show how complex structures were elaborations of simpler elements. This enterprise could be undertaken through the study of the development of the type in the animal series by means of comparative anatomy; but *embryology* enjoyed special prominence within this research programme because it allowed the direct observation of the metamorphosis of the elemental into complex structures. With the embryo, wrote Ignaz Döllinger in 1824, 'we are at the source of life'.[19]

The contrast between this and Bichat's approach to explicating organization is clear. Carl Friedrich Heusinger, who in 1822 coined the word 'histology', distinguished two departments within this science. On the one hand was 'histography', which sought the kind of static representation of tissues found in Bichat. On the other was 'histogeny', the goal of which was 'to describe the developmental stages of tissues'.[20] While histography continued after Bichat's death to attract a measure of attention, histogeny was to become the most energetic programme of research, both in Germany and later also in France itself.

Implicit in the histogenetic project was a concern with the origins of living forms. Taken to its extreme this was manifested as a desire to locate *the* 'source of life'. Unlike Bichat's general anatomy, histogeny assumed that there was an ultimate element or material from which all others derived. Phillip Franz von Walther expressed this view in 1807. While asserting that Bichat's tissues were to anatomy what elements (*Stoffe*) were to chemistry, Walther went on to declare that the differences between these tissues was only one of degree because 'all tissues contained in the texture of the organs of the animal body, are formed by the metamorphosis of one and the same original tissue'.[21] In short, while Bichat's own emphasis was upon the differences in organization and properties between the organic elements he discussed, that of Walther and others of a histogenetic bent was upon the developmental *identity* that underlay these divergences.

Walther identified the 'cellular tissue' as the common source of all others. This was a popular choice, although there was much disagreement as to just what the term connoted: while for some the cellular tissue was an areolar mesh, for others it was merely an amorphous semi-fluid. P. A. Béclard pointed out in 1823 the difficulty of finding good pictorial representations of this protean substance; indeed, it was 'impossible to represent it'.[22] This elusiveness in fact enhanced the theoretical serviceability of the cellular texture or matter; it could be all things to all men.

One conception of the *Grundstoff* was to be of special importance for future developments. This saw the origin of organization to lie in a fluid 'blastema', from which organic solids were precipitated. The structures that emerged *first* from this blastema could, therefore, be seen as elemental. At the turn of the nineteenth century these elements were usually identified as fibres, most notably by Johann Christian Reil (1759–1813) in 1795.[23] Later in the century researchers such as Johann Friedrich Meckel (1781–1833) in Germany and Henri Milne-Edwards (1800–85) in France saw minute *globules* as the elementary particle.[24] The similarities between these claims and the later postulates of the cell theory have been noted by a number of historians.[25] Although some of these analogies may be too facile, it is clear that a work like Reil's *Von der Lebenskraft* (*On the Life-force*) brings us much closer to the conceptual world of Schwann than does Bichat's *Anatomie générale*.

## THE ROMANTIC PROGRAMME AND CELL THEORY

Although Schwann presented the results of his inquiries as the result of pure empirical investigation, historians have noted the presence in his work of some of the characteristic features of the science of the Romantic period.[26] There was, in the first place, a strong commitment to synthesis: indeed, the urge to unify a large number of phenomena within a unified framework can be seen as the primary impulse behind Schwann's researches. He declared in the introduction to the *Mikroskopische Untersuchungen* that it was 'an essential advantage of our age that the individual disciplines of the natural sciences

advance in ever more intimate union with one another, and we owe a great
part of the progress which the natural sciences have recently made directly to
this fusion and completion [*Ergänzung*]'.[27] In particular, Schwann wished to
extend this process of unification by demonstrating that common principles
underlay the physiology and anatomy of plants and animals.

Schwann shared the ambition of Treviranus to advance beyond 'natural-
historical' approaches to the living world in order to attain a more profound
understanding of organisms. Traditional classifications of the elementary
parts of plants and animals had stressed the differences between them: while
plants were recognized to be uniformly composed of 'cells', a much greater
diversity of elements – fibres, tubules, globules and so on – seemed to enter
into the make-up of animals. Such, however, was merely 'a natural-historical
division', uninformed by any 'physiological' concept.[28]

Schwann's goal had been to supply just such a physiological principle.
However, he used the term 'physiological' in a particular way. For Schwann a
physiological account of tissues was equivalent to a *developmental* explanation. Despite the functional and morphological variety of elementary parts,
their origin was identical. Schwann claimed to have established the fundamental law that there was 'a common developmental principle for the most
diverse elementary parts of the organism, and that cell formation is this
developmental principle'.[29] It was because all tissue elements shared this
developmental principle that a true 'physiological' classification was
possible.[30]

Schwann therefore demonstrated the bias towards developmental studies
we have seen to emerge from the Romantic tradition. His work can, indeed, be
seen as the culmination of that tradition, the cell finally supplying the
common origin of all organic forms that earlier embryologists had sought.
The similarity is underlined by the particular *mechanism* of histogenesis that
Schwann postulated. In his view, cells arose out of a blastema by a process of
precipitation, in much the same way as the elementary fibres or globules of
previous theories had been supposed to originate.[31]

An exclusive attention to one individual – even one as important as
Schwann – cannot, however, give an adequate representation of the importance of Romanticism to the genesis and elaboration of cell theory. The
influence of Romanticism upon other significant actors, such as Rudolf
Virchow and John Goodsir, can also be demonstrated.[32] More difficult to
define, but perhaps still more important, is the role of the 'atmosphere' or
mind-set created by the prevalence of ideas such as the unity of nature, the
existence of types and the primacy of development, in biological thought after
1800. Such notions had by the 1830s become commonplace in scientific
discourse, not only in Germany but also in France and Britain.

Their wide currency generated a number of preconceptions and expectations which favoured the ready dissemination and acceptance of the cell theory once it was promulgated.[33] Cell theory satisfied a need implicit in the Romantic conception of the living world: it provided the common source – the *'Urtypus'* – of all life, the point of unity in diversity, which Romanticism supposed and demanded. The question of the origins of the organism, which had been 'posed by the philosophy of nature',[34] now possessed a putative solution.

## NOTES

1   B. Bracegirdle, *A History of Microtechnique: The Evolution of the Microtome and the Development of Tissue Preparation* (London, 1978), p. 309.
2   *Ibid.*
3   On the relations of Bichat's tissue doctrine to his physiology see T. S. Hall, *History of General Physiology 600 B.C. to A.D. 1900*, 2 vols. (Chicago, 1969), II, pp. 121–6; W. R. Albury, 'Experiment and Explanation in the Physiology of Bichat and Magendie', *Studies in History of Biology*, 1 (1977), 47–131, esp. pp. 75–7.
4   F. X. Bichat, *Anatomie générale, appliquée à la physiologie et à la médecine* (Paris, 1801), p. v. (Translations from French and German here and below are mine, unless otherwise stated.)
5   *Ibid.*, p. 1.
6   *Ibid.*, p. 3.
7   J. V. Pickstone, 'Bureaucracy, Liberalism and the Body in Post-Revolutionary France: Bichat's Physiology and the Paris School of Medicine', *History of Science*, 19 (1981), 122.
8   Bichat, *Anatomie générale*, p. 108.
9   *Ibid.*, pp. 4–8.
10  For an attempt to relate these features to the political milieu within which Bichat operated see Pickstone, 'Bureaucracy'.
11  See W. Coleman, *Biology in the Nineteenth Century: Problems of Form, Function and Transformation* (Cambridge, 1977), pp. 25–6.
12  See A. Gode-von Aesch, *Natural Science in German Romanticism* (New York, 1944; repr. New York, 1966); K. E. Rothschuh, *History of Physiology*, trans. G. B. Risse (New York, 1973), pp. 156–65.
13  G. R. Treviranus, *Biologie, oder Philosophie der lebenden Natur für Naturforscher und Aerzte*, 6 vols. (Göttingen, 1802–22), I, pp. 7–8.
14  *Ibid.*, p. 119.
15  On Goethe's notion of type see C. Sherrington, *Goethe on Nature and Science* (Cambridge, 1942), p. 22.
16  F. Tiedemann, *Zoologie. Zu seinen Vorlesungen entworfen von F. Tiedemann*, 3 vols. (Landshut, 1808–14), I, pp. 64–5.
17  See Treviranus, *Biologie*, p. 151.
18  O. Temkin, 'German Concepts of Ontogeny and History around 1800', in *The Double Face of Janus and Other Essays in the History of Medicine* (Baltimore, 1977), pp. 373–89, esp. pp. 379–80.

19  I. Döllinger, *Von den Fortschritten, welche die Physiologie seit Haller gemacht hat* (Munich, 1824), p. 10.

20  C. F. Heusinger, *System der Histologie* (Eisenach, 1822), p. 19.

21  P. F. von Walther, 'Darstellung des Bichat'schen Systemes', *Jahrbuch der Medicin*, 2 (1807), 56–7, 60–1.

22  P. A. Béclard, *Elémens d'anatomie générale, ou Description de tous les genres d'organes qui composent le corps humain* (Paris, 1823), p. 135.

23  J. C. Reil, 'Von der Lebenskraft', *Archiv für Physiologie*, 1 (1795), 8–162.

24  See J. V. Pickstone, 'Globules and Coagula: Concepts of Tissue Formation in the Early Nineteenth Century', *Journal of the History of Medicine*, 28 (1973), 336–56.

25  L. J. Rather, 'Some Relations between Eighteenth-Century Fiber Theory and Nineteenth-Century Cell Theory', *Clio Medica*, 4 (1969), 191–202, esp. pp. 199–200; E. H. Ackerknecht, *Rudolf Virchow, Doctor, Statesman, Anthropologist* (Madison, Wisc., 1953), p. 71.

26  See Temkin, 'German Concepts', p. 381.

27  T. Schwann, *Mikroskopische Untersuchungen über die Uebereinstimmungen in der Struktur und dem Wachstum der Thiere und Pflanzen* (Berlin, 1839), p. iii.

28  *Ibid.*, p. 192.

29  *Ibid.*, p. 196.

30  *Ibid.*, p. 198.

31  *Ibid.*, p. 196 and *passim*.

32  See W. Pagel, 'The Speculative Basis of Modern Pathology. Jahn, Virchow and the Philosophy of Pathology', *Bulletin of the History of Medicine*, 18 (1945), 1–43; L. S. Jacyna, 'John Goodsir and the Making of Cellular Reality', *Journal of the History of Biology*, 16 (1983), 75–99.

33  See L. S. Jacyna, 'The Romantic Programme and the Reception of Cell Theory in Britain', *Journal of the History of Biology*, 17 (1984), 13–48.

34  Balan, *L'Ordre et le temps: l'anatomie comparée et l'histoire des vivants au XIXᵉ siècle* (Paris, 1979), pp. 281–2.

# 12

# Alexander von Humboldt and the geography of vegetation

## MALCOLM NICOLSON

Alexander von Humboldt was born in Berlin in 1769.[1] As young men he and his brother Wilhelm were members of the elite German literary circles. Alexander became a personal friend of both Goethe and Schiller. He also received a scientific education at the University of Göttingen and the Freiberg School of Mines (*Bergakademie*). He later became famous for his experimental investigations, for his scientific expeditions – to the Americas from 1799 to 1804, and to Russia in 1829 – and for such major works as *Personal Narrative of Travels* and *Cosmos*. He was a remarkable man of polymathic learning and a synthetic habit of thought.

Antecedents of Humboldt's geographical concerns may be found in the work of those German scholars who pioneered the academic study of geography in the second half of the eighteenth century. Of special relevance to the present discussion are Kant's lectures on physical geography.[2] To Kant, it was artificial to arrange objects into taxonomies, according to carefully selected visible features, as had been done by Linnaeus. Such a form of classification lacked

the idea of a whole out of which the manifold character of things is being derived ... In the existing so-called system of this type, the objects are merely put beside each other and ordered in sequence one after the other.[3]

To Kant, the essential prerequisite of a knowledge of the world was a description of phenomena as they actually occur and coexist in nature. Physical geography should give 'an idea of the whole in terms of area'.[4] It was only after the task of geographical description had been undertaken that a satisfactory 'system of nature' based upon the phenomena themselves rather than upon arbitrary distinctions and aggregations would be possible.

Following Kant, German geographers assumed the existence of a functional interrelation between all of the individual phenomena of the earth's surface.[5] They postulated an underlying causal unity of nature of which the visible forms of things was only one aspect. The earth was one whole. But geographers also recognized the existence of regionality. Phenomena peculiar to a particular region were the causes of other equally regional phenomena – for example climatic and environmental conditions influenced human society so that, as Kant wrote, 'in the mountains, men are actively and continuously bold lovers of freedom and their homeland'.[6] Thus although the earth was a single organism, it contained other holistic structures which, by reason of their peculiar internal cohesive processes, had distinctive characteristics. The surface of the earth was comprised of many natural regions.

Humboldt first publicly set out his programme for a new form of plant geography in 1793.[7] In the *Florae Fribergensis Specimen*, he followed Kant in distinguishing between a true history of nature and a mere description of nature such as had been provided by the older, Linnaean system of natural history. No longer should botanists study merely individual species and their outward appearances; no longer should they be preoccupied solely with descriptive taxonomy and nomenclature. The central concern of the Humboldtian plant geographer would be, by contrast, the real phenomena of vegetation:

Observation of individual parts of trees or grass is by no means to be considered plant geography; rather plant geography traces the connections and relations by which all plants are bound together among themselves, designates in what lands they are found, in what atmospheric conditions they live . . . and describes the surface of the earth in which humus is prepared. This is what distinguishes geography from nature study, falsely called natural history . . .[8]

Humboldt identified his proposed botanical innovations with changes occurring contemporaneously in other fields of inquiry. In particular he associated his programme for plant geography with the new historical geology (geognosy) of Abraham Werner. Werner sought to transcend classical mineralogy – which had concentrated on the study of individual minerals – and to produce a unified history of the earth. Humboldt had been a student of Werner at the Freiberg School of Mines and was an enthusiastic practitioner of the Wernerian exemplar, using geognosy as the organizing principle of the geological researches he undertook on his scientific travels.[9] The gist of his remarks in the *Florae Fribergensis Specimen* was that the new programme to investigate the study of the history of the earth must encompass not only geological phenomena but biological ones as well:

Geognosy [*Erdkunde*] studies animate and inanimate nature . . . both organic and inorganic bodies. It is divided into three parts: solid rock geography, which Werner has industriously studied; zoological geography, whose foundations have been laid by [Eberhardt] Zimmermann; and the geography of plants, which our colleagues left untouched . . .[10]

This grand vision of a complete historical geography of the earth provided a central theme for Humboldt's later work. All his diverse writings were characterized by the desire to create what he termed '*la physique générale*' – the universal, synthetic science which would comprehend both the unity and the diversity of nature.[11] The geography of plants and, in particular, the geography of vegetation had a major part to play within this cosmological scheme.

The importance of geography within Humboldt's scholarly enterprise was also expressed in the priority he accorded to scientific travelling. To Humboldt exploration was an essential part of natural inquiry, a necessary condition of '*la physique générale*'. Early in his development as a natural philosopher, he had been introduced to the art of scientific travelling by an experienced practitioner, Georg Forster, who had sailed with Captain Cook on his second voyage. In 1790, the two men travelled from Germany, through the Low Countries, to France and England. Shortly after their return to Germany, Forster published a literary and scientific account of their journey.[12] *Ansichten vom Niederrhein* (*Views of the Lower Rhine*) was acclaimed in literary circles as a major achievement, particularly by Goethe, Schiller and Alexander's brother, Wilhelm. The harmonization of scientific investigation with aesthetic sensitivity which Forster had accomplished was hailed as evidence of a new maturity of attitude among natural philosophers. Here was a demonstration that scientific inquiry need not be cold and unresponsive to the beauties of nature. It could embrace and celebrate the earth in the act of studying it.[13]

As we shall see, there is much of Forster's exemplar in Humboldt's own travel writing. Both men paid particular attention to the morphology of landscape. Both favoured panoramic description. Both valued scientific accuracy and avidly collected all manner of detail and data. Their empiricism was combined with enthusiastic recording of emotional responses and subjective impressions.

Of all his travels Humboldt regarded the journey to the Americas as the most important since, in the New World, he had visited the tropics where plant and animal life displayed the greatest richness and diversity. And in his accounts of South America, Mount Chimborazo occupies a place of special symbolic importance since it was there, in 1802, that Humboldt and his companion, Aimé Bonpland, ascended from the level of the rainforest and human settlement, through the several altitudinal vegetation zones, to the region of permanent snow. On the slopes of the mountain they thus experienced, within a small compass, much of the physical and vegetational diversity of the continent. It is, likewise, a symbol of the centrality of plant geography within Humboldt's research enterprise that it was in their camp at the foot of the mountain that Humboldt began to compose a fuller articulation of the programme for plant geography he had adumbrated in the *Florae Fribergensis Specimen*.

The *Essai sur la géographie des plantes* (*Essay on the Geography of Plants*) was first published, in French, in 1807.[14] Like Chimborazo itself, the *Essai* encapsulated, for Humboldt, the totality of the scientific and aesthetic impression made upon him by the tropics of South America. Such was its unique importance that Humboldt originally intended that the *Essai* should be the introductory volume to the full scientific account of his travels in the New World.[15]

The primary purpose of the *Essai* was 'to draw natural philosophers' attention to the great phenomena which nature displays in the regions through which I have travelled. It is their whole which I have considered in this essay.'[16] Humboldt directed attention to the 'whole' because Nature could not be understood by concentrating only on particulars. Nature was one holistic unity:

This science [*la physique générale*], which without doubt is one of the most beautiful fields of human knowledge, can only progress . . . by the bringing together of all the phenomena and creations which the earth has to offer. In this great sequence of cause and effect, nothing can be considered in isolation. The general equilibrium, which reigns amongst disturbances and apparent turmoil, is the result of an infinity of mechanical forces and chemical attractions balancing each other out. Even if each series of facts must be considered separately to identify a particular law, the study of nature, which is the greatest problem of *la physique générale*, requires the bringing together of all the forms of knowledge which deal with the modifications of matter.[17]

One of Humboldt's reasons for allocating such a central position to plant geography was that the vegetation of any given region was not only a primary expression of the physical environment – it also exercised a formative influence on Mankind, both materially and spiritually. The passage with which Humboldt introduced this subject is a very characteristic one:

but the man who is sensitive to the beauties of nature will . . . find there the explanation of the influence exerted by the appearance of vegetation over Man's taste and imagination. He will take pleasure in examining what is constituted by the 'character' of the vegetation and the variety of sensation it produces in the soul of the person who contemplates it. These considerations are all the more significant because they are closely linked to the means by which the imitative arts and descriptive poetry succeed in acting upon us . . . What a marked contrast between forests in temperate zones and those of the Equator, where the bare slender trunks of the palms soar above the flowered mahogany trees and create majestical portico arches in the sky . . . How does this . . . appearance of nature, rich and pleasant to a greater or lesser degree, affect the customs and above all the sensibility of people?[18]

The vegetation was, of course, not the only feature of the environment which was morally influential. In a later essay Humboldt noted that:

The poetical works of the Greeks and the ruder songs of the primitive northern races owe much of their peculiar character to the forms of plants and animals, to the mountain valleys in which their poets dwelt, and to the air which surrounded them.[19]

But the role of vegetation in mediating between Man and the physical environment was a major one:

However much the character of different regions of the earth may depend upon a combination of all these external phenomena ... the outline of mountains and hills, the physiognomy of plants and animals, the azure of the sky, the forms of the clouds and the transparency of the atmosphere, still it cannot be denied that it is the vegetable covering of the earth's surface which chiefly conduces to the effect.[20]

Note that it is the vegetation *en masse* that is active in producing the differences in aesthetic sensibility and moral development between races and cultures. The individual plants are involved only as they contribute to the collective phenomena of vegetation.

This holistic emphasis on vegetation ran throughout Humboldt's treatment of plant geography, not only in the *Essai* but also in his later works. It constitutes one of the major novelties of his plant science and one of the principal reasons why his work must be distinguished from that of older botanists such as Johann Reinhold Forster and Karl Ludwig Willdenow.[21] Humboldt did not, of course, deny that the study of individual plants and species was an important part of botany. But this was not the main focus of his own botanical research, nor should it be, he argued, the exclusive concern of other investigators:

Botanists' research is generally directed toward objects which merely embrace a very small part of their science. They deal almost exclusively with the discovery of new plants, with the study of their exterior structure ... and of the analogies which unite them in classes and families ... it is no less important to establish Plant Geography, a science that so far exists in name only, and yet is an essential part of *la physique générale*.[22]

As he was later to put it, in his *Personal Narrative of Travels*:

... preferring the connection of facts which have been long observed to the knowledge of isolated facts, although they were new, the discovery of an unknown genus seemed to me far less interesting than an observation of the geographical relations of the vegetable world, or the migration of the social plants, and the limit of the height which their different tribes attain on the flanks of the Cordilleras.[23]

Humboldt's concern with holistic structures and the unity of landscape is well exemplified by the 'Tableau physique des Andes et pays voisins'. This is a large and elaborate engraving, folded within the pages of the *Essai*.[24] It depicts a cross-sectional profile of the Andes from the Atlantic to the Pacific at the latitude of Chimborazo. In this one figure are mapped or tabulated which plant and animal species live where, where the altitudinal zones of vegetation

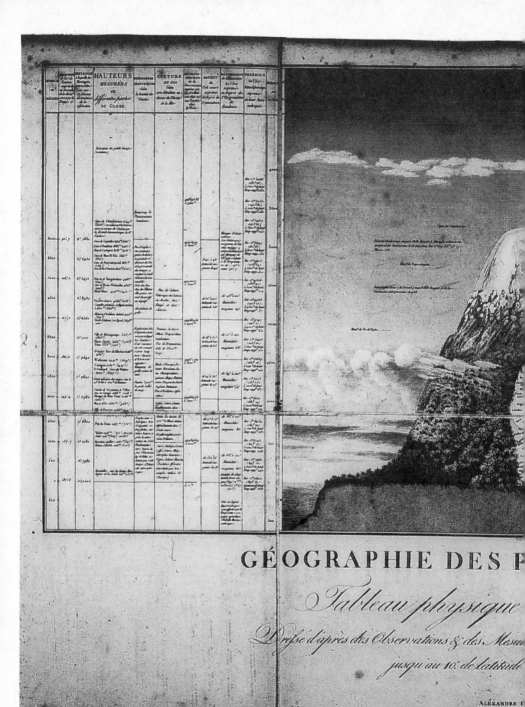

13  Profile of the Andes. 'Tableau' from Alexander von Humboldt, *Essai sur la géographie des plantes* (1807).

14 'Heights of the Old and New World, graphically compared.' Dedicated by
Goethe to Alexander von Humboldt. Note the figure of von Humboldt near
the summit of Chimborazo. From *Allgemeine geographische Ephemeriden*, 1813.

4000 Tois

9
8
7
6 áq. Luszes Luftball
5
4 Chimborazo.
3
2
1
3000 s Humboldt
Antisana
Cotopaxi
9
8
6 Gränze d. Crypto-gamen
5 Tunguragua
4 s Antecliue
3
2
1
Mieveg Antisana
2000
9
8 Micnipampa
7 Gränze der Baumarten
6
5 Quito
4
3
2 Gränze der Cinchona
Mexico
1
1000
9
8
7
6 Gränze der Palmen u. Pisangs
5
4
3
2
1

l neuen Welt

chen

begin and end, the types of agriculture pursued, the underlying geological structures and a wide variety of physical or meteorological data. The object was to give, in a single illustration, a complete impression of a natural region – the '*régions équinoctiales*' of South America.

The 'Tableau', with its holistic vision of a unified landscape, is a very typical Humboldtian production. However the concept it represents was not unique to Humboldt but rather sprang from the wider background of German Romanticism and *Naturphilosophie*. Humboldt enjoyed an enduring friendship with Goethe, to whom the German edition of the *Essai sur la géographie des plantes* was dedicated.[25] The dedication page is illustrated with an engraving which represents the genius of Poetry unveiling Nature. In the foreground lies an open copy of Goethe's great botanical work, *The Metamorphosis of Plants*. Goethe studied Humboldt's work on plant geography enthusiastically and drew an illustration for the text, 'a conventional picture of a symbolic landscape', which he, in turn, dedicated to Humboldt (fig. 14).[26] This was not the only common ground between the two men. When Humboldt sent Goethe a copy of his *Essai politique sur l'Ile de Cuba* (1826), Goethe complimented the author on not having omitted 'pointers to the incommensurable', despite the large amount of statistical information the work contained.[27]

Humboldt also had a close intellectual association with Schiller. He wrote an allegorical essay, 'The Genius of Rhodes', for Schiller's periodical *Die Horen*. Despite later being vehemently criticized by Schiller and despite having serious scientific disagreements with Goethe, Humboldt never repudiated his early intimacy with the leaders of the German Romantic movement. He reprinted 'The Genius of Rhodes' in his 'favourite' and most 'purely German' work, the compilation volume *Ansichten der Natur* (*Views of Nature*).[28] Humboldt's last and most ambitious major work, *Cosmos*, written almost forty years later still, contains many passages which give high praise to the *Naturphilosophen*.[29] Goethe's influence is acknowledged in the book's introduction and much of the text is redolent of the Romantic tradition, continuing to evince intellectual concerns seen in Humboldt's earlier works – in the *Essai* and especially in *Ansichten der Natur*.

To Humboldt, one of the principal attractions of the study of vegetation was the extent to which the plant geographer shared the interests and joys of the landscape artists. The two approaches to nature were mutually complementary. Humboldt suggested that the pictorial representation of landscape would be improved if the painter studied the classification of plant form developed by the plant geographer:

How interesting and instructive to the landscape painter would be a work that should present to the eye accurate delineation of the sixteen principal forms enumerated both individually and in collective contrast! What can be more picturesque than the arborescent Ferns, which spread their tender foliage above the Mexican laurel-oak!

15   Frontispiece to *Alexander von Humboldt und Aimé Bonplands Reise*
     (1807). The Spirit of Poetry (Apollo) unveils the mystery of Nature (Isis).

What more charming than the aspect of banana groves, shaded by those lofty grasses,
the Gaudua and Bamboo! It is particularly the privilege of the artist to separate these
into groups, and thus the beautiful images of nature . . . resolve themselves beneath his
touch . . . into a few simple elements.[30]

   Perhaps it is in the importance he attached to aesthetics within natural
inquiry that Humboldt's fraternity with the Romantic movement is most
obvious:

With the simplest statements of scientific facts there must ever mingle a certain
eloquence. Nature herself is sublimely eloquent. The stars as they sparkle in the
firmament fill us with delight and ecstasy, and yet they all move in orbits marked out
with mathematical precision.[31]

Natural science, if it was to be true to Nature, must be aesthetically satisfactory. Moreover it was not that the scientific faculty comprehended while the aesthetic faculty merely appreciated. As we have seen, to Humboldt aesthetic and emotional responses to natural phenomena counted as data about these phenomena. Aesthetic reactions to the various sorts of vegetation were indications of the particular effect of different natural environments upon human society.

It has recently been argued that different traditions within *Naturphilosophie* may be distinguished according to the role accorded to aesthetics within natural inquiry.[32] A major problem facing the philosophy of knowledge at the end of the eighteenth century was how human reason, which had only sense data to work with and was thus confined to the scrutiny of external characteristics, could ever come to comprehend the inner realities of things. The Kantian response was to argue that reason simply could never have direct access beyond the phenomena. The best one could hope for was, through establishing systematic interconnections and law-like relationships, to organize natural phenomena into synthetic holistic schemata. But the variety of *Naturphilosophen* which von Engelhardt has termed 'romantic' or 'speculative' was not prepared to accept a necessary dichotomy between the understanding of the investigator and the object being investigated. They proposed an alternative solution by which a theory of aesthetics came to the aid of the theory of rationality. Man's aesthetic sensitivities could, if suitably trained and applied, transcend the limitations of reason, penetrate beyond the surface phenomena and, sensuously and intuitively, grasp the underlying unities of Nature.

Humboldt is clearly sympathetic to this point of view:

... who is there that does not feel himself differently affected beneath the embowering shade of the beeches' grove, or on hills crowned with a few scattered pines, or in the flowering meadow where the breezes murmur through the trembling foliage of the birch? A feeling of melancholy, or solemnity, or of light buoyant animation is in turn awakened by the contemplation of our native trees. This influence of the physical on the moral world – this mysterious reaction of the sensuous on the ideal, gives to the study of nature, when considered from a higher point of view, a peculiar charm which has not hitherto been sufficiently recognized.[33]

But Humboldt, although always alive to the prerogatives of aesthetics and the appeal of the sublime, did not follow Schiller or Schelling in subordinating rationality to aesthetic sensibility. He was chastised by Schiller for his 'keen cold reason which would have all nature shamelessly exposed to scrutiny'.[34] In other words he did not repudiate empirical and experimental natural inquiry. To Humboldt, aesthetics complemented rationality; it did not make it redundant. The mathematical precision of the stars' orbits was just as valid a topic for study as their sparkle and its associated delights.

Humboldt's plant geography was a thoroughly empirical investigation of the environment of plants:

. . . it would be injurious to the advancement of science to attempt rising to general ideas, in neglecting the knowledge of particular facts.[35]

In his gathering of information, Humboldt made intensive use of instruments to measure physical parameters.[36] One of the purposes of his scientific travelling was to measure accurately, with instruments, where previous explorers had merely described. Subjective impressions were a necessary but not a sufficient part of what one might call the data-base of Humboldtian plant geography. The azure of the sky, for example, was not only to be appreciated aesthetically: it had to be quantified. Virtually everything that could be measured was measured. The readings were tabulated and compared between various sites. The physical data were then correlated with the occurrence of the various types of vegetation. Such correlations would, it was hoped, aid in the discernment of the laws which governed the distribution of vegetation. To facilitate this work, Humboldt pioneered the isoline technique of cartography.[37]

Humboldt's 'magnificent lines' enclosed areas of equal mean temperature and pressure.[38] But, in principle at least, they also marked out natural divisions of the earth's surface. Distinctive integrative processes went on within the different areas. The tables of data on the equatorial region of South America which accompany the *Essai sur la géographie des plantes* illustrate the unified interrelatedness and complexity of these natural geographical units:

The same table indicates: the vegetation; the fauna; the geological connections; the agricultural cultivations; the temperature of the air; the limits of perpetual snow; the chemical constitution of the atmosphere . . . the horizontal refraction of sunlight, and the temperature of boiling water at different altitudes.[39]

The 'regions [which] form the natural divisions of the vegetable empire' were thus real holistic entities in contrast to the artificial isolates on which herbarium practice was based.[40] They existed 'not in the greenhouses and books of botany but in Nature itself'.[41]

Regions of this sort were not, however, imagined to be topographically or vegetationally homogeneous. The 'Tableau physique des Andes et pays voisins' pictorially represented spatial differentiation within a single region. The palm might be a characteristic plant of the 'régions equitoriales' but palms were not distributed equally throughout its entire area. On the tops of the mountains one found a 'région des lichens' or, lower down, a 'région des Cinchona'. These smaller sorts of vegetational regions were distinguishable by 'physiognomy' – that is, by the life-forms, the general appearance and habit of growth, of the constituent plants.

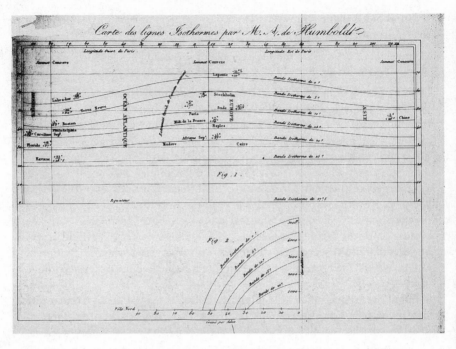

16   Map of isothermal lines by Alexander von Humboldt, from *Annales de chimie et de physique* (1817).

Classifying plant physiognomy was an important feature of Humboldt's botanical enterprise:

Among the variety of vegetation which covers the structure of our planet, one may distinguish, without difficulty, several general forms to which most of the others may be reduced . . . I have marked out by name fifteen of these physiognomic groups.[42]

Examples of the life-forms recognized by Humboldt were: the grasses, the palms, the cacti, the conifers, the lianas, the horse-tails (Equisetales), the mosses and the lichens. Thus the 'régions des lichens' were distinguishable by the obvious profusion of a number of species, all with the same lichenous life-form. It was the study of such natural physiognomies that Humboldt recommended to the landscape painter in the quotation given above.

This aspect of Humboldt's work on vegetation constitutes one of the most decisive ways in which he departed from classical taxonomic and floristic methods. Classification by life-form, although in many cases it did approximate to more orthodox arrangements, was essentially independent of floristic systems:

In determining those forms, on whose individual beauty, distribution and grouping, the physiognomy of a country's vegetation depends, we must not ground our opinion . . . on the smaller organs of propagation . . . but must be guided solely by those elements of magnitude and mass from which the total impression of a district receives its character of individuality . . . The systematizing botanist . . . separates into different groups many plants which the student of the physiognomy of nature is compelled to associate together.[43]

Similarly, species closely allied for the taxonomist might be put into different physiognomic groups by the Humboldtian plant geographer.

Humboldt's interest in reducing the diversity of plant shapes to a small number of fundamental life-forms is closely cognate with the concern of other German naturalists – such as Goethe, Lorenz Oken and the Göttingen professors, Johann Friedrich Blumenbach and G. R. Treviranus – to identify the ideal or primitive forms underlying plant and animal structure.[44] As Humboldt expressed it:

The primeval force of organization, notwithstanding a certain independence in the abnormal development of individual parts, binds all animal and vegetable structures to fixed, ever-recurring types.[45]

Thus in his study of plant physiognomy, as in much else of his plant science, Humboldt gave empirical expression to the characteristic themes and preoccupations of German Romanticism and *Naturphilosophie*.

Alexander von Humboldt may be seen as both a product of German Romanticism and an important exponent of a Romantic style within natural inquiry. He undoubtedly received formative influences from the intellectual milieu of turn-of-the-century Romanticism and from his personal acquaintance with the leading figures of the movement such as Goethe and Schiller. He combined the inputs received from these sources with the more empirical but equally holistic *Naturphilosophie* of his teachers at Freiberg and Göttingen. In Humboldt's plant geography we can see a vivid example of the Romantic commitment to a form of natural inquiry which would engage both Man's spiritual and his rational faculties, which would effortlessly combine rigorous empiricism and experimentalism with idealism and holism and which would produce a vision of nature that was both aesthetically and scientifically satisfactory.

## NOTES

1   For Humboldt's biography, see K. Bruhns (ed.), *The Life of Alexander Humboldt*, 2 vols. (London, 1873).
2   Kant's lectures on physical geography are contained in I. Kant, *Gesammelte*

*Schriften*, edited by the Royal Prussian Academy of Sciences (Berlin, 1902–), X, pp. 151–436.

3   Trans. in J. A. May, *Kant's Concept of Geography* (Toronto, 1970), p. 260.
4   Quoted in R. Hartshorne, 'The Nature of Geography', *Annals of the Association of American Geographers*, 29 (1939), 220.
5   See R. Hartshorne, 'The Concept of Geography as a Science of Space, from Kant and Humboldt to Hettner', *Annals of the Association of American Geographers*, 48 (1958), 97–108.
6   Quoted in Hartshorne, 'The Nature of Geography', 220.
7   A. von Humboldt, *Florae Fribergensis Specimen* (Berlin, 1793).
8   Quoted and trans. in Hartshorne, 'The Concept of Geography', 100.
9   For Werner's historical geology and Humboldt's use of it, see W. R. Albury and D. R. Oldroyd, 'From Renaissance Mineral Studies to Historical Geology, in the Light of Michel Foucault's *The Order of Things*', *British Journal for the History of Science*, 10 (1977), 187–215; and H. Baumgärtel, 'Alexander von Humboldt: Remarks on the Meaning of Hypothesis in his Geological Researches', in C. J. Schneer (ed.), *Toward a History of Geology* (Cambridge, Mass., 1969), 19–35.
10  Quoted and trans. in Hartshorne, 'The Concept of Geography', 100.
11  For characterizations of '*la physique générale*', see S. F. Cannon, 'Humboldtian Science', in her *Science in Culture: The Early Victorian Period* (New York, 1978), 73–110; and M. J. Bowen, 'Mind and Nature: The Physical Geography of Alexander von Humboldt', *Scottish Geographical Magazine*, 86 (1970), 222–33.
12  G. Forster, *Ansichten vom Niederrhein von Brabant, Flandern, Holland, England und Frankreich* (Berlin, 1790).
13  For the reception of *Ansichten vom Niederrhein*, see A. Meyer-Abich, *Alexander von Humboldt* (Bonn, 1969), p. 101.
14  A. von Humboldt, *Essai sur la géographie des plantes* (Paris, 1807). All translations from this work are my own.
15  *Ibid.*, p. vii.
16  *Ibid.*, p. v.
17  *Ibid.*, pp. 42–3.
18  *Ibid.*, pp. 30–1.
19  A. von Humboldt, 'Ideas for a Physiognomy of Plants', in E. C. Otté and H. G. Bohn's translation of Humboldt's *Views of Nature or Contemplations on the Sublime Phenomena of Creation* (London, 1850), p. 217.
20  *Ibid.*
21  This point is discussed in greater detail in my 'Alexander von Humboldt, Humboldtian Science, and the Origins of the Study of Vegetation', *History of Science*, 25 (1987), 167–94.
22  Humboldt, *Essai*, p. 1.
23  A. von Humboldt, *Personal Narrative of Travels to the Equinoctial Regions of the New Continent*, 6 vols. (London, 1821–5), I, p. iii.
24  Humboldt, *Essai*, between the Preface and the main text.
25  A. von Humboldt, *Ideen zu einer Geographie der Pflanzen* (Tübingen, 1807).
26  Bruhns (ed.), *The Life of Alexander Humboldt*, I, p. 176.
27  Meyer–Abich, *Alexander von Humboldt*, p. 36.
28  A. von Humboldt, *Ansichten der Natur* (Tübingen, 1808). Humboldt's comments on *Ansichten* are quoted in Bruhns (ed.), *The Life of Alexander*

*Humboldt*, I, p. 37. For an English translation of the *Ansichten*, see n. 19 above, this chapter.

29    A. von Humboldt, *Cosmos: Sketch of a Physical Description of the Universe*, 4 vols. (London, 1850).

30    Humboldt, 'Ideas for a Physiognomy of Plants', pp. 229–30.

31    Letter, 28 April 1841, in L. Assing (ed.), *Letters of Alexander von Humboldt to Varnhagen von Ense* (London, 1860), pp. 67–8.

32    See D. von Engelhardt, *Hegel und die Chemie* (Wiesbaden, 1976), and T. Lenoir, 'The Göttingen School and the Development of Transcendental *Naturphilosophie* in the Romantic era', *Studies in the History of Biology*, 5 (1981), 111–205.

33    Humboldt, 'Ideas for a Physiognomy of Plants', p. 219.

34    Letter, Schiller to Korner, 6 August 1797, in Bruhns (ed.), *The Life of Alexander Humboldt*, I, p. 188.

35    Humboldt, *Personal Narrative of Travels*, I, p. v.

36    For a full account of Humboldt's use of measuring instruments, see Cannon, 'Humboldtian Science'.

37    A. H. Robinson and H. M. Wallis, 'Humboldt's Map of Isothermal Lines: A Milestone in Thematic Cartography', *Cartographic Journal*, 5 (1967), 119–23.

38    G. Harvey, 'Meteorology', in the *Encyclopaedia Metropolitana*, quoted in Cannon, 'Humboldtian Science', p. 95.

39    Humboldt, *Essai*, p. 42.

40    Humboldt, *Personal Narrative of Travels*, I, p. 158.

41    Humboldt, *Essai*, p. 32.

42    *Ibid.*, p. 31.

43    Humboldt, 'Ideas for a Physiognomy of Plants', pp. 220–1.

44    See Lenoir, 'The Göttingen School', esp. pp. 172–3; H. B. Nisbet, 'Herder, Goethe and the Natural Type', *Publications of the English Goethe Society*, 37 (1967), 83–119; and A. G. Morton, *History of Botanical Science* (London, 1981), pp. 343–6.

45    Humboldt, 'Ideas for a Physiognomy of Plants', p. 217.

FURTHER READING

Botting, D., *Humboldt and the Cosmos* (New York, 1973)

Bowen, M., *Empiricism and Geographical Thought: From Francis Bacon to Alexander von Humboldt* (Cambridge, 1981)

Cannon, S. F., 'Humboldtian Science', in her *Science in Culture: The Early Victorian Period* (New York, 1978), 73–110

Lenoir, T., 'Kant, Blumenbach, and Vital Materialism in German Biology', *Isis*, 71 (1980), 77–108

Macpherson, A. M., 'The Human Geography of Alexander von Humboldt' (unpublished University of California, Berkeley, Ph.D thesis, 1972)

Nicolson, M., 'Alexander von Humboldt, Humboldtian Science, and the Origins of the Study of Vegetation', *History of Science*, 25 (1987), 167–94

# PART III

# SCIENCES OF THE INORGANIC

# Goethe, colour and the science of seeing

## DENNIS L. SEPPER

To someone only marginally aware of the scientific works of Johann Wolf-gang von Goethe (1749–1832) it is astonishing that he considered one of them his most important accomplishment. As Johann Peter Eckermann reports, Goethe not once but repeatedly made assertions to this effect:

I make no claims at all for what I have achieved as poet. Fine poets were my contemporaries, even finer ones lived before me, and there will be others after me. But that I alone in my century know what is right in the difficult science of colour, for that I give myself some credit, and thus I have a consciousness of superiority to many.[1]

Goethe's assessment of his achievement was shared to some degree by contemporaries who were seeking alternatives to the standard sciences of nature, for instance Schelling, Lorenz Oken, Hegel and Schopenhauer (not every *Naturphilosoph* was a supporter, however). But his *Zur Farbenlehre* (*On the Doctrine of Colours*) stirred vehement opposition among physicists, not least because he had polemicized against the colour theory of the great Isaac Newton. Although some of them admired the beauty of Goethe's descriptions of colour phenomena and granted a certain legitimacy to isolated claims, by and large they were puzzled by his attempt to reopen a closed subject, condemned his attacks on Newton and Newtonians and tried to forestall possible harm to true science from this influential poet who had decided to tread the ground of physics. Although Goethe tried personally to keep above the fray after publishing his monumental, tripartite work on colour, *Zur Farbenlehre* (1810), polemics raged between supporters and opponents for a generation and, in attenuated form, even to the present.

A basic defect in Goethe's conception seemed to be betrayed by his own words at the end of the historical part of *Zur Farbenlehre*. There he describes his first adult encounter with the fundamental phenomenon of the Newtonian theory, refraction through a prism. In an unfurnished room, freshly painted white, he took a prism in hand and looked through it. Dimly recalling youthful experiences with prisms and the Newtonian theory as he knew it, he expected to see many colours. But the walls viewed through the prism looked white. Only where there was a dark contrasting feature did colours appear. 'It did not take much deliberation for me to recognize that a boundary is necessary to produce colours, and I immediately said to myself, as if by instinct, that the Newtonian teaching is false.'[2] This 'experiment', which Goethe undertook in 1790 or 1791, led in short order to the two small volumes of the so-called *Contributions to Optics* (*Beyträge zur Optik*) (1791–92), which developed this initial insight about the importance of boundaries by presenting long series of variations on prismatic experiments.

The problem is that Goethe's expectation was erroneous: a white wall viewed through a prism will, according to Newton's theory, appear white, not multicoloured. Goethe's earliest critics explained this fact, which depends on fairly elementary geometrical considerations, in their reviews of the *Contributions to Optics*. Under the circumstances, and given the additional fact that all the phenomena Goethe adduced were already known, it seemed that Goethe's apparent anti-Newtonian animus was baseless and his contribution to optics nil. Two decades later, the reviewers of *Zur Farbenlehre* renewed these charges, basing them on Goethe's own anecdote.

Everyone knows that Newton used a prism to show that white light can be split up into component colours; he showed that there is a property of light rays (differential refractibility: that is, some rays are refracted more than others) that bears a significant relation to the colours that the rays produce (the most refractible produce violet, the least produce red, the intermediates produce blue, green, yellow and so forth). This theory can be demonstrated seemingly beyond doubt by just a few experiments.

One thing we must set aside if we are going to appreciate Goethe's colour doctrine, however, is the notion that it is refuted by the most elementary facts of optics. First, Goethe's account of his earliest prism experiment reflects not his mature understanding of Newton's theory but rather a version he derived from physical science texts he had consulted. As I have shown elsewhere, Goethe's erroneous expectation was due not to ignorance but rather to defective presentations of Newton's theory in eighteenth-century texts.[3] Within a relatively short time Goethe became aware of the differences between the theory as Newton had presented it and as Newtonians did. But through studying the history of colour science Goethe became convinced that even Newton's version was ridden with tensions, distortions and tendentious presentations of the phenomena, all directed toward the end of proving a favoured hypothesis rather than understanding colour.

An example of a tension in Newton's theory is his conception of the number of colours in the spectrum. Newton often names five, sometimes seven, the latter especially where he tries to establish a numerical analogy between the hues in the spectrum and the seven tones in the diatonic musical scale.[4]

However, in principle Newton claims that there are indefinitely fine gradations in refractibility of rays and therefore indefinitely many colours. Goethe pointed out that Newton frequently shifts between these views, depending on what suits his theory. More importantly, the primacy for Newton of the spectrum conceals other significant relationships (for instance, so-called colour temperature) that are accessible only to a more ·wide-ranging approach.

As an example of tendentiousness and distortion, one might consider that Newton constantly talks as though there is such a thing as 'the' spectrum, which is said to have a certain relationship of length to width, certain definite proportions taken up by different colours and so forth. Goethe countered that if you refract a narrow beam of sunlight and employ a mobile screen, you will find by moving the screen back and forth that 'the' spectrum is, at best, a mathematical idealization, at worst, a falsification of the phenomena. Also, by substituting different transparent materials for the prism you will find that the exact location of each colour in the spectrum varies according to the dispersive power of the material.

These objections, though legitimate, may seem rather subtle and hardly grounds for a polemical attitude toward Newton and Newtonians. The problem, however, is that the works in which Newton presented his theory of light and colours, in particular the *Opticks* (first edition 1704), were considered in the eighteenth century to be paradigms of how experimental science should be done, and the results they achieved were ordinarily taken to be proven facts rather than hypotheses or theories, thus leaving no room for disagreement. In his very earliest work on colour, therefore, Goethe was addressing not just the falsity or tendentiousness of certain theoretical claims but even more the question of how experimental science should proceed in trying to distinguish between the evidence of the phenomena (the facts) and the interpretations we make of them (hypotheses and theories).

The fundamental defect of Newton's theory, according to Goethe, was *methodological*. Newton had marshalled his evidence to prove a theory, not to become acquainted with the subject matter; he had tried to make that theory appear as the result of a perfect induction, yet only from a *few* phenomena; and he had spent decades trying to weave argument and experiments together so as to get everyone else to see things his way. What was amazing to Goethe was that Newton had by and large succeeded. Even though the physicists of the late eighteenth century followed Newton, and ultimately Francis Bacon, in professing commitment to the method of induction, Goethe concluded that they were more dedicated to looking only at colour phenomena and experiments that served the purposes of their theories.

In contrast, Goethe tried to organize the phenomena in a way that would let them as much as possible speak for themselves; and he wrote an essay, unpublished during his lifetime, that explained the advantages and difficulties of such a method, titled 'The Experiment as Mediator between Object and Subject'.[5] Ultimately Goethe came to think that no one could approach phenomena without preconceptions, without a preferred way of beginning the explanation of them, and this is one of the capital themes in the polemical and historical parts of *Zur Farbenlehre*; nevertheless, he still thought it was important to get people to look at all the phenomena relevant to a particular scientific field *before* encouraging them to develop hypotheses. In this Goethe himself was a follower of Bacon's method. But he was quite aware that a Baconian ran the risk of proliferating data without structure, to the point of being swamped by their sheer volume. He elaborated Baconianism by developing a method of analysing and comparing correlations among the experimental conditions necessary to produce, augment and diminish a given phenomenon, an analysis that undertook continuous variations of the conditions or, where continuity proved impracticable, explored degrees of similarity between affine phenomena. For example, rather than perform just a single prismatic experiment with rigidly fixed circumstances, Goethe would vary the distance of the screen, adjust the aperture to control the quantity of light, substitute prisms with different angles (he sometimes used a hinged water prism for genuinely continuous change) or made of differently dispersive materials, and so forth. Goethe's method thus aimed not at isolating particular experiments but rather at carefully controlling and varying their circumstances and noting the correlations between these circumstances and the continuously developing result.

Goethe's method was rigorous and empirical yet did not scorn (as a pure Baconian method might) the search for unities that is characteristic of a theoretical approach; the sought-for continuity is itself implicitly a principle of unification. Although he did not often take precise measurements, his comparative and contextual method nevertheless established *relative* measures. When the screen is near the prism the spectrum is round and nearly colourless; a little further on coloured fringes (yellow and red, and blue and violet) appear at opposite sides; at some more distant point the fringes begin to overlap and green appears; further yet and the blue and yellow slowly vanish to yield finally a tricoloured spectrum. Only once we have become thoroughly familiar with such phenomena does it make sense to take precise measurements.

One point to emphasize very strongly is that a major part of Goethe's positive doctrine of colours is this methodology. Not just Goethe's opponents but even most of his supporters were oblivious of this. Their immediate inclination was to look for the fundamental propositions and ask whether they were right or wrong. But Goethe was concerned even more with the question of how one arrives at propositions and relates them to the relevant field of experience. Though he recognized that theories and hypotheses are aids to articulating and unifying the variety of experience, he resisted the common tendency to let interest in theory (one's way of conceiving things) displace interest in the intrinsic phenomena of nature; in the case of colour, he feared that abstractions like ray-paths and corpuscles had illegitimately supplanted the direct experience of what they were supposed to explain. The only sensible way to judge a theory, he thought, is against the background of the totality of things in question. Goethe believed that both Newton and Newtonians had erred here. Newton had at least investigated the different classes of colour, although he raised refraction to pre-eminence and tried to force everything else into its explanatory scheme. Newton's followers, however, pronounced their fidelity to his words even when they had not experienced the phenomena themselves, and their dedication to the man and his theory made it impossible for them to see colour unprejudiced, without immediately thinking of differently refractible rays.

It is now possible to consider what Goethe hoped readers of *Zur Farbenlehre* would gain: the ability to see and experience colour afresh. Although both the polemical part, which is directed against the *Opticks*, and even more the historical part, which shows the inevitable and legitimate diversity of approaches to colour, contribute to this goal, the heart of the *Farbenlehre* is the didactic part, which presents a natural and experimental history of the known phenomena of colour organized into plausibly natural groupings through the method of the comparative analysis of phenomenal correlations (and which stands within the tradition of natural and experimental histories that were staples of seventeenth- and eighteenth-century experimental science).

The first three (out of six) sections of the didactic colour doctrine are organized according to the major categories of physiological, physical and chemical colours. (They are further subdivided within these categories.) The physiological colours are those which are due solely or chiefly to the lawful workings of the eye. Goethe calls them 'the foundation of the whole doctrine',[6] both because the laws at work in this category have analogues in the others and also because the full actuality of colour implies the activity of the seeing eye. Simultaneous contrast, afterimages and coloured shadows are examples; Goethe also discusses colourblindness as a pathological yet lawful form. The second section considers physical colours, which include those produced by refraction, reflection, diffraction and thin films; they can be defined approximately as transient colours caused by the encounter of light with physical objects. The last major category is that of the chemical colours, those that are more or less fixed in bodies; a large portion of this section treats of the ways in which chemical reactions affect or transform such colours and how colours can reveal things about the nature of the substances in which they inhere. The three sections that follow cover respectively, general principles that can be educed from the preceding natural history of colour and are intended to guide future colour researches; the relationship of the science of colour to other disciplines and crafts, like philosophy, mathematics, dyeing, physiology and pathology, natural history, general physics and music theory; and, finally, the 'sinnlich-sittliche Wirkung der Farbe', a nearly untranslatable phrase that is roughly equivalent to the psychophysiological, psychological, aesthetic and moral effects and uses of colour.

From the didactic part Goethe expected the reader to gain acquaintance with the full range of colour phenomena presented according to significant groupings and major categories that reflect the common human experience of colours and their relationships. Since the mature Goethe also recognized that there was no such thing as experience perfectly devoid of theory or conception, it was also an introduction to his way of approaching colour, which, still true to his initial insight that for prismatic colours to appear there had to be a boundary or contrast, looked for polar oppositions and dualities in the phenomena (light and dark, warm and cool, complementary hues and so on) and also for eminent phenomena in which polarities were unified (a notion Goethe called *Steigerung*, an example of which is the tendency of both spectral red and spectral violet to approach purple/magenta when intensified). The second, polemical part was included to counter the prevailing opinion that Newton had already explained the essentials about colour and to show the methodological and descriptive deficiencies of the great physicist's classic *Opticks*. The third, historical part was written to show that the understanding of colour and other natural phenomena comes through a long process in which truth is fitfully expressed, obscured and re-expressed, and in which the different conceptions corresponding to the attitudes of different ages and temperaments need to be appreciated and, where possible, reconciled. (In modern-day terms, Goethe conceived of the sciences as hermeneutical.) From the pageant of the historical part one can see that it is necessary to become many-sided and to see things from different perspectives in order to understand natural phenomena. The *Farbenlehre* thus amounts to a grand scheme for advancing the investigation of colour within the larger context of all nature and cultivating a conscious awareness of human knowing as an historical activity. In this sense the *Farbenlehre* could be called a paradigm for a science of seeing – 'seeing' both literally as sight and in an extended sense as human experience.

If we take the term 'aesthetics' in a broad sense, we can say that Goethe's doctrine of colour was training in aesthetics. This description is meant to embrace a whole range of things associated with colour: from immediate sensation, through perception, recognition and emotional affects, to symbolic overtones and ultimately intelligent reflection on all the preceding. Goethe began his physical investigations into colour because he hoped to find guidance for the artistic use of colour – he had, in the narrow sense of the term, an aesthetic purpose. But his investigation quickly broadened his purposes and interests, not just to the horizon of opposition to Newton, but more importantly to the entire range of colour phenomena. There came an explicit intention to treat colour as a whole. He eventually discovered that higher-level experiences – for instance, the artist's experience of the difference between warm and cool colours, colour harmonies and colour contrasts – had analogues, sometimes direct (like complementary colours) at the physiological, physical and chemical levels. This encouraged a conviction that he had been developing in researches in anatomy, botany and other fields: that the laws governing phenomena at their most basic level also enter into more complex, higher-level phenomena and experience. This, of course, is quite the opposite of reductionism. As a result, Goethe's 'theory' of colours (to remove the ironic quote marks one needs to remember that the Greek root refers to seeing) is a comprehensive science of aisthesis. At the very least he shows that colour scientists must include in their investigations the ways in which human beings are affected by colours and must explore the relationships these experiences have to other colour phenomena.

Goethe was not hostile to the natural science of his day and its search for quantifiable and hidden causes: Copernicus, Kepler, Galileo and even Newton the mechanist come in for praise in the historical part of *Zur Farbenlehre*, he urged others to apply mathematics and measurement to the study of colour[7] and he repeatedly asserted the need for educating naive perception. But it was his intention to situate it as just *part* of a more comprehensive science: and scientific activity and knowledge he wanted to place within the context of the whole of human life. He was sceptical of claims to definitiveness in the ongoing study of nature, and thus *Zur Farbenlehre* was not meant to be the last word on colour but rather a collation and systematization of what was known for the sake of establishing a solid foundation and basic principles for a renewed, comprehensive science of chromatics.

The ultimate goal of this chromatic science was not so much to find the (hidden) cause of colour as to identify a recognizably fundamental phenomenon that takes place wherever colour occurs – the so-called *Urphänomen*. The science therefore aims at understanding phenomena in terms of phenomenality rather than in terms of what is conceivable but not experienceable (like the infinitesimal ray or similar notions). Goethe believed he had found such a phenomenon exemplified in the setting sun. High up, the sun appears nearly white; as it approaches the horizon the light travels through more matter (air) and turns yellow, then orange, finally even red. At the same time the colour of the sky goes from bright blue to indigo. In the one case light is observed directly through translucent matter; in the other, the darkness of deep space is observed through peripherally illuminated translucent matter (diffuse light in the atmosphere). According to Goethe, these are chromatically 'Urphenomenal' encounters of light and darkness (the interaction of matter and light produces the warm colours red and yellow by the 'darkening' of light, the cool colours blue and violet by the 'lightening' of darkness). Goethe tried to adapt this 'Urphenomenal' account to explain other phenomena, but without much success. It was this failure that the first critics immediately seized upon (Goethe himself acknowledged the failure years later).[8] They were quite right that Goethe's 'Urphenomenal' explanation seemed to make little physical sense, especially in its adaptation to refractive colours, and that Newton's theory was not only more specific but also easier to imagine. Yet they ought not to have drawn the conclusion that therefore Newton was quite right and Goethe quite wrong. Goethe's theory is much more compatible with the modern wave theory that was emerging in the early decades of the nineteenth century than with Newton's, in particular because that theory (and, even more, twentieth-century quantum theory) allows one to conceive colours as produced by the interaction of light with matter rather than by a pre-existent characteristic of innumerable colour-producing rays that are materially and individually present in every beam of white light.[9] Goethe applauded the emergence of the wave theory, though he had reservations about its postulation of an ether as a medium for those waves.[10] One can also go through *Zur Farbenlehre* and point out ways in which Goethe helped stimulate further research – for instance, his systematizing of physiological colours inspired researchers like Jan Purkinje, Johannes Müller and Müller's great student, Hermann von Helmholtz, and his discussions of the colours of sun and sky stimulated the physiologist and physicist Ernst von Brücke to provide an improved explanation of the scattering of sunlight.[11] Moreover, although Goethe's attempt to array the colours on a circle in a way corresponding to principles of colour contrast and harmony was by no means the first (Moses Harris had done so a generation earlier, and the painter Philipp Otto Runge was working out the principles of his colour sphere in the first decade of the 1800s – in fact *Zur Farbenlehre*'s didactic part concludes with a letter from Runge describing his project), it was the first time that the aesthetics of colour was presented as a full-fledged branch of colour science.

Ultimately, perhaps from his reading of Kant, and certainly at the instigation of Friedrich Schiller, Goethe acknowledged that the *Urphänomen* is a kind of ideal and therefore not directly experienceable in any simple way. Rather than interpret this as a Goethean admission of failure, however, one should see it as recognition that human beings can never escape the interpenetration of experience and conception, of phenomenon and theory, of fact and perspective. Goethe's great merit, recognized clearly by none of his contemporaries except Hegel, was that he remained faithful to the way things are, while avoiding positivism, and laid the groundwork for a science that preserved objectivity (in what was commonly and basically experienced) while giving subjective variety, the diversity of human ways of conceiving things, its due.

Goethe's critics (such as Thomas Young, Helmholtz and Emil du Bois-Reymond) often attributed his 'failure' in colour science to his being a poet, or to his trying to make science romantic. Poets, they said, were at home among human truths, not scientific ones; poets cultivate the sensuous and affective life, while scientists soberly uncover the rigorous, objective laws of the foundational realms hidden from sense. Goethe not only rejected this dichotomy of poetry and science, he also tried to show that quite the opposite was true, that the more science divorced itself from the all-embracing contexts of human life and nature the more the scientific imagination became trapped in a particular and sometimes even abstractively fantastic way of conceiving things. (And so it should not seem strange that his criticisms of Newton parallel his critique of the German Romantics.)[12] He saw both science and poetry, each in its own way, as having the ultimate intention of being faithful to nature, to its actualities and its possibilities. The task of human beings was to rise to the challenge of the world, whether they conceived of themselves as physicists, poets or anything else.

Goethe thought that science could never be complete because of the incredible vastness of nature and the mortal finitude of human beings. Yet these were counterbalanced by an essential simplicity and uniformity of nature and the human capacity for insight into, if not comprehension of, its workings. Goethe had the confidence that with sustained effort and good will everyone could gain at least a glimpse of nature's complex unity, and that the fortunate and ingenious ones, including both poets and scientists, could help lead the way to a fuller experience of that nature and a deeper understanding of their role in it as human beings.

## NOTES

1  J. P. Eckermann, *Gespräche mit Goethe in den letzten Jahren seines Lebens, 1823–32*, 2 vols. (Leipzig, 1836), entry for 19 February 1829.

2  J. W. von Goethe, *Goethes Werke (Hamburger-Ausgabe)*, ed. E. Trunz *et al.*, 14 vols. (Hamburg, 1948–66), XIV, p. 259.

3  Dennis L. Sepper, *Goethe contra Newton: Polemics and the Project for a New Science of Color* (Cambridge, 1988), pp. 27–38.

4  Isaac Newton, *Opticks* (New York, 1952), pp. 125–9.

5  Goethe, *Goethes Werke (Hamburger-Ausgabe)*, XIII, pp. 10–20.

6  *Ibid.*, p. 329 (paragraph 1 of the didactic part).

198     DENNIS L. SEPPER

7  *Ibid.*, p. 484 (paragraph 727 of the didactic part).
8  J. W. von Goethe, *Goethes Werke* (*Sophien-Ausgabe* or *Weimarer-Ausgabe*), 4 divs., 133 vols. (in 143) (Weimar, 1887–1919), div. 4, XLII, p. 167.
9  Sepper, *Goethe contra Newton*, pp. 13–14, 200.
10  *Ibid.*, p. 168 and note.
11  *Ibid.*, p. 197.
12  Hans Joachim Schrimpf, 'Über die geschichtliche Bedeutung von Goethes Newton-Polemik und Romantik-Kritik', in *Gratulatio: Festschrift für Christian Wegner*, eds. M. Honeit and M. Wegner (Hamburg, 1963), pp. 63–82.

## FURTHER READING

Goethe, Johann Wolfgang von, *Goethes Werke* (*Hamburger-Ausgabe*), ed. E. Trunz *et al.*, 14 vols. (Hamburg, 1948–66). Volumes XIII and XIV contain a good selection of the scientific writings, including the entire didactic part of *Zur Farbenlehre* and the historical part abridged.
—*Die Schriften zur Naturwissenschaft* (*Leopoldina-Ausgabe*), ed. R. Matthaei *et al.*, 2 divisions, 15 vols. to date (Weimar, 1947–    ). The definitive scholarly edition of the scientific writings, now nearly complete. The second division contains excellent documentation and notes.
—*Goethe's Color Theory*, ed. R. Matthaei, trans. H. Aach (New York and London, 1971). A useful, copiously illustrated edition, which also contains the complete translation of the didactic part.
—Goethe, *Scientific Studies*, ed. and trans. D. Miller (New York, 1988). A selection representing the gamut of Goethe's scientific writings.
—*Goethe's Botanical Writings*, trans. B. Mueller (Honolulu, 1952). Includes some of Goethe's important methodological essays.
*Goethe and the Sciences: A Reappraisal*, eds. F. Amrine, F. J. Zucker and H. Wheeler (Dordrecht, 1987). Contains an excellent annotated bibliography.
Heller, Erich, 'Goethe and the Idea of Scientific Truth', in *The Disinherited Mind. Essays on Modern German Literature and Thought* (New York, 1975), pp. 3–34.
Helmholtz, Hermann von, 'The Scientific Researches of Goethe', in *Selected Writings of Hermann von Helmholtz*, ed. R. Kahl, (Middletown, Conn., 1971), pp. 56–74. The classic nineteenth-century evaluation of Goethe's science.
Newton, Sir Isaac, *Opticks* (4th edn, London, 1730). Reprinted many times since.
Sepper, Dennis L., *Goethe contra Newton: Polemics and the Project for a New Science of Color* (Cambridge, 1988)

# 14

# *Johann Wilhelm Ritter: Romantic physics in Germany*

## WALTER D. WETZELS

If there ever was a person among the German Romantics who can be and in fact was in his time considered the prototype of a Romantic physicist, it is Johann Wilhelm Ritter. And this ambiguous distinction has survived to this day. The term 'Romantic Physics' is, of course, polemical as well as programmatic. What kind of physics could there be besides Newton's comprehensive mathematical systematization of the discoveries of Copernicus, Kepler and Galileo? The ready answer to this question is usually given by citing the basic tenets of Schelling's *Naturphilosophie* discussed in chapter 2 in this volume. However, while this highly speculative system of thought about nature plays a role in the kind of physics, chemistry, physiology and optics which Ritter pursued, Romantic natural science has a strong empirical basis, and most of its speculative ventures are extrapolations from experimental data. It is also worth noticing that the natural sciences were proclaimed an integral part of early German Romanticism. The movement was not anti-science; it eagerly embraced the new discoveries in the fields of electricity and chemistry. A number of adherents of the Romantic movement in Germany actively pursued scientific research, and all of them had a keen interest in natural science. At the same time it is true that the Romantic fascination with natural science was oriented towards those new discoveries which seemed to lie outside the Newtonian system of mechanics, such as galvanism, electrochemistry and physiology. The anti-Newtonian bias is clearly there as a protest against the hegemony of the mechanical–mathematical explanation of nature. But the revolt against the tyranny of Newtonianism was not a rejection of natural science as such; rather, it was meant as a challenge to the mechanistic view of all physical reality implied in Newton's physics. A 'higher' form of physics in which the metaphor for the universe was not the cosmic clock but the cosmic organism was to be an essential part of the Romantic credo. When Friedrich Schlegel announced the need for a 'New Mythology' he proposed a grand synthesis of Spinoza's pantheism, Greek and Oriental mythology and modern physics. He concludes: 'I cannot close without once again urging you to study physics because it is from its dynamic paradoxes that the most sacred revelations of nature erupt.'[1] The source for most of these revelations was J. W. Ritter, who together with the writer and mining engineer Friedrich von Hardenberg (who wrote under the name Novalis), the brothers Friedrich and Wilhelm Schlegel, Caroline Schlegel and Dorothea Veit, the writers Ludwig Tieck and Clemens Brentano and the philosopher Schelling formed the core group of early German Romanticism. And toward the end of the year 1799, the entire group was actually living in or around the university town of Jena only a few miles from Weimar: a unique concentration of literary, scientific and philosophical talent.

Johann Wilhelm Ritter was born on 16 December 1776 in the village of Samitz, Silesia, the oldest son of a protestant minister. He went through an apprenticeship as a pharmacist to the apothecary of the nearby town of Liegnitz. Already during these years Ritter became exposed to natural science and involved with chemistry in a practical way. It is not certain whether Ritter also knew already of Galvani's and Volta's early publications, but two reasons speak for it: the general publicity which Galvani's new 'animal electricity' enjoyed throughout Europe (together with Volta's counter-proposals), and the fact that Ritter, almost immediately after enrolling at the University of Jena in 1796, began experimenting with galvanic circuits. The year is not without significance because it was in 1796 that Schelling and the Schlegel brothers made their first visits to Jena, which was to become the birthplace of the Romantic movement in Germany only two years thereafter. As a trained pharmacist, Ritter was not only more mature than the average beginning student, but he also had considerable experimental experience and extraordinary practical skills. While apparently applying himself to traditional natural scientific subjects such as chemistry and astronomy, he pursued his studies in galvanic electricity largely independently. Within a year he had already made a name for himself as an unusually knowledgeable and gifted student of Galvani's and Volta's work on 'animal' and 'contact' electricity respectively. When the young Alexander von Humboldt, therefore, wanted someone to go critically through his manuscript 'Versuche über die gereizte Muskel- und Nervenfaser nebst Vermuthungen über den chemischen Process des Lebens in der Thier und Pflanzenwelt' (Experiments about the Irritated Fibres of Muscles and Nerves, along with Conjectures on the Chemical Process of Life in the Animal and Plant Kingdoms'), it was Ritter who was the obvious person. Ritter obliged with a lengthy critique of Humboldt's work, but only some of his commentary, usually the affirmative parts, appeared in Humboldt's published version in the fall of 1797. At any rate the critical co-operation with Humboldt not only launched Ritter's full-scale research into the phenomenon of galvanism as a chemical process, it provided at the same time the grand horizon within which the investigation was to be placed: it was nothing less than to search for the principle of life in nature. Humboldt had spoken of 'conjectures on the chemical process of life in the animal and plant kingdoms' in the title of his essay; Ritter only one year later published his first book, entitled *Proof that a Continuous Galvanism Accompanies the Process of Life in the Animal Kingdom.*[2] This bold, yet for the most part painstakingly inductive, piece of research won immediate acclaim with the Romantics: Novalis and Friedrich Schlegel praised it, and Goethe was very impressed despite the fact that he otherwise kept a wary and highly critical distance from anything that was embraced by, and especially produced by, the young Romantics at Jena, be it poetry or philosophy or 'science'. But Ritter's method was different from Schelling's speculative *Naturphilosophie*; it seemed to Goethe, who prided himself on his realism, empirical yet suggestive of a general principle in nature that had the power to challenge Newton's mechanistic design of reality.

Galvani had observed the reaction of the extremities of frogs when connected to two metals. His interpretation in essence was that the biological matter was not just an electroscope which indicated some action within the circuit, it was the *source* of the observed reaction. Volta, on the other hand, proposed that the biological matter was just an electroscope, and interpreted the galvanic action as the result of the contact between three conductors, two of which were metallic, one liquid. Galvani's orientation was physiological; he wanted to prove the electric nature of whatever was the fluid in nerves. Volta's orientation was that of a physicist who saw in Galvani's combination of metals and organic matter a new form of electricity: contact-electricity between materials with an organic electrolyte as a conductor. The two interpretations of the phenomena created two camps among men of science. Since in time more and more evidence gathered by Volta and Ritter and others showed that the new contact-electricity had the same characteristics as the well-known static electricity created by friction (through the separation of positive and negative charges), especially the plus–minus polarity, Volta's theory became the dominant interpretation. The irony of the situation was that Galvani did indeed deal with, but could not possibly measure or interpret in his time, bio-electricity in muscles. On the other hand, while Volta described the results correctly in form, his concept of a mere contact (of the proper kind) as the source of electricity amounted to a *perpetuum mobile*.[3] It was Ritter, as we shall see, who introduced and experimentally developed the notion of the chemical reaction within an electrolyte as the actual source of the electrical charge at the two ends or poles of a galvanic circuit. He is, therefore, widely considered as the founder of electrochemistry.

But let us return to his first book on galvanism. Almost all the experiments described are a verification of or variations on Volta's work on galvanic circuits; they essentially support Volta's famous triplicity of individual conductors and the duplicity of classes (solid and liquid) as the necessary and sufficient condition for galvanic action. Yet the main goal of Ritter's essay is to demonstrate that such a condition can be met with organic matter, that metals are not necessary. In fact, he points out, it is the animal (or human) organism in which Volta's condition for galvanic action is omnipresent since muscle fibre, nerves and liquids are in continuous contact. A living organism is in Ritter's interpretation a complex system of galvanic circuits: a continuous galvanism 'accompanies' the process of life. Later, he equated this galvanic action in all organisms with life itself. In a circuitous fashion Ritter, it would appear, comes close to something which is not unlike Galvani's 'animal electricity' by way of Volta's law for galvanic action in a properly set up circuit between solid and liquid conductors. But, strictly speaking, the issue concerning the nature of galvanism – whether one should equate it with electricity, with a chemical reaction in organic matter, or whether it was the fluid in nerves (*Nervenfluidum*), as Humboldt thought – remains undecided for Ritter at this point. He has investigated the conditions under which galvanic action occurs.

One might wish to ask what the particularly Romantic characteristics of this early example of Romantic natural science are? Up to this point there are hardly any, if any at all, unless one considers the general orientation towards an investigation of the organism under the auspices of a new phenomenon, namely galvanism, as an indication of a new scientific paradigm to which Ritter and others contributed. However, the essay which I have been discussing does not end with the proclamation of the close affinity between galvanism and the 'process of life'. Ritter treats his reader to a concluding section which he calls 'hints', properly and very consciously separated in paragraphs 25 and 26 from the main empirical part of the book. In these two sections Ritter permits himself to speculate about possible implications of his findings and possible further experiments suggested by the ones presented before. It is probably more because of his experience as a pharmacist than because of an early infection by Schelling's *Naturphilosophie* that he is led to speculate on the possibility of influencing the degree of action in the galvanic circuits within the human body through medication, various salts, baths, alcohol or opium. His mode of thought here is really extrapolative in nature; it does not deduce phenomena from a grand schema. Ritter is obviously, albeit very tentatively and vaguely, following a train of thought which conceptually leads to the idea of a living organism as being composed of galvanic cells. Therefore it appears perfectly plausible that changing the chemical composition of the electrolyte should change the galvanic action of the cell, perhaps enhance it. Even a revival of dead parts of a body through the application of an outside source of galvanic power seemed not impossible in principle, at least not inconsistent with existent experimental evidence.

Only at the very end of the essay, after raising the interesting question of whether galvanic action might be possible in the absence of any organic matter, a question which Ritter was to address in the very near future and answer affirmatively, does the reader get a feeling for the extraordinary scope of the intellectual speculative journey he has embarked on: 'Where is the sun, where is the atom that would not be part of, that would not belong to this *organic universe, not living in any time, containing any time? –* Where then is the difference between the parts of an animal, of a plant, of a metal, and of a stone? – Are they not all members of the *cosmic-animal,* of *Nature?*'[4] The substance and tone of this message had a great appeal in Romantic circles, and Ritter almost instantly became the most respected voice in matters of Romantic natural science. While Ritter's experimental work furnished concrete evidence for the validity or at least the plausibility of the philosophical speculations of Schelling and others, indeed often seemed to give rise to further imaginative completion of the grand concept of the all-encompassing world-soul, he in turn was stimulated to no small degree by the philosophical visions generated in great profusion among the Jena circle. But the symbiosis between the speculative grandeur (and arrogance) of Schelling's *Naturphilosophie,* on the one hand, and Ritter's imaginative empiricism on the other was always volatile. He was irresistibly attracted by the overall vision of a living universe, and he felt that it was his task to demonstrate the reality of this vision. Yet at the same time he was very uneasy about the apodictical pronouncements through which speculation dictated what reality had to be like. In the introduction of the essay discussed above, Ritter states assertively: 'If, then, there is no point in piling up hypotheses, if true desire to know cannot be satisfied by a somehow, a perhaps, an it-is-possible, we can only hold on to experience. It is only at her side that we will walk happily. However, if we leave her and entrust ourselves to the wings of our imagination, we might have pleasant dreams, but wake up all that more unpleasantly.'[5] As we have seen, Ritter does allow himself such dreams occasionally, but he labels them as 'hints', clearly marks them as tentative; yet while apologizing for them to the reader, he also demonstrates his own speculative potential.

The general concept that galvanism was associated, if not yet equated, with the presence of life in nature prescribed the next major step in Ritter's investigations: a proof that galvanic action is also possible, and a reality, in inorganic nature.[6] Like his first essay, this work, too, originated from a paper which he had presented to the *Naturforschende Gesellschaft* at Jena in 1799. Again, there are extensive passages about scientific methodology, about the empirical versus the speculative approach, and although the general tone is still polemically directed against hypothesizing, certain allowances are made for the *a priori* approach. Ritter actually apologizes for not using the deductive method; the experimental, inductive approach just seems to him more convincing and convenient for the reader. One senses the growing influence of the prestige of Kant's and Schelling's axiomatic metaphysics. The question at hand was: could galvanic action, which had always been associated with organic matter, be demonstrated in circuits without such matter, which were, of course, to be built according to Volta's law? Ritter set up a galvanic circuit which consisted in its original form of a silver coin onto which he placed a drop of distilled water. He then placed a flat piece of zinc on top of the water in such a way that the zinc touched the silver coin at one point, but was separated from the silver coin at the opposite point by a piece of glass. The drop of water, somewhat flattened, did not touch the glass or the place where both metals touched. He repeated the experiment with different metals and at different temperatures. In the case of the silver–zinc combination, the zinc calcified (oxidized), and in all cases the metal which had the greater 'affinity' to oxygen calcified. Galvanic action was proved to exist in purely inorganic circuits, and it was not only 'accompanied' by the chemical process of calcification: galvanic action and chemical action were one and the same. Ritter states as the result of his experiments 'that this effect [the calcification of metals] follows the same law as those which we up to now have called galvanism. In short, galvanism, as we understood it, is not an exclusive property of organic nature, it also is present in inorganic nature, and the future will show what multitudes of chemical phenomena owe their existence to galvanism.'[7] These kinds of experimental pursuits and their interpretation are the beginning of electrochemistry. But equating electrolytical processes with galvanism and thereby giving a chemical explanation for galvanic action for the new electricity of Galvani and Volta, was only one part of Ritter's achievement. In terms of the Romantic vision of a unified picture of nature which integrated both its organic and its so-called inorganic sides, Ritter's electrochemical interpretation of galvanism was a significant experimental breakthrough: if galvanism had to be associated with life in inorganic matter and galvanic action was the result of certain chemical reactions, then these certain chemical reactions could be interpreted as processes of 'life' in a realm that had been conceived of as lifeless. In galvanism the traditional demarcation lines between organic and inorganic nature vanished. An important part of the inorganic world was capable of a form of 'life' through the process of oxidation. Galvanism has become the central phenomenon in all of nature, the phenomenon which makes inorganic and organic matter come alive, and which, therefore, is also the unifying force in nature. Thus the organic paradigm of the Romantics begins to take shape around the phenomenon of an all-pervasive galvanism.

Of course, it is apparent that Ritter's generalizations are ahead of his experiments which only covered certain non-organic circuits in which two metals and an electrolyte were the components. And it is very likely that the general Romantic idea of an organic and unified universe provided the framework and the momentum for the sweeping extrapolations which Ritter offered. Yet his generalizations never lose their ever-so-tenuous touch with empirical evidence, and the empirical findings by themselves are sound and of undeniable importance. He is very well aware of the scope and difficulty of actually providing a truly comprehensive empirical basis for galvanic action throughout nature. This task amounted to nothing less than to prove that every single piece of organic and non-organic matter was a conductor of galvanic action, or as he and others would eventually call it, electricity. (Before the invention of the Voltaic pile – that is, before the action of simple galvanic circuits could be amplified – the identical nature of static electricity and galvanic action could not be proven conclusively.) As far as organic matter was concerned, there did not seem to be any reason to doubt its conductivity. However, as experiments in electrostatics had shown, bodies seemed to fall into two classes, conductors and insulators. Both Volta and Ritter had, in fact, provided a hierarchy of metals according to their conductivity, usually with gold and silver at one end and lead and zinc at the other end of the list. Both also provided a list of insulating materials according to the degree of their insulating property. If galvanic action was to be universal, as Ritter conceived it, truly insulating materials could not exist, only less conductive ones. And Ritter attempts in the last and most ambitious scientific work of his years in Jena to construct an electric system of the earth. Under the title *Das elektrische System der Körper* (*The Electrical System of Bodies*), he organizes all metals, metal compounds, conducting liquids, then all so-called insulators (minerals, woods, precious stones, but also the resin called dragon's-blood) into one comprehensive system, the electric system of the earth. (The insulating materials were of course listed according to their behaviour in electrostatic experiments.) The first of his impossible intentions is to give a virtually comprehensive inventory of all matter on earth. The second one is to show that the organizing principle of all matter is electricity; it unifies nature and endows everything with 'life'. Although Ritter cites the experimental data which he and others have gathered, including apparent evidence from older historical accounts, in an enterprise of such ambition, he soon has to abandon the inductive method for all practical purposes. For the most part a grand thesis appears to be in search of factual evidence and finds it only sporadically and often of dubious quality. Far-fetched analogies and generous extrapolations abound and evoke more than they can possibly evince.

The same work contains extensive observations and speculations about the periodicity of nature. Ritter offers tables about the various orbital cycles in our solar system; he reflects on the periods in which meteors have been observed, thunderstorms, the seasons, the cycles in which flowers open and close, important historical events and, of course, the periodicity of acoustic and electrical vibrations. The central oscillation, the pulse rate of the 'cosmic animal' (as Ritter conceived the universe to be) is the orbit of earth around the sun in approximately 365 days. And this period of 'pulsation' repeats itself also in galvanic phenomena. Ritter finds that a Voltaic pile has its greatest effect during winter, its smallest during summer; there is also a maximum and a minimum within a day of twenty-four hours; even smaller periodic fluctuation can be observed, namely $3\frac{23}{24}$ minutes, which turns out to be just about the 365th part of twenty-four hours. Galvanic action apparently mirrors planetary cycles, particularly the fundamental period of the earth's orbiting of the sun. The important points to be made here are that Ritter's real or imaginary rhythms are for him the real pulsations of a cosmic organism, and that they are repeated not only in the more well-known natural cycles of the biosphere, but also in the central phenomenon of all nature: electricity. Of course, the book, which attempted to construct the electrical system of all bodies on earth, which claimed rhythmic correspondence between motions in our solar system and on earth, which consciously suspended all distinction between empiricism, conjecture and sheer imagination, was not received well in scientific circles.

What had become of the promising, brilliant experimental physicist who had impressed not only the Romantics in Jena, but very soon men of science all over Europe? Volta and Humboldt, to whom Ritter had dedicated his first book, thought highly of him. Between 1799 and 1805 he regularly contributed to the most prestigious journal on physics in Germany, the *Annalen der Physik*, edited by L. W. Gilbert. Summaries of his publications had appeared in the *Journal de physique* and elsewhere. In 1802 Ritter was even invited to become a candidate for Napoleon's annual prize for the best experiment on galvanism, but the prize for 1803 was not awarded because, in Ritter's case, new and solid experiments (the prototype of an accumulator or secondary Voltaic pile) were presented together with experiments which, of course, unsuccessfully claimed to prove an electric polarity of the earth (over and above the well-known magnetic one). The combined effect of these presentations, which Ritter's friend Hans Christian Oersted performed for him in Paris, was scepticism and the ultimate refusal to award the prize for that year at all.

Ritter's experimental ingenuity was unbroken, but his many valid observations, often made in extremely painful self-experiments in which he spent hours 'in' the Voltaic pile examining, enduring and recording the physiological effects of electricity on his tongue, ears, eyes and other parts of his body, became more and more buried in an increasing compulsion to force the facts he found into the grand scheme of *Naturphilosophie*, making them carry more and higher meaning than they had been set up to convey. His articles in the *Annalen der Physik* had to be edited heavily not only because of the often impenetrable style, but also because of his increasingly expansive philosophical generalizations. Ritter's personal life in Jena also became more and more frustrating. He had arrived virtually penniless, and throughout the eight years during which he established his considerable scientific reputation, and indeed fame, he was never able to secure a stable income. Aside from occasional honoraria for his writings, his means of survival was to borrow from anybody he could persuade. He was infamous for his ingenuity and frivolity in doing this, and spent lavishly what he received. Consequently, short periods of feast alternated with long times of famine, and brief intervals of sociability were followed by extended retreats and insulation from the world. At the same time, Ritter was incredibly productive in his work, and not a small amount of the money he managed to get from others either as outright gifts or as borrowed funds went for the purchase of his expensive galvanic equipment. (This was especially true for the experiments with the Voltaic pile, which had to be constructed out of several silver–zinc or later copper–zinc layers, the latter being less expensive, although less effective.) Therefore, when Ritter finally left Jena to assume his first regular academic appointment at the Royal Bavarian Academy of Arts and Sciences, his reputation in the scientific community was severely damaged, he was burdened with considerable debts and the circle of the Jena Romantics had dissolved.

Before describing Ritter's work in Munich, I should like to present two instructive examples of his work in Jena which illustrate both the illuminating and the obfuscating potential of *Naturphilosophie* when applied to concrete investigations. Ritter had read about Herschel's discovery of infra-red rays beyond the visible spectrum of normal light. Herschel had detected these invisible rays of light with the thermometer which indicated a gradual temperature rise from the violet part of the spectrum to the red part – and beyond. Before undertaking any experiment, Ritter speculates about the apparent 'polarity' of light: red and violet are two 'poles' of light; the centre of the spectrum, green, is the point of 'indifference'; it also may be compared to the equator of the earth as red and violet seemed to be analogous to the north and south poles respectively. In his diary, he jots down well-known 'polarities', like hydrogen–oxygen, negative–positive electricity, south–north magnetism, cold–warm and others. Thinking in polarities, postulating an *Urpolarität* as the fundamental dualism inherent in all of nature, had been the basic tenet of Schelling's *Naturphilosophie* as the principle of action in organic and inorganic nature, and Ritter had found polarities at work wherever he had looked. Dualism had become for him and his friends the essential structure of natural phenomena and the essential mode of thought. Herschel's extension of the visible spectrum beyond red, therefore, appeared to be a violation of polar symmetry which had to be overcome. There simply had to be a 'pole' corresponding to the infra-red rays. The question was never whether it could be found, but only how to detect these new, invisible rays beyond violet which the doctrine of polarity simply demanded. Ritter knew that the various parts of the visible spectrum had a different effect on a chemical compound (silver chloride): it blackened the compound much more in the blue–violet part of the spectrum than in the opposite part, orange–red. He therefore concluded that silver chloride could play the same role for him that the thermometer had played for Herschel. In this fashion Ritter proved on 22 February 1801 the existence of ultraviolet rays. Clearly the philosophical concept of polarity had preceded and guided his experiment.[9]

The decomposition of water into hydrogen and oxygen is a rather paradoxical case because it demonstrates how a doctrine of *Naturphilosophie* led Ritter to an abstruse misinterpretation of a crucial and important experiment when the evidence he himself produced seemed to be in conflict with that doctrine. The doctrine I am referring to was the notion of a primeval body, of a genuine 'element' in the traditional sense, from which all other material bodies developed. Earth, water, fire and air, the ancient elements, enjoyed a revival as elementary spirits in the tales of the Romantics; Goethe searched for the *Urpflanze* and celebrated granite as the primeval stone; Ritter considered iron to be the original substance primarily because of its magnetic capabilities, and because iron occupied a central position in the hierarchy of conductors for electricity. He writes: 'All matter on earth seems to be decomposed iron. Iron is the nucleus of the earth.' And he continues within the same context: 'All matter on earth, compounded, should result in iron. It is the goal of chemical action to produce this ideal iron.'[10] The general idea behind these reductions of the many to the one from which the many forms in some kind of metamorphosis derive is that of the unity of nature guaranteed by that material and structural ancestry. Nature is full of prefigurations and correspondences because of an original oneness. And one of these privileged original substances was the ancient element water. For Ritter water is 'the center of all bodies',[11] and 'everything is modified water'.[12] To be sure, these were not meant to be scientific pronouncements, but they were also not to be taken as frivolous mysticism.

It is with this background in mind that one should study Ritter's experiments in the electrolysis of water. The history of Volta's letter of 26 June 1780 to Sir Joseph Banks about an electric battery, the so-called Voltaic pile, through which the effects of simple circuits could be amplified, is well known. Under the auspices of the Royal Society, Nicholson and Carlisle demonstrated among other things the decomposition of water into hydrogen and oxygen with the Voltaic pile. The news spread to Germany, and Ritter not only repeated the famous experiment by separately collecting hydrogen at the cathode and oxygen at the anode of a cleverly designed apparatus (the configuration of which was essentially the same as the one used today), but he was also able to *measure* the results of his electrolysis of water. He arrived at a proportion of hydrogen gas to oxygen gas of approximately 2.5 to 1, which was not a bad result. Remarkable as his experimental ingenuity as well as its results were, even more remarkable was the way in which Ritter dealt with something he had proved in practice, but could not accept in theory. He could not accept the fact that water was a compound, yet he had demonstrated just that with his experiment. The dilemma was obvious and its resolution was of central importance for somebody who, as a matter of basic philosophical conviction, believed that water was an element, but had apparently just shown the opposite. Ritter's basic belief won out; something which could not be must not be. He proceeded, therefore, to develop a complicated interpretation of the electrolysis of water which made hydrogen a compound of water and positive electricity, oxygen a compound of water and negative electricity. The two gases were, according to Ritter's explanation, the result of a compositional rather than a decompositional process in which water was, as it were, galvanized to combine with the two electricities generated by the two electrodes. While the principle of polarity, central to *Naturphilosophie*, clearly guided Ritter's discovery of ultraviolet rays, that of unity blocked the obvious interpretation of his electrolysis of water.

When Ritter moved to Munich to become a member of the Royal Bavarian Academy of Arts and Sciences in 1805, he joined the company of Schelling again and of Franz von Baader, the latter having been instrumental in Ritter's securing the position. Schelling had expanded his *Naturphilosophie* into mythical dimensions, and Baader had strong theosophical leanings. It was in this general spiritual ambience that Ritter entered into a field of experimentation which may be called 'magic galvanism'. The story of his determined experimental and theoretical efforts to demonstrate the sensitivity of the human organism to water and ores in the ground with wands and water-divining pendulums, with small metal rods suspended and then mysteriously turning clockwise or counter-clockwise depending on perceived polarities of the bodies under investigation, has often been told – sometimes with sympathy, more often as the embarrassing and sad disintegration of a promising scientific mind into the morass of the occult. Ritter, of course, saw himself as still operating within natural science proper. The twitching of the muscles of the person holding a wand in order to 'feel' water or iron ore was not different from that of a frog's leg: electric or galvanic action was at work. The involuntary movements of a pendulum held over metals, fruits, eggs, over different parts of the human body, revealed hidden polarities. The dualistic structure of nature which had always been the central idea of *Naturphilosophie* now assumed more and more mystical characteristics.

Realizing that his reputation as a serious man of science had been ruined, having lost the support of the Royal Academy for any further work in these fringe areas of 'science', Ritter finally returned to solid experimental investigations, and he chose a virtually unknown area of electrical phenomena: the electrophysiology of plants. He returned, in fact, to an early idea of his, namely that the world of the plants could not be excluded from the ubiquitous actions of galvanism or electricity. His last scientific, and truly remarkable, work, published only after his death, in which he investigates the responses of the various parts of the *mimosa pudica* to various electric stimuli, is a pioneering enterprise in a field that had not existed at all as a scientific discipline.[13] The experiments are executed meticulously; the results show indeed that plants react in a complex fashion to different electrical charges – they 'responded' intelligibly. However, that was only part of Ritter's agenda. He contrasted each experiment on a mimosa with a contrasting experiment on frogs. Correspondences and parallelism became obvious in this very systematic investigation. In typical fashion, unable to resist evoking the notion of unity in nature through such correspondences, he summarizes his results by stating that mimosas obviously are '*animals in the shape of plants*', and frogs nothing but '*plants in the shape of animals*'.[14]

After years of continuous poverty, after years of abuse of his body, after having to give up his family to the care and custody of others, Ritter died in 1810, apparently from tuberculosis. His final word to friends and foes, which appeared in 1810, was entitled *Fragmente aus dem Nachlasse eines jungen Physikers. Ein Taschenbuch für Freunde der Natur (Fragments from the Literary Estate of a Young Physicist. A Pocket-Book for Friends of Nature).*[15] Its main parts are a very confessional and often moving autobiography, the 'inner' story of his life, followed by 700 entries, such as notes, aphorisms, or fragmentary thoughts that run the gamut from sheer nonsense, platitudes, playfully suggestive analogies, to scientific speculations which are sometimes tantalizingly close to discoveries made much later. The latter especially often attracted the attention of scholars who were eager to see a prophetic quality in these utterances, to read some anticipatory significance into Ritter's fanciful hypothesizing.[16] Perhaps one could collect sound waves like light rays in a focal point; perhaps magnetic and electrical telescopes were possible some time in the future; perhaps energy and matter are only two different manifestations of the same phenomena? The fragments in this collection, as Ritter claims, were taken from diaries which he had kept over many years, and he is well aware of their uneven quality, to put it mildly. Yet he is also convinced that some of them contain prophetic insights, and that all of them exhibit the true potential of human imagination when set free. Seeing himself as the last true disciple and witness of Jena Romanticism, Ritter felt that he had, with his scientific work, including these speculative sorties into the unknown, lived up to Novalis's description: 'Ritter is indeed searching for the real soul of the world in nature [*die eigentliche Weltseele der Natur*]. He wants to decipher the visible and tangible [*ponderablen*] letters and to explain the positing of the higher spiritual forces . . .'[17]

## NOTES

1  *Kritische Friedrich-Schlegel-Ausgabe*, ed. E. Behler *et al.* (Munich, Paderborn, Vienna, 1958–  ), II, pp. 319, 321. (Translations from German in this chapter are by its author.)

2  Johann Wilhelm Ritter, *Beweis, dass ein beständiger Galvanismus den Lebensprozess in dem Thierreich begleite* (Weimar, 1798).

3  *Entdeckungen zur Elektrochemie, Bioelektrochemie und Photochemie von Johann Wilhelm Ritter*, selected, introduced and interpreted by Hermann Berg and Klaus Richter, Ostwald's Classics of the Exact Sciences, 271 (Leipzig, 1986). See p. 30.

4  Johann Wilhelm Ritter, *Galvanismus*, p. 171.

5  *Ibid.*, p. 10.

6  Johann Wilhelm Ritter, 'Beweis, dass die galvanische Action oder der Galvanismus auch in der anorganischen Natur möglich und wirklich sey', *Beiträge zur näheren Kenntniss des Galvanismus und der Resultate seiner Untersuchung*, 1 (Jena, 1800).

7  H. Berg and K. Richter (eds.), *Entdeckungen zur Elektrochemie*, p. 54.

8  Johann Wilhelm Ritter, *Das elektrische System der Körper. Ein Versuch* (Leipzig, 1805).

9  Walter D. Wetzels, *Johann Wilhelm Ritter: Physik im Wirkungsfeld der deutschen Romantik* (Berlin and New York, 1973). See pp. 32–3.

10  Johann Wilhelm Ritter, *Fragmente aus dem Nachlasse eines jungen Physikers*, 2 vols. (Heidelberg, 1810), Fragment no. 51.

11  Johann Wilhelm Ritter, *Das elektrische System der Körper*, p. 85.

12  Johann Wilhelm Ritter, *Die Physik als Kunst. Ein Versuch, die Tendenz der Physik aus ihrer Geschichte zu deuten* (Munich, 1806), p. 43.

13  Johann Wilhelm Ritter, 'Elektrische Versuche an der Mimosa pudica L., in Parallele mit gleichen Versuchen an Fröschen', ed. R. L. Ruhland, in *Denkschriften der Königlichen Akademie der Wissenschaften zu München*, 1809/1810 (Munich, 1811).

14  Wetzels, *Johann Wilhelm Ritter*, p. 55.

15  Johann Wilhelm Ritter, *Fragmente aus dem Nachlasse*, Fragment no. 51.

16  Noel Deeney, 'The Romantic Science of J. W. Ritter', *The Maynooth Review*, 8 (May 1983), 43–59, esp. pp. 50–3.

17  *Novalis Schriften*, ed. Paul Kluckhohn and Richard Samuel, 4 vols. (Stuttgart, 1960–75), III, p. 655.

## FURTHER READING

Ritter, Johann Wilhelm, *Physisch-Chemische Abhandlungen in chronologischer Folge*, 3 vols. (Leipzig, 1806): a complete and unabridged collection of Ritter's work up to 1805

—*Der Siderismus* (Tübingen, 1808)

—'Versuch einer Geschichte der Schicksale der chemischen Theorie in den letzten Jahrhunderten', *Journal für die Chemie, Physik und Mineralogie*, ed. A. F. Gehlen, vol. 6 (1808), pp. 719–28

For a selection with commentaries of Ritter's contributions to natural science, there are two editions of *Ostwalds Klassiker der exakten Wissenschaften* (Ostwald's Classics of the Exact Sciences), one published in the Federal Republic of Germany, one in the German Democratic Republic.

Hermann, Armin, *Die Begründung der Elektrochemie und die Entdeckung der ultra violetten Strahlen von Johann Wilhelm Ritter. Eine Auswahl aus den Schriften des romantischen Physikers*, Ostwald's Classics of the Exact Sciences, new series, 2 (Frankfurt-on-Main, 1968)

Berg, Hermann and Klaus Richter, *Entdeckungen zur Elektrochemie, Bioelektrochemie und Photochemie*, Ostwald's Classics of the Exact Sciences, 271 (Leipzig, 1986)

Works on various aspects of Romantic natural science in Germany with a connection to J. W. Ritter include:

Burwick, Frederick, *The Damnation of Newton: Goethe's Color Theory and Romantic Perception* (Berlin and New York, 1986)

Neubauer, John, *Bifocal Vision. Novalis' Philosophy of Nature and Disease* (Chapel Hill, 1971)

Snelders, H. A. M., 'Romanticism and *Naturphilosophie* and the Inorganic Natural Sciences 1797–1840: An Introductory Survey', *Studies in Romanticism*, 9 (1970)

Tatar, Maria, *Spellbound. Studies on Mesmerism and Literature* (Princeton, 1978)

Williams, L. Pearce, *Michael Faraday, a Biography* (New York, 1965)

# 15

## The power and the glory: Humphry Davy and Romanticism

### CHRISTOPHER LAWRENCE

Romanticism is one of the many threads, though at times an extremely fragile one, on which it is possible to hang the career of one of the nineteenth century's most famous chemists, Sir Humphry Davy (1778–1829). At first sight Davy's relatively short working life seems to fall into three periods, and it is in the middle one of these that the Romantic link appears the most tenuous. In the late 1790s, the young Davy lived and worked in Bristol where he came to know Coleridge and Southey. At this time the Romantic connection seems undeniable. In Davy's last years too, a knowledge of Romanticism appears to be the key to understanding a man who devoted his time to writing philo-sophical works, ostensibly about the consolations of travel and fly-fishing. In the middle years, however, when Davy was a lecturer on chemistry at the Royal Institution and hobnobbed with the great in London society, the Romantic relation seems thin indeed. The difficulty of synthesizing these three periods of Davy's life is perhaps one of the reasons why there is no substantial modern biography. Historians drawn to his life or work have usually concentrated on one of the three periods or focused on Davy either as chemist or as Romantic. The life of Davy written by his brother John Davy (1790–1868) still remains a valuable source, quoting from many of Humphry's letters and reproducing many of his poems.

If Romanticism is a continuous thread in Davy's life, it is a Romanticism which has to be understood in two rather different senses. In the first sense, Romanticism comprises a cluster of ideas, assumptions and practices which Davy drew on, often quite unreflectingly, throughout his life. Thus if Roman-ticism encompasses, for instance, a belief in the unity of nature, an active universe and the sublimity of light, Davy was indeed a Romantic from birth to death. In the second sense, however, Romanticism comprises a number of ideas and practices which Davy employed self-consciously and self-creatively. In this sense Davy's career and his presentation of himself were shaped by his use of a Romantic idea of genius. The two senses are of course related. Davy's idea of genius was built from and incorporated his Romantic assumptions about nature. The bridge between these two Romanticisms is Davy's concept of power. Anyone who reads Davy's works cannot fail to be struck by his repeated use of the word 'power' in every conceivable context, whether in his discussions of chemistry, poetry, the mind or society. It is power, too, which suggests that there are concrete social dimensions to Davy's thought. Davy's Romanticism, or certainly that of his last works, was an endorsement of industrialization, of the unequal division of property and labour and of European racial ascendency.

Humphry Davy was born in 1778 at Penzance, Cornwall, still, in the late eighteenth century, a relatively isolated Celtic community. Throughout his life his native county, as Michael Neve has noted, 'retained a pull on his poetic imagination'. Davy had a regular school education, during which he culti-vated his enduring passion for fishing, writing poetry and jotting down his most exalted and most commonplace thoughts into notebooks. In 1795 Davy was bound apprentice to a Penzance surgeon Bingham Borlase (1752–1813). During his apprenticeship Davy undertook a mammoth project of self-education. He also wrote verses and romances and jotted down his metaphysi-cal speculations. A characteristic Romantic concern, which was to preoccupy Davy all his life, crops up in the notebooks at this time: genius. In or around 1796 Davy composed a poem 'The Sons of Genius', which was later published in Southey's *Annual Anthology* of 1799. In the poem Davy describes how the moon inspires that small but select group of men, the sons of genius, who can 'rise above all earthly thoughts' and, 'enraptured', partake of 'a thousand varying joys'. Such joys were both sensual and intellectual – for instance, imbibing the 'charms of nature', understanding nature's laws and knowing the 'fair sublime immortal hopes of man'.

By 1797 Davy had developed an interest in chemistry, a subject relevant to his projected medical career. To throw oneself into chemical studies in the mid-1790s, however, was not simply to engage in an arcane professional pursuit. Quite the reverse: chemical theory lay in the heartland of the contem-porary European revolution. Chemical pronouncements about the nature of matter and spirit when made by such men as the political radical, theologian and chemist Joseph Priestley (1733–1804) were recognized by contemporaries as utterances that questioned the structure of society itself. As Coleridge put it in 1798, 'I regard every experiment that Priestley made in chemistry as giving *wings* to his more sublime theological works.' Davy plunged himself into the centre of this world, experimenting on the most dramatic and political substances: heat, light and gases.

Underlying Davy's apparently uncomplicated early career was a complex knot of social and intellectual relationships. The undoing of this knot starts, curiously, not in Cornwall but in Edinburgh. In the 1780s the composure of

Edinburgh society and, in particular, the self-confidence of Edinburgh medicine, was rocked by the political storms raised by the physician John Brown (1735–88). Brown attacked the Edinburgh establishment and proposed what he and his followers construed as a new theory of life based on the excitability of the animal body. Brunonianism, as Brown's medical theory was called, was widely adopted abroad, especially in Germany, where it was used as the basis of Romantic medicine. In England Brunonianism seems to have been less well received. It was, however, taken on board by a number of radical doctors who used it as a stick with which to beat the establishment and what they saw as the sterile mechanist medicine of the Enlightenment (see chapter 7 of this volume). One of these radical doctors was the Bristol physician Thomas Beddoes (1760–1808). Beddoes was a prominent figure in English medicine. He had been a Reader in chemistry at Oxford and, in the late 1790s, he was installed at Bristol, experimenting with gases and using them as therapeutic agents. Beddoes was the regular intimate of a variety of powerful and often slightly dangerous social groups. His connections stretched to the Midlands' famous Lunar Circle, which included industrialists such as Matthew Boulton (1728–1809) and James Watt (1736–1819), the radical doctor, poet and Brunonian of sorts Erasmus Darwin (1731–1802) and Joseph Priestley who, in the heat of Tory reaction in 1794, had left for America. Beddoes's connections, however, were not merely scientific and medical. This immensely learned man, who was probably more familiar with German culture than anyone else in England, had friends throughout the literary and political world, most notably Coleridge and Southey, both of whom lived in the West Country in the late 1790s. Beddoes's lines of communication, however, extended not only to Scotland and the Midlands but deep into Cornwall where he geologized and corresponded with doctors, surgeons and influential men of affairs. Cementing these friendships was a shared intellectual interest in natural philosophy, particularly chemistry, as a tool for investigating the system of nature, the forces of life, the powers of the atmosphere and the properties of electricity and light. It is not surprising then that Davy, who was dabbling with these socially dangerous commodities, became known to Beddoes. By 1798 Davy was communicating the results of his chemical experiments to the Bristol doctor.

   Davy's experiments on light so excited Beddoes that, in 1799, he published Davy's work as *Essays on Heat and Light*. These early experiments and speculations, which Davy later repudiated, in fact contain various general ideas which he seems to have stuck to and employed in different ways throughout his life. In the *Essays*, Davy seems to have been using Brunonian concepts to develop an unorthodox view of matter, life and the mind. In orthodox or dualist Enlightenment thought matter and spirit (or soul) were entirely different categories. Davy, however, rather like a number of other thinkers, collapsed these categories. At the end of the eighteenth century various authors, for instance, Erasmus Darwin and a Unitarian circle centred on Priestley and including Coleridge, had developed a concept of the 'active universe'. Each of these authors used this concept differently, but they all employed the idea that there was a single active power or principle underlying the natural world which was itself organized so as to be capable of intelligent purpose. For some, such as Darwin, this new philosophy spelled the death of Christianity. For others, such as Priestley, it meant a vigorous reinterpretation of biblical teaching. To the establishment and the established Church all those who dabbled with these ideas seemed hostile radicals. Employing these concepts Davy developed the idea of power as the fundamental building block of nature. This concept was not for Davy, any more than for Priestley, materialist in the sense of atheistic and mortalist. Rather it included an insistence on the durability after death of the power that makes life; an 'energy of mutation', as he called it. Similarly, mind was not a different substance but was only the highest expression of a fundamental 'law of animation'.

In another essay written at about the same time Davy suggested that genius was some sort of universal self-comprehension by this power. 'Consciousness of life', 'strong feelings', 'great powers' and genius were, for Davy, all synonymous. These idealist speculations of Davy's, in which spirit or power was the ultimate reality, are similar to views which Coleridge held at about the same time. The roots of Coleridge's views have been traced by Robert Schofield to the Cambridge Platonists of the seventeenth century and from them through a Neoplatonic tradition which, paradoxically, takes in aspects of the work of John Locke, David Hartley and Joseph Priestley (see Further Reading section). For Davy there were no fundamental breaks in the universe; rather it was a continuous, hierarchical, increasingly organized scale of perceptive existence in which light was the most fundamental power.

In the *Essays* Davy gave a concrete explanation of the operations of light. Sensibility, he argued, was the fundamental property of the brain and nerves, and was the basis of all perception and volition. Oxygen, he further suggested, was, in the state in which we breathe it, actually composed of two parts – oxygen (the gas) and light, a combination Davy called 'phosoxygen'. Davy explained how during respiration this substance enters the blood and is dissociated, the light then fuelling the sensibility of the nerves. He concluded, 'Thus essential then is light to perceptive existence.' And, in a paragraph which nicely illustrates the central role that chemistry could occupy in contemporary philosophy, he declared:

We may consider the sun and the fixed stars, the suns of other worlds, as immense reservoirs of light destined by the great ORGANIZER to diffuse over the universe organization and animation. And thus will the laws of gravitation, as well as the chemical laws, be considered as subservient to one grand end, PERCEPTION.

It is hardly surprising that, in October 1798, the young surgical apprentice, poet, imaginative chemist and speculator on life and nature took the long journey to Bristol to become Beddoes's assistant.

Beddoes had first come to Hotwells at the fashionable Clifton end of Bristol in 1793. Here he intended to open an institution devoted to the treatment of disease by the inhalation of gases. While working towards this objective the jovial Beddoes experimented on airs, built up a large private practice and married Anna Maria Edgeworth, the sister of the novelist Maria Edgeworth. Davy moved into the Beddoes' household, which was a stopping-off point for a train of intellectuals, notably Coleridge.

Coleridge had known of Beddoes's writing before he knew him personally. In 1792 Beddoes had published a moral tale, *Isaac Jenkins and Sarah his Wife*. It was a story set in a pastoral context about the evils of drunkenness, and Coleridge had written admiringly of 'the genius of him who wrote Isaac Jenkins'. Coleridge first came to Bristol in 1794 on the invitation of Southey, in order to further their famous 'American plan', a project to set up an ideal community, a 'Pantisocracy', on the banks of the Susquehanna. Coleridge returned again in 1795 and in January 1797 he moved to Nether Stowey. In the same year Wordsworth, who was living in the West Country at Racedown, moved to nearby Alfoxden, and here they composed the *Lyrical Ballads*. Coleridge had first met Beddoes in 1795. Not long after, Coleridge was reading Beddoes's medical works and those of John Brown. Among the many attractions of Beddoes for Coleridge must have been Beddoes's comparatively prodigious knowledge of German literature. In 1793 Beddoes had published an essay which was an early, if not the first, English account of Kant's *Critique of Pure Reason*, a work which Coleridge later read in German. In the *Monthly Review*, for which he wrote extensively, Beddoes reviewed German editions of the lesser-known works of Kant and, in 1798, Goethe's *Wilhelm Meister's Apprenticeship*. Besides their interest in German literature, Coleridge and Beddoes also shared radical political sympathies. Soon after they met they were jointly engaged in political agitation and, in 1796, Beddoes assisted Coleridge with the publication of his democratic journal *The Watchman*.

In 1798, with William and Dorothy Wordsworth, Coleridge departed for Germany. Davy arrived in Bristol shortly after they had left. Coleridge returned in July of the following year, and probably met Davy in the October. Coleridge first mentions the young chemist in a letter to Southey in December 1799. Coleridge and Davy became lifelong mutual admirers. Coleridge found much that was exciting in the young man, and no doubt fuelled Davy's sense of his own genius. 'Every subject in Davy's mind', Coleridge wrote, 'has the principle of vitality. Living thoughts spring up like turf under his feet.' Davy also excited the approval of the other Romantics. Southey, for instance, wrote that Davy had 'the feeling and habit of seeing all things with a poet's eye'. It is not exactly clear just when Davy met Wordsworth. When the poet returned from Germany he went directly to the Lakes. Wordsworth undoubtedly first heard of Davy from Coleridge. Later Wordsworth and the chemist became long-term friends. There was, of course, much in Davy's poetry and his cosmic speculations that the Romantics would have welcomed. Roger Sharrock has observed that the

subject-matter of his poetry is nature – moonlight, rivers and rocks – and nature viewed in terms of the one vital life shared between man and the universe . . . as with Coleridge and Wordsworth in the period 1797–9, his interest in the beauty of the moon amounts almost to an obsession.

Davy's notebooks at this time are filled with dozens of projects, verses, stories and essays. There is also a fragment on genius which ponders on the difference between men who are 'insignificant in their powers' and those who 'are full of energy in life, and leave behind them monuments of thought capable of perpetuating their existence'. The notebooks are also filled with Davy's ruminations on the consolations of philosophy, his 'sympathy with nature', his hatred of materialism and atheism and his conjectures on imagination and feeling.

At Bristol the main business of the day for Davy was not of course poetry, but gas chemistry, producing and experimenting with different kinds of airs. In the spring of 1799 Beddoes (with the financial support of the Wedgwoods) had opened a Pneumatic Institution, or hospital for the treatment of the sick by the inhalation of gases. Davy was made superintendent in residence. Clearly, from Davy's notes, the project which the Institution's members were engaged on was a Brunonian one – determining whether diseases caused by 'deficient energy' could be cured by gaseous inhalation. The work of the Institution, however, was not simply directed to the sick. Davy, Beddoes and their associates experimented on each other and on themselves. In his laboratory in April 1799 Davy first prepared pure samples of nitrous oxide ('laughing gas'), a substance which he had encountered when performing his first experiments in 1798. Nitrous oxide was an instantaneous hit with the Bristol circle. This is hardly surprising. Davy and the others, although ostensibly engaged in a scientific inquiry into the physiological effects of a gas, employed it as a poetic device for exploring the senses. Nor indeed were a science of life and poetry separate enterprises. According to Brunonian theory life depended on stimulation, and the more the stimulation the higher the states of perception. Davy recorded one of his earliest experiences with the gas as 'a highly pleasurable thrilling . . . The objects around me became dazzling . . .' Although occasionally the gas produced 'little pleasure', at other times it generated 'sublime emotions' and 'vivid and agreeable dreams'.

Unlike many other eighteenth-century doctors the Brunonians were unusual in regarding opium as a stimulant, and thus a valuable, almost universal, remedy. The Brunonians also argued, however, that excessive use of stimulants could eventually exhaust the system and produce death. It is clear that Davy viewed the inhalation of nitrous oxide from this perspective. He recorded that he was interested in discovering whether its stimulating powers destroyed life by producing 'the highest possible excitement', and he resolved 'to breathe the gas for such a time and in such quantities as to produce excitement equal in duration and superior in intensity to that occasioned by high intoxication from opium'. The transports produced by the gas were in fact so overwhelming that Davy briefly became an addict. He observed 'that a desire to breathe the gas is always awakened in me by the sight of a person breathing, or even by that of an air-bag or an air holder'. In his private notebooks Davy transformed his nitrous experiences into poetry:

> Not in the ideal dreams of wild desire
> Have I beheld a rapture-wakening form;
> My bosom burns with no unhallow'd fire,
> Yet is my cheek with rosý blushes warm;
> Yet are my eyes with sparkling lustre fill'd;
> Yet is my mouth replete with murmuring sound;
> Yet are my limbs with inward transports fill'd,
> And clad with new-born mightiness around.

It was not only Davy who inhaled the gas. The poets, too, queued up for a fix. Coleridge, already an opium addict, tried the new drug. Thirty years later he recalled 'the voluptuous sensation' it had produced. Southey wrote that in the inhalation of nitrous oxide Davy had discovered a pleasure for which language had no name.

Both Coleridge and Wordsworth were attracted to Davy, not only as a poet, but also as a natural philosopher. They did not, however, see chemistry in quite the same way that Davy did. Sharrock has argued that, at this time at least, in Wordsworth's and Coleridge's view chemistry, like the other natural sciences, involved an analysis of nature which was useful but superficial. As H. W. Piper has described it, they saw the sciences as ministering 'in the cause of order and distinctness, to the higher faculties'. In February 1800, Davy received a letter from Coleridge, who was now in the Lakes, about a project to build a laboratory where Coleridge and the Wordsworths might study chemistry under Davy's tutelage. Davy, however, turned the offer down. The results of his chemical researches on nitrous oxide had been published in 1800 and his name had reached the ears of the founders of the Royal Institution in London. They were seeking a lecturer and Davy was offered the job. In March 1801 he left Bristol to take up the appointment. Davy was to find, however, that the chemistry he was expected to pursue at the R. I. was quite different from the wild intellectual explorations of the Pneumatic Institution. At the

R. I. he was obliged to engage in a chemistry of the most economically practical sort, investigating tanning and agriculture at the behest of the aristocratic interest which funded the Institution. At first, at least, Davy found this sort of chemistry to be irksome and superficial. In August 1801 Coleridge wrote to Southey, 'I had one very affecting letter from Davy soon after his arrival in London – and in this he complained in a deep tone of the ill effect which perpetual analysis had on his mind.' Both Coleridge and Southey were worried that Davy's integrity might be wrecked by his brilliance: Coleridge feared that 'Chemistry turns its priests into sacrifices.' Davy, however, was soon to argue that chemistry was the equal of poetry.

In July 1800 Davy had been requested by Coleridge to proof-read the sheets of the second edition of his and Wordsworth's *Lyrical Ballads*. The book was to be published in Bristol and Davy was on hand to do the work. Although Davy at this time had not met Wordsworth, it has been suggested by Sharrock that shortly after Davy's proof-reading activities, there began to appear in his letters and poetry sentiments and terminology similar to those of Wordsworth. For instance, in July 1801 Davy wrote to a friend about 'That part of almighty God which resides in the rocks and woods . . .', a sentiment to be found in 'Tintern Abbey'.

Although Davy shared many of the sentiments of his poetical friends, and in spite of his initial unhappiness at the R. I., he did not concur with the Romantics' estimation of the role of chemistry. On 21 January 1802 Davy gave an introductory lecture to a course in chemistry. Coleridge was in the audience. In this lecture Davy directly confronted the Romantic view that the mind was engaged in a lesser activity when dealing with chemical affairs. Chemistry, Davy argued, takes 'the beings and substances of the external world' and 'explains their active powers'. Chemistry, however, was only arbitrarily distinguished from other sciences. 'The son of true genius . . . in the search of discovery . . . will rather pursue the plans of his own mind than be limited by the artificial divisions of language.' In doing so, he will be able to ascertain nature's 'hidden operations'. The fruit of such knowledge was both practical – it could give men power to invent and improve their lot – and moral, for the 'active powers' were also the basis of life and thought. Thus chemistry could 'exhibit to men' that system of knowledge 'which relates so intimately to their own physical and moral constitution'. When engaging with nature in this way, the mind of the natural philosopher must be active and creative. Where Coleridge and Wordsworth saw chemistry as a lower-order activity, Davy regarded it as an active pursuit 'connected with the love of the beautiful and sublime . . . eminently calculated to gratify and keep alive the more powerful passions and ambitions of the soul'. It was, in other words, a pursuit worthy enough for genius to show itself. Davy's description of chemistry, in fact, was similar to Wordsworth's account of poetry in the preface to those very same *Lyrical Ballads* which Davy had corrected. Indeed, it has been suggested by Piper that Davy's views on creativity 'echo Wordsworth's beliefs on the formation and nourishment of the mind'.

According to Sharrock, the 1802 edition of the *Lyrical Ballads* contains a favourable response by Wordsworth to Davy's lecture. Although Wordsworth still distinguished the poet's gifts he described poets and 'Men of Science' as partners in the amelioration of the human lot. This latter is a theme which appears repeatedly in Davy's works. Chemistry not only leads to understanding of nature but confers power over it, which in turn leads to power to transform society:

By means of this science man has employed almost all the substances in nature either for the satisfaction of his wants or the gratification of his luxuries. Not contented with what is found upon the surface of the earth, he has penetrated into her bosom, and has even searched the bottom of the ocean for the purpose of allaying the restlessness of his desires, or of extending and increasing his power.

The transformations that Davy envisaged the sciences bringing about related to material comforts and did not entail any reordering of society. Davy argued:

The unequal division of property and of labour, the difference of rank and condition amongst mankind, are the sources of power in civilized life, its moving causes, and even its very soul.

In ten years at the Royal Institution Davy transformed himself from an unknown provincial into a metropolitan lion. He was a successful lecturer, and the results of his research were pleasing to his aristocratic masters as well as being hailed internationally. In spite of his success Davy protested the banality of his everyday tasks. For instance, in a letter of 1803 he wrote, 'I am a lover of nature with an ungratified imagination. I shall continue to search for untasted charms, for hidden beauties . . . Common amusements and enjoyments are necessary to me only as dreams to interrupt the flow of thoughts.' It is equally clear, however, that throughout his life Davy's idea of genius was pre-eminently social. Genius was not raving or mad or betrayed in the solitary researches of the alchemist or the frenzied composing of a delirious poet. The mark of genius, of having tapped the active powers, was public recognition and acclaim. In 1801 in a letter to his Bristol friend John King (1766–1846), Davy admitted, 'The voice of fame is murmuring in my ear . . . I dream of greatness and utility.' 'Genius', 'hero', 'great man' and 'public figure' were effectively synonymous in Davy's writings. Thus, for Davy, ambition and a desire for fame were not base sentiments, but noble and necessary ingredients of genius. As a youth Davy had declared his own genius and, according to his own criteria, his later fame should have demonstrated to him the truth of his declarations.

Much of the work which made Davy famous was done during the first decade of the new century and was based on the use of the new Voltaic battery, invented in 1800. Its invention Davy described as 'an alarm bell to experimenters in every part of Europe'. In 1806 Davy used the current electricity produced by the battery to decompose water into 'oxygene and hydrogene'. In the following year he isolated potassium from potash by the same method. At first sight, Davy's work at the R. I. was centred on what might seem to have been either obscure chemical problems or on the application of chemistry to agricultural and industrial matters. However, as David Knight has shown in his *Atoms and Elements* (see Further Reading section), Davy's chemistry still had a Romantic core. Davy used chemistry to ask and answer questions about the material and spiritual constituents of the universe and thus about the nature of life and mind. Although many of Davy's chemical views changed during the R. I. years his Romantic assumptions did not. In spite of some fairly broad shifts of opinion, he remained committed to unity in nature and to powers rather than matter as the fundamental agency of order and change. Davy's work, although it used many of the concepts developed by the great French chemist Antoine Lavoisier (1743–94), was in many ways positively anti-French. The basis of Lavoisierian chemistry was the property-bearing element. Thus oxygen, for example, was regarded as the acidifying principle. In Davy's chemistry, however, such 'elemental' behaviour was simply the consequence of the activity of more fundamental powers of nature. In this respect, Knight has shown that Davy's chemistry drew on a British tradition originating with Newton, a figure Davy probably regarded as the greatest historical genius. Thus, Knight has argued, Davy's work on the Voltaic pile incorporated the presupposition that electricity was not some extra substance which combined with other things but that chemical affinity and electricity were manifestations of a single power. It is important to see therefore that Romanticism was not a hindrance to chemistry, nor yet something that enabled Davy to discover truths about nature. Rather the nature that Davy discovered was itself Romantic. This bears on a rather more general point. Davy's Romantic chemistry was not idiosyncratic nor his Romanticism unique. Davy's chemistry was congenial to his audience which, by its approval, validated it as a true picture of nature. Such approval was also an endorsement of Davy's account of his own genius.

Coleridge therefore was vocalizing a widespread assumption when he proclaimed Davy 'the Father and Founder of philosophic alchemy, the Man who *born* a Poet first converted Poetry into Science and *realized* what few men possessed genius enough to *fancy*'. Chemistry of this sort, as Coleridge saw it, was 'a pursuit of unity of principle, through a diversity of forms'. In this sense, Coleridge held, it was 'poetry, substantiated and realized'. During these years Davy, like Coleridge, also speculated on the nature of life and, also like

Coleridge, developed a vitalist philosophy. That is, Davy and Coleridge were committed to the view that the mere organization of matter could not give rise to life. The organized living body was the instrument of some more fundamental, perhaps undiscoverable, hidden power.

Argued in this way Davy's chemical and physiological ideas appear to follow from a number of Romantic assumptions about nature. Such an argument in turn reinforces the practice of representing Romanticism solely as a cluster of *ideas* about nature, or God and the mind. But Davy's lectures on matter, life and mind were also political practices which can only be understood in the context of the dangerous years of the French Wars, the years of conservative reaction. Davy, who had possibly flirted with radicalism in his Bristol years, increasingly became, like Coleridge, a Tory. During the first two decades of the new century any account of life or mind which so much as hinted that they were merely the products of corporeal organization was condemned as atheistical and politically subversive, which usually meant French-inspired. Conversely vitalism of the sort espoused by Davy and Coleridge was used to underwrite the existing social order. This was achieved by the identification of the vital power with the Creator worshipped by the Church of England. In 1817 the surgeon William Lawrence (1783–1867) was to find out the cost of ignoring this orthodox identification. In a series of lectures given in 1817 Lawrence argued for an autonomous science of physiology and suggested that the mind should be investigated as an action of the brain. In doing so Lawrence attacked the teachings of the eminent London surgeon, John Abernethy (1764–1831), who had earlier used the works of Davy to defend the existence of a separate vital principle. In 1817 Abernethy effectively denounced Lawrence as a materialist, a dangerous charge in those years. Coleridge took Abernethy's side in the controversy.

During the second decade of the century Davy, to borrow the poet Philip Larkin's words, 'burst into fulfilment's desolate attic'. He had become a figure of national and international importance. By his own lights this was nothing but entirely consonant with the order of things. The man of genius, by reason of his capacity to see further, had become the man of power. In April 1812 he left the Royal Institution, was knighted and was married, all in the space of four days. His wife was the wealthy socialite Jane Apreece (1780–1855). Although England was at war he travelled to France in 1813 and was fêted by French savants. He then journeyed on to Italy, studying geology and experimenting on the combustion of diamonds. He finally returned to London in 1815. In 1818 Davy set out for Italy again, and returned in 1820 when he was elected President of the Royal Society.

It was in this second decade of the century that Davy, on the basis of a number of experiments, devised a safety lamp for miners. This invention, so often singled out as a blessing, epitomizes so much of Davy's work. On the one hand it was, apparently, a practical demonstration of his lifelong rhetoric that natural science could be used for the relief of man's estate. On the other hand, it demonstrates his continual enslavement to the industrial and aristocratic interests which he served so well. The lamp could well be used to symbolize the new web of social relations which were generated during the Industrial Revolution. Far from saving lives, the lamp was used to effect the exploration of deeper and more dangerous seams which in turn resulted in an increased death toll. The point is not that the lamp was a good invention badly used; rather, as Morris Berman has argued, the invention itself was the product of an order, of which Davy was part, based on division and exploitation (see Further Reading section, below). In 1817, John Davy recorded that his brother thanked the colliery owners for a gift of plate given to him for the invention of the lamp. In giving thanks Davy took the opportunity to discuss invention. In doing so he represented the mind of the inventor as an equal partner with capital in the creation and maintenance of the social order. He reminded the owners that it was natural science which had been the means of their raising their 'subterraneous wealth'. The extraction of coal, however, had not only increased the 'wealth and power' of the country, for it also drove the steam engine which

has produced even a moral effect, in rendering capital necessary for the perfection of labour, credit essential to capital, and ingenuity and mental energy a secure and dignified species of property.

Davy's lamp, which, like the steam engine, he portrayed as an embodiment of the utility of chemical genius, caused him much strife. He was accused of having stolen the idea.

These years of fulfilment were indeed the years of desolation. Davy's marriage was desperately unhappy. His most brilliant experimental work was behind him. Davy, however, continued to experiment and to travel. After he became President he ran the Royal Society as an autocrat. In 1826 he fell ill and much of the remainder of his life was spent in retreat in Europe, fly-fishing in Austria and travelling across Italy and Switzerland. During these years he wrote *Salmonia* and, finally, *Consolations in Travel*. In his biography, John Davy remarked on the similarities of these last philosophical works to Davy's youthful sketches. It was as if, he said, in later life, 'there was a renewal of former ideas, a revival of former intentions'. More recently, Michael Neve has argued that in these works Davy attempted to bring together and organize the lost strains of a Romantic consciousness that had 'been disconnected and diffused' (see Further Reading section). Such observations seem just, since the similarity with the speculations of the Bristol years are marked. Yet such idealist accounts should not be allowed to obscure the material social relations to which these texts address themselves.

The *Salmonia* is a work, ostensibly about fly-fishing, modelled on Isaak Walton's *The Compleat Angler*. It is, however, also a poetic hymn to the 'wild and beautiful scenery of nature' and a meditation on death and the fear of dying. It is the *Consolations in Travel*, however, which is most reminiscent of the musings of the young Davy. The *Consolations* is a series of dialogues and is a florid example of the historical–evolutionary thought of the early nineteenth century. It is also, as are so many Romantic Tory texts of the time, a deeply religious work musing on the Christian God and his world.

In the first dialogue the narrator (Philalethes) describes a vision conjured up as he sits among the ruins of the Colosseum in Rome. The vision is stimulated by the full moon which, the narrator says, 'has always a peculiar effect on these moods of feeling in my mind, giving to them a wildness and a kind of indefinite sensation, such as I suppose belong at all times to the true poetical temperament'. In the vision a 'voice' which the narrator calls 'that of the Genius', shows Philalethes the 'origin and progress of human society' and how improvement had been wrought by 'a few superior minds'. The narrator then describes the most recent and significant revolution in human affairs, that produced by the experimental sciences and invention. It is at this point that the imagery of power, so important in Davy's account of chemistry and the mind, can be seen to have its roots in industrial society. The steam engine is praised by the narrator as one of the greatest inventions in human history. Davy describes it as a 'power' able to drive 'the machinery of active life' in a manner which 'could hardly have been imagined' by the old philosophers. It is also a 'power placed in human hands which seems almost unlimited'. Such a power is itself the product of another, 'the exertions of one man of genius, aided by the resources of chemistry'. Similarly, it is the 'intellectual power' of the 'Caucasian stock' and those 'powers . . . acquired by cultivation' which are the source of European superiority. In time the narrator says this civilizing process will lead to the extinction of the 'negro race, and the red men' as the Caucasians absorb the 'power belonging to a ruder tribe'. This progressive motion results in an increase in the complexity of life and a proliferation of living forms.

Following this the narrator discloses a philosophy of life and mind which rests on the familiar assumptions that there was unity in nature and a hierarchy of spirit from the least perceptive to the omniscient. According to this philosophy, our spiritual natures are 'parts more or less inferior of the infinite mind'. As such, they are 'continually aiming at, and generally rising to a higher state of existence'. Thus beyond this earthly world is, for instance, the 'monad or spirit' that was Newton, which 'is now drinking intellectual light from a power source and approaching nearer to the infinite and divine mind'. In this elated sphere monads have 'modes of perception' of which we are ignorant. Like men in earthly society, these higher monads have modified their material world but 'with far superior results'. Through all its stages of being the monad carries with it one 'sentiment or passion'; this is 'the love of knowledge or of intellectual power, which is . . . the love of . . . unbounded power, or the love of God'. One of the keys to this cosmic speculation is to recognize in it the Romantic assumptions that had been used by the young Davy over twenty-five years previously, for the narrator remarks that 'power and knowledge . . . and . . . intellectual life . . . depend more or less upon the influence of light'.

In another dialogue in the *Consolations* the narrator engages with the origins of genius, describing his own career in a way that leaves the reader in no doubt that Davy saw his own life as having followed the ideal trajectory, both glorious and tragic. Philalethes describes how as a young man he came to the metropolis and how his 'ambition was satisfied', but how the life of the metropolis 'without losing its power . . . had become bitter'. He records how he was seized with the desire to travel, and how the English burn out their limited 'human powers' and how 'distinguished men . . . poets, and even philosophers . . . before the period of youth is passed' pass over into 'premature winter time'. (Davy had left the R. I. in his early thirties.) In the course of this dialogue Davy's characters are led to speculate on the nature of life. As in Davy's youth, one of the characters conjectures that during respiration 'some sensible matter' reaches the blood and stimulates the vital property of sensibility. Sensibility, and at a higher level intellectual activity, were not to be explained simply as the product of organized matter – the optic nerve and brain 'are but the instruments of a power which has nothing in common with them'. This power, the sentient principle or monad, is connected to the body 'by kinds of ethereal matter'. Indeed, although at death the gross organs are destroyed, 'some of the refined machinery of thought' mediated by this 'ethereal nature' might survive. This argument seems to be the intellectual face of Davy's self-confessed horror of being buried alive. At this point in the dialogue, in a revealing sentence, Davy couples together the two Romanticisms which are woven through his life, justifying the pursuit of glory – the exhibition of genius – by drawing on assumptions about the unity of nature and its hierarchical and ultimately spiritual reality:

The desire of glory, of honour, of immortal fame and of constant knowledge, so usual in young persons of well-constituted minds, cannot, I think, be other than symptoms of the infinite and progressive nature of intellect – hopes, which as they cannot be gratified here, belong to a frame of mind suited to a nobler state of existence.

Davy died on the European continent, as had, within the previous ten years, Keats, Shelley and Byron ('powerful spirit', Davy called him in a poem). Ironically Davy, the Celt, was buried in Protestant Geneva and, contrary to his wishes, he was interred only three days after his death was deemed to occur.

(Davy, however, had asked for ten days' grace.) The great Englishman (for that was how Davy saw himself) had been a major figure in the shaping of European thought and industrial society. Davy had seen power as the key to the universe. He had himself sought to discover it in chemistry and to exercise it in society. Romanticism was indeed the thread which ran through his life. Romanticism, however, was not a body of ideas in the light of which Davy helped shape the modern world – quite the reverse. Davy fashioned Romanticism to bring about a new order of things.

## FURTHER READING

Berman, Morris, *Social Change and Scientific Organization. The Royal Institution (1799–1844)* (London, 1978)

Cartwright, F. F., *The English Pioneers of Anaesthesia: Beddoes, Davy, Hickman* (Bristol, 1952)

Davy, Humphry, *The Collected Works*, ed. John Davy, 9 vols. (London, 1839–40)

Davy, John, *Memoirs of the Life of Sir Humphry Davy* (London, 1839)

Knight, David M., *Atoms and Elements. A Study of Theories of Matter in England in the Nineteenth Century* (London, 1967)

—*The Transcendental Part of Chemistry* (Folkestone, Kent, 1978)

Neve, Michael, 'The Young Humphry Davy: or John Tonkin's Lament', in Sophie Forgan (ed.), *Science and the Sons of Genius. Studies on Humphry Davy* (London, 1980)

Piper, H. W., *The Active Universe. Pantheism and the Concept of Imagination in the English Romantic Poets* (London, 1962)

Schofield, Robert, 'Joseph Priestley, Eighteenth-Century British Neoplatonism, and S. T. Coleridge', in Everett Mendelsohn (ed.), *Transformation and Tradition in the Sciences* (Cambridge, 1984)

Sharrock, Roger, 'The Chemist and the Poet: Sir Humphry Davy and the Preface to *Lyrical Ballads*', *Notes and Records of the Royal Society*, 16 (1961), 57–76

Stansfield, Dorothy A., *Thomas Beddoes M.D. 1760–1808. Chemist, Physician, Democrat* (Dordrecht, Holland, 1984)

Temkin, Owsei, 'Basic Science, Medicine and the Romantic Era', *Bulletin of the History of Medicine*, 37 (1963), 97–129

# 16

## Oersted's discovery of electromagnetism

### H. A. M. SNELDERS

On 21 July 1820 the Danish physicist and chemist Hans Christian Oersted (1777–1851) announced in a four-page Latin pamphlet his discovery of the effect of an electric current on a magnetic needle suspended in the earth's magnetic field. He sent his publication to a large number of prominent men of science and scientific societies. It was headed *Experimenta circa effectum conflictus electrici in acum magneticam*. Before long it was translated into Danish, Dutch, English, French, German and Italian. The English version appeared in the October issue of the *Annals of Philosophy* as 'Experiments on the Effect of a Current of Electricity on the Magnetic Needle'.[1]

In the spring of 1820 Oersted found out that a pivoted magnetic needle placed parallel to a wire carrying an electric current made a great oscillation. The needle deflected one way for one direction of the current and the opposite way for the other direction. Oersted made use of a large battery of low internal resistance, which he had constructed together with his friend, the lawyer Lauritz Esmarch (1765–1842).[2] The apparatus consisted of twenty copper troughs, twelve inches long and equally high and two and a half inches broad (fig. 18). The trough formed the positive pole of the cell. In each cell a zinc plate was fastened to a hoop which protruded from the copper trough of the adjoining cell. The troughs were filled with water containing 1/60th of its weight of sulphuric acid and an equal weight of nitric acid:

Let the straight part of this wire [carrying the current] be placed horizontally above the magnetic needle, properly suspended, and parallel to it. If necessary, the uniting wire is bent so as to assume a proper position for the experiment. Things being in this state, the needle will be moved, and the end of it next the negative side of the battery will go westward.[3]

If the wire was held under the magnetic needle, it moved in the opposite direction. The deflection of the magnetic needle depended on the distance between the needle and the wire and on the 'power of the battery'. In Oersted's experiments the declination of the needle made an angle of about 45°.

17　A 'Romantic' representation of Oersted's discovery of electromagnetism.
From Hans Kraemer, *Weltall und Menschheit* (1902–5).

In his *Experimenta* Oersted described various experiments. He found that it made no difference whether the wire was made of platinum, gold, silver, brass, iron, ribbons of lead and tin or a mass of mercury. The conductor did not even lose its effect when interrupted by a little water. Moreover Oersted found that a deflection also took place when glass, metal, wood, water, resin, stoneware or stones were placed between the magnetic needle and the wire. He clearly realized the new aspects of his discovery:

It is needless to observe that the transmission of effects through all these matters has never before been observed in electricity and galvanism. The effects, therefore, which take place in the conflict of electricity are very different from the effects of either of the electricities.[4]

As a simple rule for finding the direction of the force upon the magnetic needle, Oersted went on to propose:

That these facts may be the more easily retained, we may use this formula – the pole *above* which the *negative* electricity enters is turned to the *west*; *under* which, to the *east*.

It came as a surprise that the direction of the force was perpendicular to a plane through the wire and the needle. This observation was more remarkable than the fact that an electric current acted on a magnetic needle. For a long time scientists had assumed some connection between electricity and magnet-

18   Oersted's apparatus, now in the Danmarks Tekniske Museum, Elsinore.

ism. Oersted's discovery, however, seemed to contradict Newton's law that forces of gravitational attraction act along the line connecting the two gravitational bodies. Charles Coulomb's law of the interaction of electrically charged bodies and of magnetized bodies, respectively, implied that the attractive and repulsive forces between magnetic bodies as well as between electrified bodies also lay along the connecting lines. Oersted, however, had found that the force between a current and the pole of an adjacent magnet was *not* along the line connecting the two, but *perpendicular* to that line.

At first sight Oersted's explanation of his discovery seems somewhat strange. Already in the title of the Latin pamphlet he spoke of a 'conflictus electricus', an electric conflict, and not of an electric current. The Italian, the German and the Danish translations also used this expression. The English translation, however, spoke about 'a Current of Electricity', and the Dutch about 'de werking van het galvanismus op de magneetnaald' (the action of galvanism on the magnetic needle). But in the texts of these translations we also meet the term 'electric conflict':

The opposite ends of the galvanic battery were joined by a metallic wire, which, for shortness sake, we shall call the *uniting conductor*, or the *uniting wire*. To the effect which takes place in this conductor and in the surrounding space, we shall give the name of the *conflict of electricity*.[5]

The expression 'electric current' was already used before Oersted's time. In the *Dictionnaire raisonné de physique* (1781) of Mathurin Jacques Bisson (1723–1806) we find the entry 'courant électrique'. The reason for Oersted's use of the concept of an electric conflict is to be found in his philosophical background.

## OERSTED'S EARLY LIFE

Oersted, son of a pharmacist at Rudkøbing on the Danish island of Langeland, was born in 1777. At the age of eleven he began to assist in his father's apothecary shop. He developed an increasing interest in chemistry and physics. In 1794 Oersted started his studies in natural sciences at the University of Copenhagen. Here he also followed courses on Kant and the critical philosophy, given by the Kantian philosopher Børne Riisbrigh (1722–1809). In 1797 Oersted passed his examination in pharmacy. Two years later he wrote a paper on Kant's *Metaphysische Anfangsgründe der Naturwissenschaft (Metaphysical Foundations of Natural Science)* (1786) for a new Danish journal which was started for the purpose of promoting Kant's philosophy. Oersted treated the same subject more elaborately in his thesis for the doctorate from the same year. The *Dissertatio de forma metaphysices elementaris naturae externae* (1799) was a critical account of the content of

Kant's ideas on natural philosophy, in particular on such questions as: what is the *a priori* basis of science, what are the necessary presuppositions of all experience and what laws of matter and its motion may be deduced *a priori*? Kant considered the concept of matter as 'the moveable insofar as it fills a space'.[6] He immediately added that matter does not fill a space 'by its mere existence, but by a special moving force'.[7] Matter fills its space by the antagonism between two forces: attracting and repulsing. Both forces determine the dynamical nature of matter. Kant used these concepts for the explanation of chemical and physical processes: solution, separation, friction and elasticity. In his dynamical theory he restricted himself to a general foundation on which matter is built up, namely the two basic forces. That there are two basic forces is essential: by mere attraction, matter would 'coalesce in a mathematical point and the space would be empty and hence without any matter.'[8] With only repulsive forces matter 'would be held within no limits of extension, i.e. would disperse itself to infinity, and no assignable quantity of matter would be found in any assignable space.'[9] According to Kant the physical divisibility of a substance which fills a space continues as far as the mathematical divisibility of that space. Therefore he rejected an atomistic structure of matter: 'Matter is divisible to infinity, and indeed into parts each of which is again matter.'[10]

Oersted became an ardent exponent of Kant's new critical philosophy, which was to be of fundamental importance to his scientific development. He was less affected by the speculative elaboration of the Kantian philosophy by Schelling and his adherents. Kant saw the world as an equilibrium between the opposing forces of attraction and repulsion, while Schelling believed in a conflict in which these forces constantly strove to overcome each other, an attractive force constantly battling with a repulsive force producing a basic polarity in matter and, ultimately, in the whole universe.

In his thesis Oersted showed his reserve against the rising *Naturphilosophie*. He was critical of the way Schelling used empirical theses in his books *Ideen zu einer Philosophie der Natur (Ideas of a Philosophy of Nature:* 1797) and *Von der Weltseele (On the World Soul:* 1798):

These two books no doubt deserve attention for the beautiful and great ideas we find in them, but on account of the not very rigorous method by which the author intermingles empirical propositions without sufficiently distinguishing them from *a priori* propositions these books are robbed of much of their value, especially as the empirical propositions adduced are often utterly false.[11]

Oersted's study of Kant's critical philosophy gave him an excellent background for his later scientific work. It led him to the realization that for a law of nature which is absolutely valid an *a priori* foundation is necessary. But it also made him an opponent of the atomic theory for the greater part of his life.

After his studies at Copenhagen University Oersted became manager of the Lion Apothecary in Copenhagen. In 1801 he was appointed an assistant professor, without pay. He began experimental studies with the Galvanic pile, which was described by Alessandro Volta in 1800. With this new source of electric current a continuous flow of electric fluid could be produced.

From the summer of 1801 until the end of 1803 Oersted made an educational trip (*Wanderjahr*) through Germany, France and the Netherlands. In September 1801 he met the German Romantic physicist Johann Wilhelm Ritter (1776–1810) in Oberweimar. Ritter was first and foremost an experimental physicist, but was also strongly influenced by the thoughts of the Romantic movement. Time and again his publications illustrate excellent experimental methods based upon fanciful speculations. His interest was mainly focused on the nature of galvanism. For three weeks Oersted worked together with Ritter on the subject of galvanism.

The latter's belief in a connection between electricity and magnetism had taken a firm hold of Oersted's mind during his stay with Ritter. It was this Romantic belief in and search for unity and polarity in nature that became the guiding principle in Oersted's scientific researches. Oersted and Ritter met for the second time in Jena (August to September 1802). The two meetings led to an animated and interesting correspondence lasting until Ritter's untimely death in 1810.

In a letter to Oersted of 22 May 1803, Ritter prophesied a remarkable discovery in 1819 or 1820. He meant that the years of maximum inclination of the ecliptic (1745, 1764, 1782, 1801) coincided with outstanding discoveries in the field of electricity ('$\frac{1}{3}$' means in the first four months of the year, and so forth):

$1745\frac{1}{3}$ Invention of the Leiden jar by Kleist (1745)

1764 Invention of the electrophorous by Wilcke (1764)

$1782\frac{1}{3}$ Invention of the condenser by Volta (1783)

$1801\frac{1}{3}$ Invention of the Voltaic pile (1800).

Ritter went on: 'You will not have to reckon with a new epoch or its start any earlier than the year $1819\frac{2}{3}$ or 1820. This we might well witness.'[12] And then lo and behold Oersted discovered in 1820 the effect of the electric current upon the magnet!

Ritter was not the only one to influence Oersted. During a six-months' stay in Berlin, he not only studied chemistry and physics, but also attended lectures by Fichte and the brothers Schlegel; he got acquainted with Franz von Baader and studied intensively, but not uncritically, the writings of Schelling. He found a complement to these philosophical pursuits during a stay in Paris (1802), where he met French savants and was impressed by their highly developed use of careful experiments in the study of nature. In Germany he was confronted with scientists who – often through mere speculation – were seeking the unity of nature they so firmly believed in. This influence of the speculative German spirit on Oersted was not, however, superseded by French experimentation, notwithstanding his somewhat critical attitude towards *Naturphilosophie*. From a philosophical point of view he was able to accept without question the more speculative views not only of Ritter, but also of the Hungarian Jakob Joseph Winterl (1739–1809). This professor of chemistry and botany at the University of Buda believed that he had found compounds which were still more simple than chemical elements and from which in the end all matter was built up.[13] Winterl called these hypothetical substances 'Andronia' (the principle of acidity) and 'Thelycke' (the principle of alkalinity). Oersted became acquainted with Winterl's speculations during his stay with Ritter. He was especially enthusiastic about Winterl's meta-physical idea that all the forces of nature arise from the same fundamental principles of heat and light, acid and alkali, electricity and magnetism, namely the positive and negative principles of electricity. It was this concept which gave unity and connection to all experience. Because of this interest in Winterl's views, Oersted came to be considered, around 1805, as a *Naturphilosoph*. However, his faith in Winterl's two principles sustained a severe shock with the failure of a number of chemists to isolate the 'Andronia'. By 1807 his confidence in Winterl's theory had dissipated, and at the same time Ritter's influence on him decreased considerably. Oersted's belief in Romantic natural philosophy declined in the following years, even though his faith in the unity of all the forces of nature remained unchanged.

In 1806 Oersted became Extraordinary Professor of physics at the University of Copenhagen. (Only in 1817 was he promoted to Ordinary Professor.) He made experimental researches on the acoustical figures of Ernst Chladni (the geometrical figures formed when a sand-covered plate is subjected to acoustical vibration), to which he was led by the hope of finding electric effects accompanying the oscillations (1807). Oersted held the belief that electricity, galvanism and magnetism were not imponderable substances, as was generally assumed, but mere modifications of the general primitive forces (*Grundkräfte*) of attraction and repulsion under different circumstances. At the end of a book entitled *Materialen zu einer Chemie des neunzehnten Jahrhunderts* (*Materials for a Chemistry of the Nineteenth Century*) (1803), Oersted stated:

The constituent principles of heat which play their role in the alkalis and acids, in electricity, and in light are also the principles of magnetism, and thus we have the unity of all forces which, working on each other, govern the whole cosmic system, and the former physical sciences thus combine into one united physics.[14]

In 1806 Oersted suggested that all natural phenomena are produced by the same principle, which 'appears in most different forms as, for example, light, heat, electricity, magnetism, and so on'.[15]

In the same year he published an article in which he stated that the electric conflict in a wire is caused by the contrary effect of the two kinds of electricities which are accumulated in the poles of the Galvanic battery.[16] The propagation of electricity in a wire is a continuous disturbance and restoration of an equilibrium and therefore not to be considered as a continuous current.

## OERSTED'S DISCOVERY OF ELECTROMAGNETISM

During a stay in Berlin (May 1812 – Summer 1813) Oersted wrote a book entitled *Ansicht der chemischen Naturgesetze durch die neuern Entdeckungen gewonnen* (*Consideration of the Physical Laws of Chemistry Deduced from the New Discoveries*) (1812), in which – in his own words –

he proved that not only chemical affinities, but also heat and light are produced by the same two powers, which probably might be only two different forms of one primordial power. He stated also, that the magnetical effects were produced by the same powers, but he was well aware, that nothing in the whole work was less satisfactory, than the reasons he alleged for this. His researches upon this subject, were still fruitless, until the year 1820.[17]

An important reason for these claims was the action of the column of Volta and the chemical action of an electric current. A description of the background to the discovery of electromagnetism was given by Oersted in an article on thermo-electricity, which he wrote for David Brewster's *Edinburgh Encyclopaedia* in 1827:

In the winter of 1819–20, he delivered a course of lectures upon electricity, galvanism, and magnetism, before an audience that had been previously acquainted with the principles of natural philosophy. In composing the lecture, in which he was to treat of the analogy between magnetism and electricity, he conjectured, that if it were possible to produce any magnetical effect by electricity, this could not be in the direction of the current, since this had been so often tried in vain, but that it must be produced by a lateral action. This was strictly connected with his other ideas; for he did not consider the transmission of electricity through a conductor as a uniform stream, but as a succession of interruptions and re-establishments of equilibrium, in such a manner, that the electrical powers in the current were not in quiet equilibrium, but in a state of continual conflict. As the luminous and heating effect of the electrical current goes out in all directions from a conductor, which transmits a great quantity of electricity, so he thought it possible that the magnetical effect could likewise irradiate. The observations . . . of magnetical effects produced by lightning, in steel-needles not immediately struck, confirmed him in his opinion. He was nevertheless far from expecting a great magnetical effect of the galvanic pile; and he still supposed that a power, sufficient to make the conducting wire glowing, might be required.[18]

The electric conflict in the conductor between the opposite electricities produces many effects: chemical effects in the conductor in the direction of the current, and heat and light effects which radiate in all directions from the conductor. Might it therefore not be possible that the magnetic effect is a special action of the same forces that are found in heat and light? In his *Ansicht der chemischen Naturgesetze* Oersted remarked that one must try to show the action of magnetism on electricity. 'One feels that magnetic forces are as general as electric forces. An attempt should be made to see if electricity, in its most latent stage, has any action on the magnet as such.'[19]

It was not until 1820 that Oersted succeeded in proving the connection between electricity and magnetism with a very simple experiment. In April 1820, during an evening lecture on electricity, galvanism and magnetism, he found that the current in a wire caused a magnetic needle, placed at a distance from it, to move. The effect, however, was very feeble. At the beginning of July 1820 the experiments were resumed using a stronger galvanic apparatus. The effects were still feeble and irregular because Oersted employed very thin wires, supposing that the magnetic effect would not take place when heat and light were not produced by the galvanic current. Soon he found that wires of a greater diameter give a much stronger effect. The experiments were done in the presence of a number of witnesses: Lauritz Esmarch, the mathematician Peter Wleugel (1766–1835), the physicist Adam Hauch (1755–1838), the biologist Johannes Reinhardt (1776–1845), the physician and biologist Ludwig Jacobson (1783–1843) and the chemist William Zeise (1789–1847). Soon after he published a preliminary report of his results. This report caused a sensation, but the verdict of the French chemist Pierre Louis Dulong (1785–1838) was significant. On 2 October 1820 Dulong wrote to the Swedish chemist Jöns Jacob Berzelius (1779–1848): 'At first the news is received very coolly here. One believes that it is again a German dream.'[20] The German physicist Ludwig Wilhelm Gilbert (1769–1824), editor of the *Annalen der Physik*, who published attacks on *Naturphilosophie*, as well as appending critical notes to articles with a speculative tendency, declared Oersted's discovery to be a mere accident (1820). His colleague Christian Heinrich Pfaff (1773–1852) also held the same opinion (1824). Of course Kantian and Schellingian scientists and philosophers were very enthusiastic. On 2 August 1820 the Kantian mineralogist Christian Samuel Weiss (1780–1856) wrote to Oersted that his discovery was a confirmation of Kantian dynamics.

Despite Oersted's fundamental experiments, his explanation was clearly speculative. The electric conflict exerts only an influence 'on the magnetic particles of matter'. All non-magnetic bodies appear permeable by the electrical conflict, while magnetic bodies (or rather their magnetic particles) resist the passage of this conflict. 'Hence they can moved by the impetus of the contending powers.'[21] Oersted himself realized that 'the electric conflict is not confined to the conductor, but dispersed pretty widely in the circumjacent space' and must be assumed to traverse circles whose planes are at right angles to the conductor:

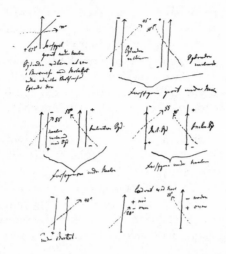

19  A page from Oersted's laboratory notebook, July 1820, showing the action of electricity on a magnet.

From the preceding facts we may likewise collect that this conflict performs circles; for without this condition, it seems impossible that the one part of the uniting wire, when placed below the magnetic pole, should drive it to the east, and when placed above it towards the west; for it is the nature of a circle that the motions in opposite parts should have an opposite direction.

The centre of force does not act attractively or repulsively on the magnetic poles, but it drives the poles in a circle around it. Oersted was well aware that this whirling was a new action of force and that it was not analogous to the central forces (forces acting in straight lines between points) of attraction and repulsion that underlie the phenomena of gravitation, electricity and magnetism. His notion of circular magnetic lines of force, which cut a current-carrying wire in planes perpendicular to the conductor, seems very modern, but has nothing to do with the later concepts of field theory.

Immediately after the foregoing, Oersted suggested that the ideas of a circular movement in the medium surrounding the electric conflict would be of significance for the theory of the nature of light:

I shall merely add to the above that I have demonstrated in a book published five years ago that heat and light consist of the conflict of the electricities. From the observations now stated, we may conclude that a circular motion likewise occurs in these effects. This I think will contribute very much to illustrate the phenomena to which the appellation of polarization of light has been given.

Oersted's Latin pamphlet was a preliminary communication. It was rather too brief to be perfectly intelligible. In the July issue of the *Journal für Chemie*

238      H. A. M. SNELDERS

*und Physik*, the same number in which the German translation of the Latin pamphlet was printed, we find another paper with Oersted's extension of his experiments. This article was translated into English as 'New Electromagnetic Experiments' and printed in the *Annals of Philosophy* of November 1820.[22] Oersted reported that a Galvanic pile composed of a hundred discs of two square inches of zinc and copper, and a paper moistened with salt water to serve as the fluid conductor, has hardly any effect upon the magnetic needle. On the other hand a single galvanic cell of zinc and copper, with for the conductor a liquid of one part sulphuric acid and one part of nitric acid in sixty parts of water, gave a clear deflection of the needle. All the effects of the experiments are made more powerful by using a cell with large plates instead of smaller ones, and also by using a large cell instead of a battery of smaller cells: 'The magnetic effects do not seem to depend upon the intensity of the electricity, but solely on its quantity.'[23] Oersted also showed the effect of a stationary magnet upon a moveable part of an electric current (the so-called reversed Oersted effect).

Oersted's discovery of the action of an electric current on a magnetic needle was a direct consequence of his metaphysical belief in the unity of all natural forces. In the circle of the German Romantics, he was held in great respect. But in contrast to certain speculative scientists of his time, Oersted was a man of such solid learning that he was apparently but little susceptible to purely speculative thinking. From the beginning of his scientific career we can distinguish in Oersted two main streams: a speculative one and an empirical one. Both aspects divided his work and his interests. Sometimes one of the two dominated, but never suppressed the other totally. Oersted was an adherent of the critical philosophy of Kant, but he was also influenced by the speculative *Naturphilosophie*. He took a liking to vaguely formulated hypotheses, but differed sharply from Schelling in his acceptance of the fundamental importance of careful observation. His attitude towards *Naturphilosophie* changed in the course of his life, but the idea of unity and polarity in nature remained prominent. Until the end of his life, Oersted believed in a philosophy of the unity of all forces of nature:

The laws of nature in the material world, are laws of reason, revelations of a rational will; but when we thus consider all material nature, as the constant work of eternal reason, our contemplation cannot remain at this point, but leads us by thought to view the laws of the universal nature. In other words, soul and nature are one, seen from two different sides: thus we cease to wonder at their harmony.[24]

NOTES

1   H. C. Oersted, 'Experiments on the Effect of a Current of Electricity on the Magnetic Needle', *Annals of Philosophy*, 16 (1820), 273–6.

2  H. C. Oersted, 'Bemerkungen hinsichtlich auf Contactelectricität', *Journal für Chemie und Physik*, 20 (1817), 205–12.

3  H. C. Oersted, 'Experiments on the Effect of a Current', p. 274.

4  *Ibid.*, p. 275.

5  *Ibid.*, p. 274.

6  I. Kant, *Metaphysical Foundations of Natural Science*, trans. with Introduction and Essay, by James Ellington (Indianapolis and New York, 1970), p. 40.

7  *Ibid.*, p. 41.

8  *Ibid.*, p. 60.

9  *Ibid.*, p. 57.

10  *Ibid.*, p. 49.

11  H. C. Oersted, *Naturvidenskabelige Skrifter*, ed. Kirstine Meyer (Copenhagen, 1920), I, p. 77.

12  H. C. Oersted, *Correspondance avec divers savants*, ed. H. C. Harding (Copenhagen, 1920), II, p. 36.

13  H. A. M. Snelders, 'The Influence of the Dualistic System of Jakob Joseph Winterl (1739–1809) on the German Romantic Era', *Isis*, 61 (1970), 231–40.

14  H. C. Oersted, *Naturvidenskabelige Skrifter*, I, p. 209.

15  H. C. Oersted, 'Die Reihe der Säuren und Basen', *Journal für Chemie und Physik*, II (1806), 509–47; quotation is on p. 538.

16  H. C. Oersted, 'Ueber die Art, wie sich die Electricität fortpflanzt', *Neues allgemeines Journal der Chemie*, 6 (1806), 292–302.

17  H. C. Oersted, 'Thermo-Electricity', *The Edinburgh Encyclopaedia*, XVIII (1830), 573–89; *Naturvidenskabelige Skrifter*, II, p. 356.

18  *Naturvidenskabelige Skrifter*, II, p. 356–7.

19  *Ibid.*, p. 148.

20  J. J. Berzelius, *Bref*, ed. H. G. Söderbaum (Uppsala, 1915), IV, p. 18 (my translation).

21  H. C. Oersted, 'Experiment on the Effect of a Current', p. 276.

22  H. C. Oersted, 'New Electromagnetic Experiments', *Annals of Philosophy*, 16 (1820), 375–7.

23  *Ibid.*, p. 375.

24  H. C. Oersted, *The Soul in Nature* (London, 1852), p. 384.

## FURTHER READING

The primary sources are in H. C. Oersted, *Naturvidenskabelige Skrifter*, ed. Kirstine Meyer, 3 vols. (Copenhagen, 1920), and *Correspondance de H. C. Oersted*, ed. H. C. Harding, 2 vols. (Copenhagen, 1920).

Secondary sources on Oersted:

K. Meyer, 'The Scientific Life and Work of H. C. Oersted', in *Naturvidenskabelige Skrifter*, I, pp. XIII–CLXVI

B. Dibner, *Oersted and the Discovery of Electromagnetism* (New York and London, 1962)

J. Rud Nielsen, 'Hans Christian Oersted – Scientist, Humanist and Teacher', *American Physics Teacher*, 7 (1939), 10–22

O. I. Franken, *H. C. Oersted. A Man of the Two Cultures* (Birkerød, 1981)

On Romanticism and *Naturphilosophie*:

R. C. Stauffer, 'Persistent Errors Regarding Oersted's Discovery of Electromagnetism', *Isis*, 44 (1953), 307–10, and 'Speculation and Experiment in the Background of Oersted's Discovery of Electromagnetism', *Isis*, 48 (1957), 33–50

H. A. M. Snelders, 'Romanticism and *Naturphilosophie* and the Inorganic Natural Sciences 1797–1840: An Introductory Survey', *Studies in Romanticism*, 9 (1970), 193–215

B. Gower, 'Speculation in Physics: The History and Practice of *Naturphilosophie*', *Studies in History and Philosophy of Science*, 3 (1973), 301–56

# I7

## Caves, fossils and the history of the earth

### NICHOLAS A. RUPKE

### INTRODUCTION

This essay focuses on Germany, from the 1770s till the 1810s, when the study of rocks and fossils expanded the horizon of history and was made into one of the corner-stones of Romantic *Naturphilosophie*. During this period the scope of history was widened by a more sympathetic investigation of those past ages which the Enlightenment had treated as unenlightened or barbaric and had left in obscurity. The religious scepticism of the French *philosophes* was replaced by a new love of Christianity and a form of philo-Catholicism which expressed itself in an idolizing of the Middle Ages, a period which had been denigrated or ignored by the previous generation. The new philosophy of history was characterized also by what may be called organicism. Human civilization was examined in relation to its natural environment, and the successive periods of its history were studied as part of a connected development. Herder depicted human life as closely related to its milieu in the natural world, and the first volume of his *Ideen zur Philosophie der Geschichte der Menschheit (Ideas on the Philosophy of the History of Mankind)* (1784–91) discussed the earth as a planet, as the theatre of geological upheavals and changes and as a surface with an atmosphere, oceans and continents fit to support plant and animal life and eventually man.

Human society, in this view of history, is a product of historical change and so is man himself. This developmental view focused initially on linguistics: the question of the origin of language was turned into a debate on social anthropology. But in the course of this period the testing ground for different and opposing views of the pattern of history was changed and broadened to include geology. Organic fossils and minerals were no longer just curious stones, but archival evidence of a pre-Adamic world. The animals to which large bones and teeth, now found in caves and gravel, had belonged, were interpreted as the inhabitants of the earth before man had acquired civilization. Thus the study of fossils extended the range of history, dramatically so. Its implications for the historicity of nature produced a wave of real enthusiasm which swept Germany from the 1780s on. During these early years of the modern study of fossils and minerals, the subject was readily accessible and its results available in the intellectual market place, before geology was withdrawn to a more confined shop as a result of academic departmentalization.

During the 1770s and 1780s the new ideas were in part borrowed from the French: from Buffon, Rousseau or Jean-Louis Giraud Soulavie. The German contributions, however, became increasingly significant – contributions to anatomy and mineralogy, to paleontology and stratigraphy. In no other European country were these subjects so extensively brought to bear on the philosophy of history and nature as in Germany.

## THE FASCINATION WITH CAVES

In the course of the eighteenth century, especially during its last few decades, a number of caves in the area of the Harz Mountains and the French Alps were systematically explored. Enormous quantities of fossil teeth and bones were found in these. An early classic of cave paleontology was the *Ausführliche Nachricht von neuentdeckten Zoolithen (Detailed account of Newly Discovered Zoolites)* (1774) by Johann Friedrich Esper (1732–81).[1] The fossils were distributed to German centres for the study of anatomy, but also to Buffon and Daubenton and, later, to Cuvier in France, to Petrus Camper in the Netherlands and to John Hunter in England. This turned out to be a major stimulus to the development of vertebrate, mammalian paleontology and to the study of the history of the earth. The Universities of Göttingen and Erlangen, located not far from the caves, figured prominently in this development, although Halle and Leipzig, on the other side of the Harz Mountains, were also involved. The main fossils were those of 'bear', 'elephant' and 'rhinoceros', and the story of their discovery and identification is in part a tale of cave studies.

At the time, caves occupied a much more central position in geological theory than they do today. They were not regarded as minor and accidental perforations in the uppermost skin of the earth, as we regard them now, but they were believed to be pervasive and primordial features, present since the birth of the earth as a planet, providing essential information as to the manner of its origin. Caves were imagined as corridors to the deep recesses of our globe in which the archives of its history were stored, and where the secrets of its past could be discovered, including those of its antediluvian inhabitants, a mighty race of giants and monsters.

Consider, for example, the major two-volume work, *Beschreibung merkwürdiger Höhlen (Description of Curious Caves)* (1799, 1805), written

20  Contemporary illustration showing the wonders of the cave at Gaylenreuth in Bavaria. From William Buckland, *Reliquiae Diluvianae*, 1823.

by two young doctors from the University of Leipzig, Johann Christian Rosenmüller (1771–1820) and Wilhelm Gottfried Tilesius (1769–1859). It contained over 110 instances of caves, drawn mostly from the published literature, and in part also from personal observations of a number of German examples. Some of these they believed to be primordial, but in addition they emphasized a theory of the origin of caves, taken from Tobern Bergman's *Physicalische Beschreibung der Erdkugel* (*Description of the Physics of the Globe*) (1769), which envisaged dissolution of limestone in groundwater percolating through faults and fissures. Rosenmüller and Tilesius were deeply convinced of the importance of caves for a proper understanding of the earth and its history. They referred to cave systems as the bowels of our earth, corridors to the very workshop of Mother Nature. The examination of caves was especially significant for paleontology: the first and best-known examples of extinct animals came from caves; caves could be regarded as the mother of vertebrate paleontology, a subject which had become enhanced in interest as a result of recent contributions by Cuvier and Faujas de Saint-Fond.

Praise of caves, no less lyrical, was sung by Friedrich von Hardenberg, alias

Novalis (1772–1801), probably Germany's purest Romantic. Novalis particularly romanticized mining and the exploration of caves, more so than, for
example, the ascent of a lofty mountain peak. He enlisted as a student at the
Mining Academy of Freiberg, where the most famous teacher was Abraham
Gottlob Werner, who lectured on mineralogy and stratigraphy (see below, pp.
250–52). Novalis was deeply impressed by Werner with whom he continued a
correspondence after having left the Academy. The deep recesses of the earth
did not possess for Novalis the demonic quality they held for other Romantics
such as Ludwig Tieck and E. T. A. Hoffmann. To him the mines and caves
preserved the peaceful perfection of a golden past when harmony reigned. The
subterranean vault is to the miner an instructive and wise teacher:

| | |
|---|---|
| Die mächtigen Geschichten | The vault is happy to tell him |
| Der längst verflossnen Zeit | about the great events of ages |
| Ist sie ihm zu berichten | long gone by. |
| Mit Freundlichkeit bereit. | |
| | |
| Der Vorwelt heilge Lüfte | The sacred breezes of this pre-historic |
| Umwehn sein Angesicht, | world waft around his face |
| Und in die Nacht der Klüfte | and in the night of the abyss |
| Strahlt ihm ein ewges Licht.[2] | shines for him an eternal light. |

Novalis's romanticized description of caves occurs in his *Bildungsroman*
(psychological novel) *Heinrich von Ofterdingen* (1802), which remained
fragmentary because of the author's early and tragic death from tuberculosis.
The novel describes the journey from Eisenach to Augsburg by young
Heinrich. On the way he encounters an old miner who had been taught by a
wise teacher named Werner (in honour of Novalis's own teacher). In the
company of some fellow travellers and led by the old miner, Heinrich explores
a major cave system. In its deep recesses they come upon a recluse, Count of
Hohenzollern, who in his subterranean vault initiates young Heinrich in the
true meaning of history. Novalis's cave description is remarkably vivid – a
cold draught at the narrow entrance, smooth walls and the presence of large
fossil bones:

Everyone's attention was in particular occupied by the countless mass of bones and
teeth which covered the floor. Many of these were fully preserved, others showed signs
of decay, and those which jutted out from the walls seemed to have turned to stone.
Most of the specimens were unusually large and sturdy.[3]

The company is excited to find these 'Überbleibsel einer uralten Zeit' (remains
of very ancient times). Whereas to astronomers the heavens are the book of the
future, to them the earth is an archive which reveals the existence of an *Urwelt*
(a prehistoric world). The large bones point to an early period in the history of
the world when Nature still displayed the vigour of youth. But although
Nature has lost its gargantuan powers to uplift mountains, to generate

precious metals and stones or outsized plants and animals, it has become more delicate, diverse, impressionable and imitative of man. Thus the primeval golden age, which has disappeared in discord and strife, will return in the millennium taking the form, as Novalis says elsewhere, of the perfection in man of his inner quality. The old miner says:

> It may be that Nature is no longer as fertile as it used to be, that today no metals and precious stones originate any longer, no masses of rock and mountains, that plants and animals no longer acquire such astonishing size and strength; to the extent that the generative force has diminished, the forces which shape, ennoble and produce social activity have increased; nature's inner constitution has become more sensitive and soft, her imaginative faculty diverse and capable of dealing with symbols, her hand more delicate and artistic. She comes closer to man, and if formerly she was a wild rock, she now is a calm, blossoming plant.[4]

## DEGENERATION AND EXTINCTION

From the time of Leibniz's *Protogea* (1748), it was widely believed that mammalian fossils were to be referred to unknown species still alive in unexplored regions of the globe. Towards the end of the eighteenth century, however, this notion lost its currency. For example, in his doctoral dissertation, *Beiträge zur Geschichte und nähern Kenntniss fossiler Knochen* (*Contributions to the History and More Detailed Knowledge of Fossil Bones*) (1795), the above-mentioned Rosenmüller presented a precise comparison, side by side, of the entire skull of a cave bear, a brown bear (*Ursus arctus*) and a polar bear (*Ursus maritimus*). The latter had been studied in recent years by both Camper and Peter Simon Pallas (1741–1811). With respect to dentition, size and proportions, the fossil skull differed as much from the skull of brown and polar bear as the latter two between themselves and the conclusion seemed inevitable that the fossil skull represented a separate species which no longer exists:

> Because we are unable to find among the known and described species of bear the type of which the bones occur in caves, the conclusion seems entirely appropriate that this type of bear no longer exists.[5]

However, Rosenmüller did not regard extinction as the only possibility. The large fossil bones might have belonged to animal types which are still in existence today, but have degenerated into relatively small forms as a result of environmental changes. The two possibilities, 'ausgerottet' ('exterminated') and 'ausgeartet' ('degenerated'), were joined to become the popular alliteration of paleontological speculation. Rosenmüller regarded degenerative variability as a fact and he guessed that as a result of changes of climate, food and habitat the once mighty forebear from the caves could have diminished to become our familiar polar or brown bear:

Would it therefore not be possible that the type of bear, to which the fossil bones belong, has degenerated into the polar bear species? Or, what seems rather more likely, could we not look upon the brown bear as the degenerated offspring of our cave bear?[6]

Unlike certain of his contemporaries, Rosenmüller did not believe that the cave fossils had been washed in by deluges or introduced by men. He suggested that the caves had been used by the bears as their living quarters, possibly as recently as the Celtic inhabitation of the region. However, he conceded that further proof was required, and this was adduced much later when William Buckland, who was familiar with Rosenmüller's work, developed it into a fully fledged ecological theory.

Although specimens from caves constituted the vast majority of mammalian fossils, the cave bear was not the only animal that conditioned people's conceptions of the possibly extinct and gigantic inhabitants of a previous world. With anatomical skill and fantasy, another outsized animal, the mammoth, was reconstructed from its disjointed remains of bones, teeth and tusks. These were not only known from caves, but also from superficial sand and gravel deposits which we now call Pleistocene. Attention was drawn to the mammoth by Pallas. His father was a professor of medicine in Berlin, and Peter Simon followed in his footsteps studying medicine at the Universities of Göttingen, Halle and Leiden. In 1767 he was invited to work at the St Petersburg Academy of Sciences, and for more than forty years he devoted himself to the study of Russian natural history.

Pallas's early expeditions, in the years 1768–73, were described in his widely read *Reise durch verschiedene Provinzen des russischen Reichs (Journey through several Provinces of the Russian Empire)* (1773–6; translated into Russian, French, English, Italian). The expeditions covered both European and Asiatic Russia, and included western Siberia. Pallas reported the astonishing occurrence, in a fair number of places along his itinerary, of elephant and rhinoceros bones, even at arctic latitudes. Similar occurrences were reported from North America and became widely known through Buffon's *Epoques de la nature (Epochs of Nature)* (1778). The description of a modern rhinoceros skull by Camper added further fuel to a debate about the origin of such fossils and what climatological revolutions their present locations might prove. Inspired by all this, similar finds of elephant and rhinoceros bones were made in German soil and described by the remarkable Johann Heinrich Merck (1741–91), an exact contemporary of Pallas. He initially followed an unfulfilling career in the military; an unhappy marriage added further to a need for compensation which he found in literary activity and in a circle of *literati* to which Herder, Goethe and Christoph Martin Wieland belonged. A journey to St Petersburg in 1773 was the start of a deep interest in natural history. Merck was an autodidact whose correspondence with Camper,

Samuel Thomas Sömmerring and other anatomists of first rank gave him a real competence in the comparative osteology of mammalian fossils.

Merck obtained a collection of elephant and rhinoceros bones, not only from such caves as the Baumannshöhle, but also from the Pleistocene sediments along the Rhine between Mainz and Mannheim. In a number of published letters (1782–6) he argued that the fossil species were different from the related ones of today, and Merck speculated that the differences in size and other characteristics might have been produced by a process of degeneration, operating over a period of centuries:

You know, Sir, that the fossil bones of elephants, discovered in America and Siberia, are very much larger than the skeletons exhibited in Europe's museums. Let us therefore not leave this matter without concluding that those bones belong to an animal of enormous size, but that Mr Buffon's rule is once more confirmed, namely that the animals of long ago, the remains of which are found in many places, possessed an astonishing strength and size, entirely degenerated in the course of the past centuries.[7]

Merck's paleontology inspired Goethe to an enthusiastic, almost feverish, participation. In a letter of 1782 he asked Merck to forward the letters by Camper, and with a near-childish pride he mentioned that he knew his osteology by heart and was able to identify the individual parts in the skeletons of all animals by the names of the human bones. Goethe was not only interested in identification of fossils, but also in the information they give us as to the history and periodization of the geological past: 'Soon now the time will come when fossils will no longer be lumped together, but will be arranged according to the different epochs of the history of the world.'[8] It is likely that the awareness of geological history and time which Herder demonstrated in the first volume of his *Ideen* of 1784 was deepened by Merck and the interest which he raised in his literary circle for the nascent subject of paleontology.

Goethe's early prediction that fossils would soon be classified according to geological age came true in the work of Johann Friedrich Blumenbach (1752–1840). He dominated German paleontology during the decades 1790–1810, till Cuvier's monumental *Recherches sur les ossemens fossiles* (*Researches on Fossil Bones*) (1812) eclipsed all previous work in vertebrate paleontology. Blumenbach's belief in both extinction and degeneration, and his definition of a metaphysical life force (*nisus formativus*) provided points of contact with the emergent Romantic concept of history and *Naturphilosophie*.

Blumenbach studied medicine at the Universities of Jena and Göttingen, and was named full professor of medicine at the latter university in 1778. Already the previous year he had given lectures on comparative osteology which he subsequently expanded into a full course of '*anatome comparata*'. In the preface of his *Handbuch der vergleichenden Anatomie* (*Textbook of Comparative Anatomy*) (1805) he claimed to have been the first to lecture on the subject at a German university. Among the men with whom he corresponded were Camper, Albrecht von Haller and Charles Bonnet; and he counted Jean-André Deluc and Georg Christoph Lichtenberg among his friends and patrons at Göttingen. Blumenbach's insistence on the fact of extinction, his subdivision of earth history according to fossils and the wide distribution of his ideas through the popular *Handbuch der Naturgeschichte* (*Textbook of Natural History*) (1779; many times reprinted) were valuable contributions to geology, later largely fallen into oblivion.

Blumenbach's expertise in comparative anatomy made it possible for him to identify fossils at the species level and not to be satisfied with a classification based on a general similarity with living types. In a paper, 'Beyträge zur Naturgeschichte der Vorwelt' ('Contributions to the Natural History of the Prehistoric World') (1790), he argued that only on the basis of precise identification could fossils become archives of earth history proving the fact of extinction and the reality of a pre-Adamic world. He condemned the opinion of men such as James Hutton who believed that fossil invertebrates from the distant geological past still have living representatives in today's oceans. Blumenbach described an encrinite, argued that it was extinct and suggested that the degree of similarity of fossils with the living fauna might constitute a basis for the subdivision of earth history:

If one looks at fossils from the grand viewpoint that they are the most infallible documents in Nature's archives, demonstrating that our planet has gone through several upheavals, showing even the manner and to some extent the times of these major changes, thus making it possible to determine the relative ages of the various major formations – it is obvious that their history must be regarded as one of the most important and instructive parts of all of natural history, and especially of scientific mineralogy.[9]

In the first volume of his most readable booklet, *Beyträge zur Naturgeschichte* (*Contributions to Natural History*) (1790; second edition, 1806), Blumenbach maintained that neither Providence nor any supposed chain of being disallowed the belief in extinction. Its reality could be demonstrated even in the present Adamic creation: the dodo, a large bird from Mauritius, had disappeared in recent times. From a pre-Adamic world not only many species of ammonites and belemnites had vanished, but even entire genera and families. Blumenbach left open the question whether or not the cave bear and the fossil elephant and rhinoceros were extinct species. In his *Handbuch der Naturgeschichte* he classified them as dubious fossils, naming them respectively as '*Ursus spelaeus?*', '*Elephas primigenius?*' and '*Rhinoceros antiquitatis?*'. This uncertainty gave Rosenmüller grounds to interpret the cave bear as the original type (*Urtypus*) from which other bear species had originated by degeneration. Blumenbach himself limited degenerative variability to change within a particular *Typus* or *Urbild* ('species').

Thus three periods of earth history could be distinguished. In his 'Specimen archaeologiae telluris' ('A Case-Study of the Archaeology of the Earth') (1801) Blumenbach proposed the following subdivision: (1) the most recent deposits, characterized by fossils identical to today's species, such as the Oeningen marl (they were formed by local inundations); (2) deposits of intermediary age, with fossils similar to, but not identical with, present-day species now living in tropical regions far away from the northern caves and gravel (the change was caused by a climatological revolution); (3) the oldest deposits which contain entirely unknown fossil organic remains, made extinct by a major global upheaval.[10]

Blumenbach ascribed the regeneration of new forms of life, after the global catastrophe, to a metaphysical life force or *Bildungstrieb*. He defined this force in a booklet, *Über den Bildungstrieb* (1789), which was his contribution to the debate between preformationists and epigenesists. Are germs, or are sperm cells preformed little organisms? Initially, Blumenbach had followed Haller's preformational view. But he changed his mind and attributed the growth of an organism into the form of its species, and also its regenerative power to heal wounds, to a specific growth-urge activated at the moment of conception. The *Bildungstrieb* would have repopulated the surface of the earth after the destructive catastrophe. But because physical conditions had been altered by the upheaval, the new organisms were not identical to those of the previous world. As an example Blumenbach cited a marine snail, *Murex despectus*, which looks identical to a fossil snail, *Murex contrarius*, except for the fact that the former has a shell which winds rightwards, the latter leftwards. Such organic changes were not caused by a mere degeneration of early forms into later ones, but by 'transformation through an altered direction of the *Bildungstrieb*'.

Blumenbach saw a meaningful parallelism between his three geological periods and the three epochs of human history: 'three divisions which to a certain extent can be compared with the three periods of secular history, namely the *historic*, the *heroic* and the *mythical* period'.[11] Stated more specifically: the microcosm of human history recapitulates the history of the earth. The idea was not uttered in isolation; Blumenbach's older colleague Lichtenberg, the professor of physics, expressed a similar idea in his *Sudelbücher* (*Notebooks*) (1789–93). Such speculations were near to the Romantic *Naturphilosophie* with its fascination for the parallelism between man's embryological development and the lower rungs of the animal kingdom. Establishing the order of a developmental succession, whether in the crust of the earth or in the human embryo, was not considered profound enough. What was the logic behind the sequence? How might one get to the principle controlling the development? Such questions fascinated German philosophers of this period, and found amplification in the mushrooming speculative *Naturphilosophie* of around 1800 (see pp. 252–7).

## NEPTUNIST STRATIGRAPHY

Blumenbach recognized that a proper periodization of the history of the earth had to be based not only on a study of fossils, but also on that of the succession of rock formations in space and time (stratigraphy, in modern parlance). In the course of the 1780s and 1790s, when the fact of extinction was placed on the firm foundation of comparative anatomy, and fossils were interpreted as archives of the history of life, a related development took place in mineralogy. During the last two decades of the eighteenth century, the mineral composition of rock types and formations was given a time significance: a particular mineralogical composition was believed to indicate the relative age of a rock formation and its place in the succession of geological history. However, paleontology and stratigraphy, two parts of what later became known as historical geology, developed in different institutional contexts. The former had an intellectual affinity to anatomy and was fostered in medical faculties; the latter was of use to mining and was given a modern form within the mining school of Freiberg.

Traditionally, the study of both organic fossils and minerals had been part of medical education. For example, during the early part of the eighteenth century a student at the University of Tübingen received his MD under Rudolph Jakob Camerarius with a thesis *De lapidum figuratorum usu medico* (*On the Medical Use of Figured Stones*). Similar topics were given space in the *Acta physio-medica*. Before cave fossils became objects of anatomical study they were collected and ground into a powder to be sold in pharmacies as an effective medicine. When through the late eighteenth century the study of organic fossils began to take the form of modern paleontology, it did not leave its institutional context but remained part of the study of medicine.

In contrast to organic fossils, minerals and rocks were gradually removed from the institutional embrace of medicine, and in Germany received a new home at the Mining Academy of Freiberg. The first lecture courses at this school were given in the year 1766. From 1769 till 1771, Abraham Gottlob Werner (1749–1817) was a student there. On the basis of a brilliant study of mineralogical classification, *Von den äusserlichen Kennzeichen der Fossilien* (*On the External Characteristics of Fossils*) (1774) he was appointed at his *alma mater* as inspector and lecturer of mining and mineralogy. Over the years he diversified the topics of his lectures, and from the year 1786 he gave annual courses in oryktognosy (nearly synonymous with mineralogy) and geognosy (nearly synonymous with geology). His *Kurze Klassification und Beschreibung der verschiedenen Gebirgsarten* (*Short Classification and Description of the Different Formations*) (1787) contained a stratigraphic classification of rock formations. Although Werner published some twenty-five articles and books, his popular lectures were the main outlet for his ideas.

Through the 1790s Werner perfected his stratigraphical classification. His courses were attended by an eager and international student audience. The list of his pupils included the names of such later luminaries as Leopold von Buch (from 1790 to 1793), Alexander von Humboldt (from 1791 to 1792), E. F. von Schlotheim (from 1791 to 1793) and J. F. d'Aubuisson (from 1797 to 1802); Novalis entered in 1797 and Robert Jameson came to Freiberg in 1800. A fairly large proportion of the students at the Academy belonged to the aristocracy. Werner's impact as a teacher can be measured by the fact that at the time of his death more than twenty of his pupils occupied professorial positions at institutions of higher education in Europe.[12]

Werner speculated that initially, some one million years ago, the earth had been enveloped by a universal ocean. The oldest crystalline rocks, such as granite, had originated from its waters by chemical precipitation. As the ocean subsided, more and more of the new rocks originated mechanically, by erosion from older formations exposed to the air. Among the chemical precipitates Werner counted basalt. Others believed that it was igneous in origin. This led in the course of the 1780s to a heated controversy between 'Neptunists' who adhered to Werner's view and 'Vulcanists' who insisted on an igneous origin, the latter party headed by Werner's former pupil J. C. W. Voigt. Feelings ran high and the debate continued well into the nineteenth century.

More fundamental was the difference between Werner's view and James Hutton's 'Plutonism', the latter also formulated in the course of the 1780s. In his 'Theory of the Earth' (read 1785; published 1788) Hutton proposed that most rocks are not chemical in origin, but detrital – eroded from dry land, deposited on the sea floor, lithified by heat from below and uplifted again to become new dry land. Thus a new cycle of erosion, deposition, lithification and uplift started, *ad infinitum*. To Hutton not only basalt but also granite was of igneous origin. This type of rock therefore did not represent the oldest layer of the crust of the earth, the foundation of all other formations, but originated during any geological period or cycle. The clash between the Neptunist and Plutonist view of earth history was very fierce indeed. The protracted nature of the debate and its ferocity cannot be explained in purely geological terms, but must be viewed at the level of clashing philosophies of history. Hutton's Plutonism was part and parcel of Enlightenment history with its penchant for eternalism. Werner's Neptunism was not merely a system of chemical lithogenesis, but had a consanguinity with both the

traditional, biblical cosmogony and the more secularized view of history held
by many Romantics. Goethe saw the controversy in this light, and so did men
such as Blumenbach and his friend Jean-André Deluc.

Neptunist stratigraphy represented a directional, orthogenetic view of
history which incorporated the notions of both degeneration and progress.
Huttonian Plutonism with its implications of a steady state, even in the
organic realm, was fundamentally uncongenial to the Romantic philosophy
of history. The directionalism of Wernerian stratigraphy was spelt out by von
Buch, in a (for him) rare instance of metaphysical speculation, when he chose
as the topic of his inaugural lecture to the Berlin Academy of Sciences 'Ueber
das Fortschreiten der Bildungen in der Natur' ('On the Progressiveness of
Forms in Nature' (1806; published 1808).[13] In a language reminiscent of that of
speculative *Naturphilosophie*, von Buch interpreted the stratigraphic succession as a process of gradual progress. The history of nature has been a process
of struggling forces, directed towards the highest form, the human *Gestalt*, in
which the eternal desire for freedom is striving in vain to loosen itself from the
material world. Von Buch depicted the stratigraphic succession as a progressive drive towards individuality (*Trieb zur Selbständigkeit*). The first manifestation of this drive was the formation of crystalline rocks, of the granitic
triplet of quartz, feldspar and mica. To the inner force of chemical crystallization was added the external force of mechanical erosion. Forces and their
products became diverse and multiple. Crystalline rocks were followed by
detrital ones, and an impulse (*Triebkraft*) generated primitive life. The
organic *Triebkraft* produced ever higher forms of plants and animals, at an
accelerating pace.

## ROMANTIC HISTORY AND *NATURPHILOSOPHIE*

In the middle part of the eighteenth century the question of the origin of
language had in part been the battle-ground of opposing anthropologies and
of conflicting philosophies of history. Lord Monboddo's (James Stewart's)
*On the Origin and Progress of Language* (first volume, 1773) was an example
of an outspoken social anthropology along the lines of Enlightenment progressivism. The Berlin Academy of Sciences promoted a number of prize
questions on the origin of language. Among the answers which attracted
attention was that by Johann Peter Süssmilch (1707–67), also known for his
pioneering work in population statistics. The gist of his argument was
summarized in the title of his essay of 1756 (published 1766), namely *Versuch
eines Beweises, dass die erste Sprache ihren Ursprung nicht vom Menschen,
sondern allein vom Schöpfer erhalten habe* (*Attempt to Prove that the First
Language Had its Origin not in Man, but Came Exclusively from the Creator*). The prize question posed in the year 1771 read: 'Imagine man left to his
natural faculties; would he be capable of inventing language? And how would

he arrive at that invention?' The award went to Johann Gottfried Herder (1744–1803), whose answer was entitled *Abhandlung über den Ursprung der Sprache* (*Treatise on the Origin of Language*) (1772). The first part of the question was answered with an unequivocal 'yes'. Not the language, only the linguistic facility is of divine origin.[14] This disagreement with Süssmilch did not imply that Herder stood on the side of rationalist progressivism. Language might well have developed by the action of reason shaping the early imprecise vocal signs into clear concepts; but to characterize this process as progressive was something altogether different. Like the other Romantics, Herder idolized early language and literature, from a time when these had still been direct expressions of inner feeling, not yet spoiled by the sophistication of reason and reflection.

However, during the last few decades of the eighteenth century, when the study of fossils and rocks widened the scope of nature's historicity, paleontology and stratigraphy became a new testing ground for philosophies of history. It is no coincidence that such early stratigraphers as Georg Christian Füchsel and Werner were interested in linguistics and the problem of the origin of language.

The new and vast expanse of geological history was not interpreted in purely progressivist terms (as is commonly supposed). A number of the Romantics saw it just as much as a process of degeneration, and as an illustration of the magnificence of the very early past. The history of the earth could be interpreted as analogous to and supportive of their cherished pattern of human history and linguistic development.

Novalis, in his politico-religious essay *Die Christenheit oder Europa* (*Christianity or Europe*) (1799), venerated medieval times when no split had yet broken the religious community apart, and when a naive–harmonious *Weltbild* had not yet been corrupted by rational contemplation. To Novalis, the truth about nature is not to be discovered along the road of reason, but by an intuitive–mystical unity with the world around us; priests and prophets are closer to it than *philosophes* and our hope for a future of progress lies in taking up where the Middle Ages left off. Accordingly, in his *Die Lehrlinge zu Sais* (*The Novices at Sais*) (1802) the pupils of the title sought the essence of nature under the guidance of a wise master. As a devoted Neptunist, Novalis attributed the terrestrial origins to water, to a universal ocean or *Weltmeer*. As I have mentioned before, in his *Heinrich von Ofterdingen* (1802) Novalis used the new geological discoveries to give substance to the belief in a former existence of giants, and in a primeval voluptuousness of nature in comparison to which today's physical reality is a weak residue. Novalis mourned our loss of original innocence and strength, and the corruption which has occurred since early times, but he also believed in the possibility of improvement, in the coming of a golden age, a millennium to which he often referred.

Not only to the Romantic *literati* but to contemporary painters, too, rocks were of interest for what they communicated about the history of the earth and for what they could express about the artist's search for the lost unity with nature. Pre-eminent among these painters was Caspar David Friedrich (1774–1840) who was friendly with Wernerian naturalists and whose work was characterized by a close attention to geological features. His many representations of rock formations and mountain views are among the most characteristic expressions of Romantic feeling on canvas.

The Romantic *Naturphilosophen* Lorenz Oken and Gotthilf Heinrich von Schubert made explicit and substantial use of the new geological material, although in support of rather different and conflicting systems of nature. Schubert stayed close to such Romantics as Novalis, with an emphasis on the degenerative aspect of earth history. Progress is little more than a repeatedly failed upward tendency, although it may be sustained in the perfecting of human morality. Oken's view was much more secular, less attached to the traditional view of history of Lutheran orthodoxy and of Pietism. He proposed an organistic progressivism closely linked to physiology of which the *Naturphilosophen* were so very fond.

Schubert (1780–1860) was the son of a clergyman, and in his younger days was influenced by Herder, Schelling and Werner. He obtained an MD at the University of Jena and not long after went to Freiberg to attend the lectures by Werner. Subsequently he was invited to give a series of lectures to a gathering of members from the upper classes in Dresden and in this way originated his *Ansichten von der Nachtseite der Naturwissenschaften* (*Views of the Dark Side of the Natural Sciences*) (1808), a book warmly applauded by traditionalist Romantics. Schubert depicted the physical history of the world, the history of life and human history as analogous processes, each characterized by a trend of degeneration and corruption, but with an upturn towards the end: a decline from a former paradisiacal state, but with a glimmer of hope that paradise might be regained.

In the *Ansichten* Schubert maintained that the crust of the Earth had originated by chemical precipitation, and that the earliest formation consisted of granite. The later formation indicated a decline from this primitive purity: the crystals were smaller and so were the extent and thickness of the formations. 'The most recent river deposits, from the alluvial period, appear almost entirely characterless, merely minor hills; their feeble endeavour to rise seems flattened at the very moment of their origin.'[15] However, at the depth of anorganic decline, nature was resurrected in the appearance of organic life:

In the form of the organic world nature rises again from the grave of decay, and the cause of the organic inceptions has been simultaneously that of the decline of the inorganic world. Thus a new period is merrily built on top of the ruins of the old submerged one, in the hope of establishing its handiwork more firmly on the deep foundations of the most remote times, not as a result of the permanence of corporeal mass, but through spiritual strength.[16]

21  Caspar David Friedrich, *Kreidefelsen auf Rügen* (*Chalk Cliffs on Rügen*) (1820).

In the organic world a similar pattern of initial grandeur followed by decline could be discerned. Originally, plants and animals could reach truly gigantic heights: lofty trees and mighty elephants had populated the earth. Animal populations had been more numerous and plant growth more luxuriant than they are today. Innocence had reigned, all early animal species supposedly having been herbivorous.

Schubert believed that the same pattern could be seen in the history of mankind. Man's original state had been one of innocence and pure knowledge, acquired through inspiration and harmony with nature. Man's golden childhood was described by Schubert as follows:

It was a time of childhood, but higher than this helpless infancy of today. Mortal mothers now give birth, but the original infancy was nursed by an immortal mother, and man has departed from the direct observation of an eternal idea, has been unaware in the midst of the highest knowledge and forces, which now the later generation must regain by way of lofty but hard struggles.[17]

Corruption had set in when the priest-king had become a conqueror-king, no longer in harmony with nature, but trying to impose his will on nature instead. Degeneration was visible in language, and even in man's dentition which supposedly had changed from a purely herbivorous to an omnivorous type. The Middle Ages, however, had shown a return to a state of harmony and a *Weltbild* based on alchemy and astrology. This improvement of man's relationship to the world around him had been temporarily halted by the mechanistic view of nature, present during the time of the Enlightenment, but was now taken up again and continued with an interest in the invisible forces behind the empirical reality, exemplified not only by electricity and magnetism, but also by such phenomena as precognition, dreams and other psychical experiences, the *Nachtseite* of the natural sciences.

Schubert gave a more factual account of the Wernerian theory of the earth in his *Handbuch der Naturgeschichte* (*Textbook of Natural History*) (1813, 1816). And in the continuation of his Dresden lectures, published under the title *Die Urwelt und die Fixsterne* (*The Pre-historic World and the Fixed Stars*) (1822), he returned to the question of geological progressivism, taking the opportunity to argue against it. Schubert's use of the analogy of earth history and human history was akin to the idea that man is a microcosm in which the world around him is summarized. This was a beloved notion among Romantics and *Naturphilosophen* because it stressed the unity of man and nature, the oneness of the individual with the totality of existence. The new periodization of earth history provided a further area where the microcosm-macrocosm analogy could be applied. Human history was a recapitulation of the whole history of the earth and life.

The application of this view to the geological past, in a very general and all-encompassing manner, was the work of Lorenz Oken (1779–1851), who carried it out with a near-maniacal gusto. Like Schubert, Oken was influenced by Schelling. Much of his own system was contained in two early booklets, *Abriss der Naturphilosophie* (*Outline of Nature Philosophy*) (1805) and *Die Zeugung* (*Generation*) (1805), which he wrote as a young man in his middle twenties. He left the empirical ground on which Johann Friedrich Meckel (1781–1833) had founded his idea of embryological recapitulation, and he opened his *Abriss* with the following rhetorical question:

What else does the animal kingdom represent but the anatomized state of man, the macrozoon of the microzoon? In the former is most beautifully displayed and un-wrapped what in the latter has come together in the form of small organs, according to the same beautiful order.[18]

To Oken nature is merely the dissected display of man, or, conversely, the history of the earth and of life are in essence nothing more or less than the birth of man. Wherever man looks he encounters himself, 'and in happy exultation now calls out: you have found the organs of the universe in yourself, and yours in the universe!'[19] Thus the counterpart in nature to man's eyes (or vision) was the insects; when nature considered man's hand (or tactility) it made snails; the ear appeared early on as birds; worms were merely unconnected lips; and nose and tongue can be referred back to the appearance of fishes and amphibians. Accordingly, in his *Lehrbuch der Naturphilosophie* (*Textbook of Nature Philosophy*) (1809) and his *Lehrbuch der Naturgeschichte* (*Textbook of Natural History*) (1815) Oken classified animals using the criterion of human organs (*Aderthiere*, 'vein-animals'; *Darmthiere*, 'intestinal animals'; *Lungenthiere*, 'lung-animals', etc.).

The last two books contained also Oken's exaggerated version of Wernerian Neptunism. Like all speculative *Naturphilosophen* he was not interested in the stratigraphical succession as such, but in the metaphysical logic behind it. This was his teleological progressivism which made chemical lithogenesis, and in particular crystallization, the logical starting point of a process from simple to complex and advanced. 'Crystallization is part of the essence of the earth, just as the spherical form belongs to the essence of water, down to its smallest particles. The earth's life consists in the crystalline form. Earth and crystal are one and the same reality.'[20] The earth is therefore not a ball, not even a spheroid, but a polyhedron. Granite formed by chemical precipitation, and must be regarded as the *Urgebirge* (oldest formation).

In Oken's teleological model of the past man is the final product of nature. However, like Blumenbach and Cuvier, Oken rejected the notion of a single and linear chain of being. Instead he postulated a series of levels of differentiation. Although he held that at each level in her progressive unfolding nature had to rest and gather energy for the next step, he did not use the notions of degeneration and transformation.

# NOTES

This Chapter is a revised version of one which originally appeared in *History of Science* in 1983. Science History Publications have kindly agreed to its inclusion here.

1  See Florian Heller, 'Die Forschungen in der Zoolithenhöhle bei Burggaillenreuth von ESPER bis zur Gegenwart', *Erlanger Forschungen*, 5 (1972), 7–56. See also Armin Geus's 'Einführung' ('Introduction') to the facsimile reprint of Esper's *Ausführliche Nachricht von neuentdeckten Zoolithen unbekannter vierfüssiger Thiere* (Wiesbaden, 1978), and J. F. Esper, 'Kurze Beschreibung der in den Osteolithen bey Gailenreuth ohnweit Muggendorf im Baireuthischen neuerlich entdeckten Merkwürdigkeiten', *Frankisches Archiv*, ed. Heinrich Christoph Büttner *et al.*, 1 (1790), 77–105 (these were posthumously published notes made by Esper in the year 1778). See also S. T. Sömmerring, 'Über die in Leibnitz Protogaea abgebildeten fossilen Thierknochen', *Magazin für die Naturgeschichte des Menschen*, ed. C. Gross, 3 (1790), 60–74.

2  *Novalis Werke*, ed. Gerhard Schulz (Munich, 1981), p. 186 (my translation).

3  *Ibid.*, p. 192.

4  *Ibid.*, pp. 201–2. See also Gerhard Schulz, 'Novalis und der Bergbau', *Freiberger Forschungshefte*, series D, 11 (1955), 242–55.

5  J. C. Rosenmüller, *Beiträge zur Geschichte und nähern Kenntniss fossiler Knochen* (Leipzig, 1795), p. 60.

6  *Ibid.*, p. 65.

7  J. H. Merck, *Lettre à monsieur de Cruse: sur les os fossiles d'éléphans et de rhinocéros qui se trouvent dans le pays de Hesse-Darmstadt* (Darmstadt, 1782), pp. 20–1. See also *Johann Heinrich Merck, Werke und Briefe*, ed. Arthur Henkel and Herbert Kraft (Frankfurt-on-Main, 1968).

8  Goethe to Merck, 27 October 1782: Weimar edition of *Goethes Werke*, IV 6, p. 75 (my translation). See also W. von Engelhardt, 'Goethes Beschäftigung mit Gesteinen und Erdgeschichte im ersten Weimarer Jahrzehnt', in *Genio huius loci: Dank an Leiva Peterson*, ed. D. Kuhn and B. Zeller (Vienna, 1982), pp. 169–204.

9  J. F. Blumenbach, 'Beyträge zur Naturgeschichte der Vorwelt', *Magazin für das Neueste aus der Physik und Naturgeschichte* (ed. Georg Christoph Lichtenberg and continued by Johann Heinrich Voigt), 6 (1790), 1–27. The quotation is on pp. 1–2 (my translation).

10 J. F. Blumenbach, 'Specimen archaeologiae telluris, terrarumque inprimis Hannoveranarum', *Göttingische Anzeigen von gelehrten Sachen*, vol. for the year 1801, 1977–84.

11 J. F. Blumenbach, *Beyträge zur Naturgeschichte*, vol. I (Göttingen, 1806), pp. 115–16.

12 See C. G. Gottschalk, 'Verzeichniss Derer, welche seit Eröffnung der Bergakademie und bis Schluss des ersten Säculum's auf ihr studirt haben', *Festschrift zum hundertjährigen Jubiläum der Königlichen Sächsischen Bergakademie zu Freiberg* (Dresden, 1866), 221–95; Otfried Wagenbreth, 'Abraham Gottlob Werner und der Höhepunkt des Neptunistenstreites um 1790', *Freiberger Forschungshefte*, series D, 11 (1955), 183–241, and 'Werner–Schüler als Geologen und Bergleute und ihre Bedeutung für die Geologie und den Bergbau des 19. Jahrhunderts', *Freiberger Forschungshefte*, series C, 223

(1967), 163–78; W. von Engelhardt, 'Die Entwicklung der geologischen Ideen seit der Goethe-Zeit', *Abhandlungen der Braunschweigischen wissenschaftlichen Gesellschaft*, (1979), 1–23, and 'Neptunismus und Plutonismus', *Fortschritt der Mineralogie*, 60 (1982), 21–43.
13  *Leopold von Buch's gesammelte Schriften*, ed. J. Ewald *et al.* (Berlin, 1870), II, 4–12.
14  See Hans Aarsleff, 'The Tradition of Condillac: The Problem of the Origin of Language in the Eighteenth Century and the Debate in the Berlin Academy before Herder', in *Studies in the History of Linguistics*, ed. Dell Hymes (Bloomington, Ind., 1974), 93–156.
15  G. H. von Schubert, *Ansichten von der Nachtseite der Naturwissenschaften* (Dresden, 1808), p. 198.
16  *Ibid.*, p. 201.
17  *Ibid.*, pp. 7–8. See also Carl Beck, 'E. T. A. Hoffmanns Erzählung "Die Bergwerke von Falun"', *Freiberger Forschungshefte*, series D, 11 (1955), 264–72, for a discussion of Schubert's influence on Hoffmann.
18  L. Oken, *Abriss der Naturphilosophie* (Göttingen, 1805), p. iii (my translation).
19  *Ibid.*, p. v.
20  L. Oken, *Lehrbuch der Naturphilosophie*, vol. I (Jena, 1809), p. 149.

## FURTHER READING

Laudan, Rachel, *From Mineralogy to Geology: The Foundations of a Science, 1650–1830* (Chicago, 1987)
Levere, Trevor H., *Poetry Realized in Nature: Samuel Taylor Coleridge and Early Nineteenth-Century Science* (Cambridge, 1981)
Oldroyd, David, 'Historicism and the Rise of Historical Geology', *History of Science*, 17 (1979), 191–213, 227–57
Ospovat, Alexander M., 'Romanticism and German Geology: Five Students of Abraham Gottlob Werner', *Eighteenth-Century Life*, 7 (1982), 105–17
Porter, Roy S., *The Making of Geology: Earth Science in Britain, 1660–1815* (Cambridge, 1977)
Rupke, Nicholas A., *The Great Chain of History: William Buckland and the English School of Geology, 1814–1849* (Oxford, 1983)
—'The Apocalyptic Denominator in English culture in the Early Nineteenth Century', in M. R. Pollock (ed.), *Common Denominators in Art and Science* (Aberdeen, 1983), pp. 30–45

# PART IV

# LITERATURE AND THE SCIENCES

# 18

## Goethe's use of chemical theory in his Elective Affinities

### JEREMY ADLER

By 1800, the gulf between natural philosophy and the arts had already developed into a cultural problem which some of the best minds determined to resolve. Ever since the return from his Italian journey of 1786–7, Goethe was intent on linking literature with the sciences: 'Nowhere', he protested, when commenting on the poetic version of his *Metamorphosis of Plants* (1798), 'would anyone admit that science [*Wissenschaft*] and poetry could be united. People forgot that science had developed out of poetry, and failed to consider that after a revolution of the ages [*Umschwung von Zeiten*], the two could cheerfully meet again on a higher level and to mutual advantage' (*LA*, I, 9, p. 67).[1] The work of Goethe's which most powerfully combines science – that is, natural philosophy – with literature is his novel of 1809, *Die Wahlverwandtschaften (Elective Affinities)*. The very title translates a chemical term, *attractio electiva*, and the process which this term designates provides a basic pattern for the plot: a so-called 'double affinity' (AB + CD→AC + BD) takes place between a married couple, Eduard and Charlotte, and their two friends, the Captain and Ottilie. The novel pursues this chemical analogy into the characters' innermost thoughts. In the central episode, the 'spiritual adultery' (*Ehebruch im Ehebett*), a 'double affinity' alienates spirit from flesh. Eduard and Charlotte come together, supposedly in the act of love, yet each imagines a different partner: Eduard sees only Ottilie before him, and the Captain's image floats before Charlotte's mind. To some considerable extent, the action externalizes this psychological 'affinity'. The erotic entanglement increasingly dominates and eventually ruins the lives of all four characters, resulting in the tragic death of Ottilie, and that of her beloved Eduard. Thus radically, and for some of his contemporaries scandalously, did Goethe confront chemistry with life, seeming thereby to inquire whether human life is governed by chemical laws. By extending the reference of an established chemical theory to encompass social interaction, the novel provides the basis for a universal theory of affinity.

No other novel of world literature both makes such detailed and extensive use of a scientific doctrine, and actually contributes to it, too, by developing unresolved theoretical issues. The boldness of this enterprise lay not just in Goethe's uniting of chemistry with literature, but in his choice of a theory which many contemporaries regarded as central to the study of natural philosophy as a whole. No less a figure than Newton had emphasized the theoretical importance of chemical 'affinity' or 'attraction', i.e. the propensity of chemical substances to combine.[2] In the Introduction to his *Philosophiae naturalis principia mathematica* (*Mathematical Principles of Natural Philosophy*), published in 1687, Newton announced his success in discovering the mathematical rules governing the force of gravity 'with which bodies tend to the sun and the several planets', and according to which he could 'deduce the motions of the planets, the comets, the moon, and the sea'. However, the motion of smaller bodies, and specifically the laws of physical cohesion and chemical attraction, eluded him:

I wish we could derive the rest of the phenomena of Nature by the same kind of reasoning from mechanical principles, for I am induced by many reasons to suspect that they may all depend upon certain forces by which the particles of bodies, by some causes hitherto unknown, are either mutually impelled towards one another, and cohere in regular figures, or are repelled and recede from one another. These forces being unknown, philosophers have hitherto attempted the search of nature in vain; but I hope the principles here laid down will afford some light either to this or some truer method of philosophy. (pp. xvii–xviii)

Newton's implicit suggestion that natural philosophy should study chemical 'affinity' became explicit in the celebrated 31st Query of his *Opticks* (1718): 'We must learn from the Phaenomena of Nature what bodies attract one another, and what are the Laws and Properties of the Attraction' (p. 376). From Newton to Laplace, studying chemical affinities provided the hope for a theory of attractions which could unite celestial physics with terrestrial chemistry under a single, physical law.[3] Thus, Laplace asserted in his *Exposition du système du monde* of 1796 that 'all chemical combinations are the result of attractive forces' (II, p. 196), and that their study was a central task of physics. If only the 'impossible' could be achieved, he argued, and the *form* of the molecules constituting matter could be demonstrated, *all* terrestrial and celestial phenomena would be explained by a single law: 'from the variety of these forms, one would be able to explain all the varieties of attractive forces, and could bring together every phenomenon of physics and astronomy under a single, general law' (II, p. 197). For Laplace, chemical affinity was potentially nothing less than the key to the system of the universe. From this perspective, one can begin to assess the full significance in Goethe's choice of theme: it alludes to a hitherto incomplete *chemical* theory that holds out the promise of a universal *physical* law, and transfers it to the *human* sphere.

Thus mankind is seen in terms of a hitherto unexplained, because insufficiently researched, universal order, stretching from chemical matter to the stars. Significantly, Goethe placed his novel into an open-ended context, in which a literary work might yet make an original contribution to philosophic debate.

In so doing, Goethe built on and combined conflicting elements of a tradition which can in fact be traced from the pre-Socratics to the Stoa and to Plotinus, and thence to the Renaissance theory of a universal *sympathia* linking all parts of the cosmos. In the works of Ficino and della Porta,[4] *sympathia* captured the Renaissance imagination as an 'occult quality' that caused a host of otherwise inexplicable phenomena, from human love to the movement of the planets. Descartes's critique of 'occult qualities' in his *Principia philosophiae* of 1644 questioned the very basis for such theories, and *sympathia*, now understood as 'attraction', was transmuted into – or replaced by – a physical concept in Newton's physics. Though Goethe recovers many of these older layers, the framework he chose was decidedly contemporary and – perhaps surprisingly – Newtonian. In a field where he did not pretend to be a specialist, he outwardly accepted the prevailing doctrine, yet, by subtle changes, accommodated it to his own views.

Several key episodes in the eighteenth-century study of 'affinity' played a part in shaping Goethe's novel.[5] The theory took its characteristic form in 1718 with the work of E. F. Geoffroy. It was Geoffroy who formulated as a 'law' of affinity an observation that had previously been made by authorities such as Geber, Paracelsus, Bacon and Glauber:

Whenever two substances are united that have a disposition to combine and a third is added that has a greater affinity with one of them, these two will unite, and drive out the other. (p. 203)

This law provided the mainstay of affinity studies until the end of the century. It could, theoretically, be used to work out the relative affinity of all known substances, to be drawn up in so-called 'affinity-tables'. If, for example, a body, E, is driven out of the combination AE by D, D is driven out by C and C by B, then D has a stronger 'affinity' for A than E, C has a stronger one than D, and B has a stronger one than C. The order of 'affinities' is then: ABCDE. It was on this principle that Geoffroy constructed his tables (see fig. 22). However, the law was questionable since certain substances violated its predictions. He had no real successor until, in 1749, P. J. Macquer republished the tables in his own *Elémens de chymie-théorique* of 1749, which is probably the first chemical textbook to use 'affinity' as a consistent conceptual tool to unify the whole of chemistry. As Macquer claims, 'almost all of the phenomena that appear in chemistry are based on the affinities which exist between different substances' (p. 256). The title-page of the second edition (see fig. 23)

22   E. F. Geoffroy's table of affinities, 1718, published in the *Mémoires de l'Académie royale des sciences* (Paris, 1719).

signals the theory's new importance: on the left and right of a small engraving stand the *practical* tools of chemistry, whilst at the centre three putti are studying the *theoretical* centre, namely Geoffroy's tables. Macquer's major contribution lay in his bringing a geometrical order to affinity by his use of a Linnaean system of classification, first presented as six theses in 1749, refined as seven theses in 1753 and finally formulated as a typology of seven types of affinity in his *Dictionnaire de chymie* of 1766 and 1778. This typology provides the basis for the chemical theory in Goethe's novel.

Remarkably, the central reaction upon which the theory from Geoffroy to Macquer depended $(AC + B \rightarrow AB + C)$ had no name until the mid-eighteenth century. Various terms were then proposed, but the most widespread was 'elective affinity' or 'elective attraction', perhaps first used by William Cullen in 1748, and certainly used in 1755–6 by Cullen's pupil Joseph Black.[6] In its Latin form, *attractio electiva*, the term became widely known through the work of Torbern Bergman, who was generally thought to have originated it in his famous study, the *Disquisitio de attractionibus electivis* (*A Dissertation on Elective Attractions*) of 1775.

Bergman's work became the most famous in the field, and gave affinity studies their definitive shape for the next twenty-five years. The *Disquisitio* presented a thorough analysis of all the known facts of chemistry, and included the fullest tables ever prepared, with fifty columns for reactions 'in the dry way' (i.e. using fire) and thirty-six 'in the wet way' (i.e. using water as a solvent). The extended second edition contained fifty-nine and forty-three

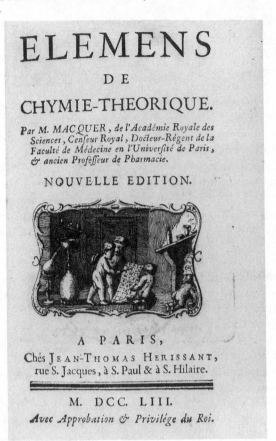

ELEMENS

DE

CHYMIE-THEORIQUE.

Par M. MACQUER, de l'Académie Royale des
Sciences, Censeur Royal, Docteur-Régent de la
Faculté de Médecine en l'Université de Paris,
& ancien Professeur de Pharmacie.

NOUVELLE EDITION.

A PARIS,

Chés JEAN-THOMAS HERISSANT,
rue S. Jacques, à S. Paul & à S. Hilaire.

M. DCC. LIII.

Avec Approbation & Privilége du Roi.

23   Title-page of P.J. Macquer's *Elémens* (2nd edn, 1753).

columns respectively. Bergman's achievements were many: his introduction
of algebraic letter-symbolism (e.g. AC + B→AB + C) gave the laws of affinity
a new, mathematical precision; his diagrams made possible a stereochemical
representation of reactions; his tables enshrined practically all known chemical facts in tabular form; and, finally, his treatment of 'apparent anomalies' to
the laws of affinity went a long way towards bridging the gap between
observation and theory. His classification built on Macquer and, in its turn,
probably influenced the final form of Macquer's typology. What might be
called the Macquer–Bergman theory dominated chemistry till around 1800,
though there were significant competitors and problems, the most serious
among which was the fact that numerous reactions appeared to be reversible,
or depended on seemingly arbitrary variables (such as the effects of mass) that

were not yet fully recognized as chemically significant. This inability to verify the theory in strictly empirical terms led Lavoisier to exclude it from his *Traité* of 1789 that founded the New Chemistry,[7] and stimulated Berthollet, Lavoisier's collaborator, to disprove the central 'law' of affinity. In place of 'simple elective attraction' $(AC + B \rightarrow AB + C)$, Berthollet asserted that a substance (A) was divided between two others $(AC + B \rightarrow A1B + A2C)$. Moreover, reactions were determined by a host of factors, such as quantity, solubility or cohesion. Thus Berthollet replaced the single attractive force by a multiplicity of phenomena.

Goethe himself referred to both Bergman and Berthollet in connection with 'affinity'. In a conversation with Riemer of 24 July 1809, he names Bergman as a source for the idea of 'elective affinity', using the early German translation of that term, namely '*Wahlverwandtschaft*', that became an inspiration for his novel. Then, almost twenty years later, in a letter of 26 September 1826, he mentions the fact that 'for decades' he has been struggling 'with Berthollet in the matter of the affinities'. As to detailed knowledge, Goethe had access to a variety of possible sources.[8] Berthollet's refutation of the standard theory appeared in 1801 and was quickly publicized in Germany. Goethe could have known it from at least four sources.[9]

Goethe's first detailed presentation of 'affinity' occurs in his third *Lecture on Anatomy* of 1796 (*LA*, I, 9, pp. 202f.). Brief though this treatment is, it reveals how thoroughly Goethe had absorbed the current theory, and how radically he was beginning to transform it. Interestingly, Goethe steers a path between the existing views by granting that 'affinities' operate 'according to fixed laws', whilst denying that they have a definable 'limit'. Thus he accepted the general theory, yet rejected the possibility of establishing a fixed 'order' that could be summarized in a table. Then, Goethe implicitly argues against chemical method by focusing not on experiments but on nature: a new combination can tear the 'parts of a body' 'out of [their] earlier combination' and create 'a new substance', the 'constituents' of which 'appear' 'to wait for a new, or, under certain circumstances, a re-combination'. The concept of 'affinity' as used by Goethe provides a tool for the observer. Significantly, he makes no reference to the tables which had hitherto been such a point of interest. As he later put it in the Foreword to his colour theory with respect to diagrams in optics, such tabular 'hieroglyphics' actually hinder the interpretation of nature since observers tend to see and explain their own diagrams in place of actual phenomena (*LA*, I, 4, pp. 9f.). Goethe accordingly rejects the experiments that can be summarized in an abstract, tabular form, and envisages a sequence of chemical processes as they actually occur, or might occur, in nature. Experience, not experiment, provides the touchstone of his theory.

The notion that 'affinity' might be broadened further still to include not just natural but human relations recalls the older idea of 'sympathy' and reflects the Romantic tendency to innovate with a received doctrine:[10] Goethe himself considered using affinity theory as a model for his colour theory; Schelling incorporated it into his *Naturphilosophie*; whilst Ritter perceived a parallel between chemical and electrical attraction which supported his belief in a universal 'sidereal' force. This desire to innovate imaginatively with natural philosophy provided the general context for Goethe's procedure in his novel.

In *Elective Affinities*, Goethe presents 'affinity' through the ironic distancing device of a dialogue between his characters. The Captain and Eduard initiate Charlotte into the doctrine in the realistically plausible context of an educated conversation. Various elements of the theory are distributed among the three characters in a masterly transformation of what is – quite literally – textbook chemistry, as each character speaks according to his own knowledge, insight and personality. However, the discussion begins with a significant caveat: the Captain will explain 'affinity' as he had 'learnt, as [he] had read it' 'about ten years ago' (*Wv*, p. 34). If we take him literally, this would date the Captain's knowledge to the time just before Berthollet's main findings appeared, i.e. to the years around 1799. 'Whether it still fits the newer doctrines', the Captain adds, 'I am unable to say.' This proviso signals a divergence between the Captain's knowledge and the narrator's awareness of a newer, changed theory.

Accordingly, the conversation closely follows the theories of Macquer and Bergman, as they were propagated in numerous eighteenth-century chemical handbooks, dictionaries and compendia. Indeed, the dialogue's structure derives from Macquer's seven types of 'affinity'. The order of examples that the characters give adapts Macquer's typology:

1   *Affinity of aggregation* or *cohesion*. Macquer's first type also provides the starting point in the novel. The Captain explains it in abstract terms as a thing's 'relation to itself' ('Bezug auf sich selbst'). Eduard gives examples: 'Just imagine water, oil or quicksilver: you will find that they display a unity, a cohesion of parts' (*Wv*, p. 35). Similarly, Macquer writes of the 'tendency to combine which two drops of water, or oil, or quicksilver . . . display towards one another' (*Elémens*, V, p. 435). The example was in fact a commonplace in the eighteenth century. However, after Macquer's *Elémens* textbooks increasingly distinguished cohesion from chemical 'affinity' proper.

2   *Affinity of composition*, which involves chemical change, does not appear at this point in the novel; instead, a new type features, barely distinguished as such in the chemical literature, namely the cohesion of different but similar substances, like wine and water:

'Let me hurry on', said Charlotte, 'to see whether I understand what you mean. Just as everything must have a relation [*Bezug*] to itself, so it must also have a relationship [*Verhältnis*] to others.'

'And that will vary according to the difference in their natures', Eduard quickly continued. 'Some will meet like old friends and acquaintances, who quickly come together and unite, without changing each other in any way, as wine mixes with water . . .' (*Wv*, p. 36)

3 *Compound affinity of composition* is also omitted, being replaced by an example of two substances which do *not* mix, like oil and water:

whilst on the other hand, some substances will coexist like strangers, and cannot even be combined by mechanical mixing and rubbing; like oil and water which, after being mixed together, immediately separate again. (*Wv*, p. 36)

By introducing these new examples, the novel creates a gradual transition to:

4 *Mediating affinity*, illustrated in Macquer by sulphur and water, that can be united through an alkaline salt (the 'mediator'). Most later writers give Macquer's example, but also include a second one, which is that adopted in the novel. Thus, J. S. T. Gehler writes in his *Physikalisches Wörterbuch*:

Oils and sulphur are insoluble in water, but by combining with the alkaline salts they change into soaps and liver of sulphur, and as such dissolve in water, whereby the alkaline salt acts as the intermediary. (IV, pp. 475f.)

The Captain says: 'Thus we combine . . . oil with water by means of an alkaline salt.' (*Wv*, p. 37)

5 *Affinity of decomposition*, i.e. *simple elective affinity*, which may be represented thus: $AC + B \rightarrow AB + C$. Macquer illustrates this with an example involving acids, alkalis and metals. The novel instead uses a standard test for limestone, namely the acid + limestone → gypsum + water + aerial acid reaction:

'Let me continue', said the Captain. '. . . For example, what we call limestone is a more or less pure earth, intimately joined up with a delicate acid, which has become known to us in the form of air. If one places a piece of this stone into dilute sulphuric acid, the latter takes hold of the lime and together they will appear as gypsum; but on the other hand, that delicate, aerial acid will disappear. A division has taken place, and a new combination has emerged. One even considers it justified to apply the word elective affinity here, since it really looks as if one relation had been chosen in preference to another, as if one had been selected instead of another.' (*Wv*, p. 38)

J. F. A. Göttling includes the same example in both his *Versuch einer physischen Chemie* of 1792 and *Der physische-chemische Hausfreund* of 1804.

6 *Reciprocal affinity*, involving a substance which enters a relation first with one, then with another substance, does not occur in the novel; instead, the

characters treat the phenomenon of *replacement*, not isolated as such in the chemical literature, whereby the separated substance (here, aerial acid) recombines with another (here, water). This effectively provides a bridge to:

7   *Double affinity*, which the novel illustrates with the letter-symbolism that became well-nigh universal after Bergman:

> Imagine an A, which is intimately united with a B, and cannot be separated from it by various means or by force; imagine a C, united to a D in the same way; now bring the two pairs into contact: A will throw itself to D, and C to B, so that one cannot say which left the other one first, and which was the first to re-unite with another. (W*v*, pp, 40f.)

Of the innumerable analogues, the briefest may be found in Göttling's *Handbuch der theoretischen und praktischen Chemie* (1798–1800):

> If a compound A + B comes together with another, C + D, they will exchange parts, and create two new bodies, A + C and B + D. (p. 71)

As can be seen, *Elective Affinities* largely adopts Macquer's seven-part system, and even includes four of Macquer's types of affinity in the same numerical position (1, 4, 5 and 7) whilst simultaneously making modifications which effect a transition between types: e.g. 6 (*replacement*) links 5 (*simple elective affinity*) and 7 (*double affinity*). Consequently, Macquer's typology becomes a morphology: the former is an analytic abstraction; the latter approximates to the natural development from simpler to more complex forms, as this might occur in nature. Thus Goethe transformed an analytic theory of 'affinity' into an organic one.

The very title of the novel implies a fundamental analogy between human and chemical behaviour, and the characters' names seem to support the view that some chemical 'affinity' exists between them: both Eduard and the Captain turn out to be called OTTO, and the two women, CharloTTe and oTTilie, have related names. However, the actual workings of these affinities prove difficult to disentangle. Whereas the chemical discussion offers a lucid typology (in the manner of Macquer), the actual novel reveals the complexities of natural life (as suggested by Goethe's own view in his *Lectures on Anatomy*).

The interpretation of the novel's chemical import largely hinges on the analogies with 'simple elective affinity' and 'double affinity'. Ironically, it is Eduard – a partial witness given to hasty judgements and misprisions – who explicitly introduces these comparisons. In so doing, he somewhat overlooks the possibility of 'double affinity', and asserts a similarity between the limestone–acid reaction and the main characters. He does so to symbolize his immediate situation: the Captain has come to stay and thereby (temporarily) has drawn Eduard away from Charlotte. Eduard sees this reflected in the reaction, as he tells Charlotte:

> In the end, in your eyes, I am the limestone, that has been taken hold of by the Captain as an acid, removed from your charming company, and turned into refractory gypsum. (*Wv*, p. 39)

The situation may be remedied, Eduard argues, by bringing Ottilie, Charlotte's young niece, to the estate. He now re-interprets the situation in terms of the letter-symbolism that the captain used for 'double affinity':

> You represent the A, Charlotte, and I am your B: for actually I only depend on you and follow you as the B does the A. The C is quite obviously the Captain, who for the moment somewhat draws me away. Now it is only right, if you are not to depart into uncertainty, that a D should be found for you, and that, without any question, must be the charming little lady Ottilie . . . (*Wv*, p. 41)

Eduard's playful *ad hoc* analogizing entirely neglects the possibility that real human emotions might be involved. Crucially, he envisages no more than a 'replacement', remaining blithely unaware that the letter–symbolism implies 'double affinity' and an actual exchange of partners. Hence he fails to recognize that his ultimate separation from Charlotte might be wrought not by the Captain, but by Ottilie. Later, he compounds the muddle when addressing the Captain:

> 'Beware of the D . . . for what should B do, if C were taken away from him?'
> 'Well, I would have thought', said Charlotte, 'that that would be self-evident.'
> 'Of course' cried Eduard, 'he would return to his A, to his A and O!', he exclaimed, jumping up and pressing Charlotte firmly to his bosom. (*Wv*, p. 46)

Thus Eduard bungles his analogy, repeatedly referring to a form of 'simple elective affinity' without entertaining the possibility of a 'double affinity'. However, after Ottilie's arrival, the couple, Eduard and Charlotte, really does begin to break up, and the two recombine with Ottilie and the Captain respectively. Indeed, Charlotte and the Captain embrace for the first (and only) time at the very moment when Eduard and Ottilie first declare their love. This seems to correspond to the simultaneous exchange which the Captain emphasizes – introducing a feature not found in the textbooks – when discussing 'double affinity'. Later on, as chemical substances exhibit an attractive force so, too, Eduard and Ottilie seem to exert a similar power over each other, as described by the narrator towards the end of the novel:

> As before, they exerted an indescribable, almost magical power of attraction over each other. They lived under the same roof; but even when not thinking of each other, preoccupied by other things, pulled hither and thither by society, they drew near to each other. If they were in the same room, it did not take long and they were standing, sitting down beside each other. . . . Yes, if you had held one of them fast at one end of the house, the other would, by and by, and wholly without intent, have moved to him. (*Wv*, p. 274)

For the first time in the novel, the narrator introduces a standard eighteenth-century term for 'attraction' or 'attractive force', *Anziehungskraft*, albeit

24  'Häuslicher Verein' ('Family Group'): an illustration by H. A. Dähling for
an 1811 German edition of *Elective Affinities*. From *Urania. Taschenbuch
für Damen auf das Jahr 1812.*

qualified by the phrase 'almost magical', and does so in connection with a
passion which constantly manifests itself as a form of 'elective affinity': even
when 'attracted' or 'torn' ('hin und her gezogen') by society, they seek each
other out. The force appears to be absolute, and inexplicable; and by introducing it the text appears to imply that 'attractions' comparable to those in the
natural world may determine human relationships.

To interpret the novel in this way seems to suggest that one need only
correct Eduard's reading of the chemical analogy to arrive at a correct
formula. If one were to adjust the attributions of the chemical substances and
the letter-symbolism, the following, more accurate, analogy might result:

| AB | + | CD | → | AC | + | BD |
|---|---|---|---|---|---|---|
| Limestone (calcareous earth + aerial acid) | + | Dilute sulphuric acid (acid + water) | → | Gypsum | + | Aerated water |
| Eduard/Charlotte | + | Ottilie/Captain | → | Eduard/ Ottilie | + | Charlotte/ Captain |

The analogy is in fact multivalent, and invites other possible interpretations, too.[11] However, such a reading (and any rival interpretation in the same terms), which attempts to locate an unchanging, quasi-algebraic formula for the whole action, only raises further problems. Most obvious among them is the fact that, in terms of 'double affinity', the Captain and Ottilie never form a couple, and that Charlotte and the Captain renounce their love almost as soon as they become aware of it. Does this mean, as has been argued, that the novel actually disavows the concept of 'elective affinity'? This whole mode of argument in fact does no more than take issue with Eduard's view of chemistry and its application to human life, since it is only Eduard who proposes the formulaic approach. To debate whether or not the novel can be interpreted in terms of the formula ignores the position of the narrator, and ulimately, that of Goethe himself. Eduard's formula implies a strictly empiricist method, but in fact the novel, whilst making use of an empirical framework, resurrects the Neoplatonic implications of 'affinity' associated with Plotinus' concept of 'sympathy'. Specifically, it is to Plotinus' concept of an 'idea' that the novel returns.

In one of his many, often difficult, remarks about the novel, Goethe said that it was his only longer work written according to a single 'idea', and compared it in this respect to a poem like *The Metamorphosis of Plants* (*GA*, XXIV, p. 636). Goethe had long previously adopted the Neoplatonic concept of an 'idea' as 'the law of all phenomena' (*HA*, XII, p. 366) and used the term to describe his own *Urpflanze* (*WA*, IV, 27, p. 144). The 'idea' had a characteristically Goethean dimension as a scientific concept. It was to be understood both abstractly, and concretely as a generative model for all plants (*WA*, IV, 8, pp. 232f.). In *The Metamorphosis of Plants*, Goethe indicates how the ultimately inexpressible (because abstract) 'idea' manifests itself differently in every plant:

> Alle Gestalten sind ähnlich, und keine gleichet der andern;
> Und so deutet das Chor auf ein geheimes Gesetz,
> Auf ein heiliges Rätsel. O könnt' ich dir, liebliche Freundin,
> Überliefern sogleich glücklich das lösende Wort! (*LA*, I, 9, p. 67)

All figures are similar, and none is like the other; / And so the chorus points to a secret law, / To a sacred mystery. Oh, if only I could, dearest friend, / Immediately and happily give you the answer in a word.

Likewise, introducing his concept of 'morphology' in 1807, Goethe asserts the difference between manifestations of the 'idea':

The flexible life of Nature resides in the fact that phenomena, which appear the same according to the Idea, can in experience appear either as the same, or as similar, or even as completely unlike or dissimilar. (*LA*, I, 9, pp. 236f.)

Just as the *Urpflanze* provides a generative model of *all* plants, so 'elective affinity' may be interpreted as a model of all possible chemical and human relationships: it should be understood not as a formula, but as an 'idea'. The whole of Macquer's typology in Goethe's redaction may therefore be interpreted as a manifestation of this 'idea': 'cohesion' is its simplest, 'double elective affinity' its most complex form.

This reading is borne out by the action; it cannot be fitted to a single formula, because the narrator presents it as the continuously changing manifestation of the central 'idea': first, Eduard and Charlotte are alone together on their estate ('simple affinity'); then the Captain draws Eduard away from Charlotte ('simple elective affinity'); Ottilie joins the company, at first spending her days with Charlotte ('replacement'); and then the new couples, Eduard and Ottilie and Charlotte and the Captain, begin to form ('double affinity'). This relation comes to a head in the 'adultery in the marriage bed', when Eduard and Charlotte make love, each imagining a different partner. The use of linguistic parallelism and words like 'attract' (*anziehen*) and 'repel' (*abstossen*) evoke the analogy: 'Eduard only held Ottilie in his arms; the Captain wafted nearer or further before Charlotte's soul . . .' (*Wv*, p. 93). This scene offers one of the closest points of analogy with 'double affinity'; but soon Charlotte and the Captain renounce their love, and the actual exchange of partners – so desired by Eduard – never takes place. To an extraordinary degree, the action manifests the 'idea' of 'double affinity', yet the empirical reality, by definition, differs considerably from the ideal pattern.

The manner in which the central exchange of partners in *Die Wahlverwandtschaften* differs from Bergman's scheme of 'double affinity' suggests an awareness of Berthollet's theory, as if Goethe had intentionally drawn on two contradictory views. The procedure would be typical of Goethe's 'theory of modes of understanding' (*Theorie der Vorstellungsarten*): 'Since it is my aim, to place in a clearer light a number of relations and effects in Nature, I cannot be concerned solely with *one* hypothesis' (*WA*, II, 7, p. 8). To grasp the phenomena, which lie beyond language, Goethe aimed to exploit *all* available hypotheses. Thus two hypotheses might be played off against each other:

If we can get used to considering both hypotheses as problematic, and weigh them up against one another, combine one with the other, or dispel one by means of the other, the mind will perhaps become accustomed to grasping both at once, and then it will be possible to go further than I am at present able to conceive.

(*WA*, II, 6, pp. 316f.)

In accord with this kind of dual vision, the action of Goethe's novel bears witness *both* to Bergman's view of 'double affinity', *and* to Berthollet's thesis that no such reaction ever occurs. Thus, while it is true that Eduard falls in love with Ottilie, he never divorces Charlotte, and at the end of the novel, all four friends come together on Eduard's estate. This coincides with a formula that Goethe could have known from J. B. Trommsdorff's *Systematisches Handbuch der gesammten Chemie* (1805–15) (I, p. 51f.) and that expresses Berthollet's theory of double affinity, namely AB + CD→ABCD. To the extent that the four friends divide and reunite, Bergman's view holds (AB + CD→AC + BD); but to the extent that the old relationships continue until the end of the action and the four characters remain friends, Berthollet's view applies (AB + CD→ABCD). No one formula can account for what occurs. Affinity, the novel appears to assert, proves to be a more complex phenomenon than any single theory would allow. It emerges that the imaginative handling of affinities in human life not only extends natural philosophy by exposing it to moral reflection, but creates a serious intellectual forum which can actually refine chemical doctrine.

The very factors which interfere with 'double affinity' are similar to the kind that Berthollet envisaged. To begin with, it is Charlotte's exercising of free will and her moral sense – her renunciation of the Captain and her correlative demand that Eduard exercise the same degree of self-control – which prevents a divorce and an exchange of partners; and later on, when Charlotte finally gives in, it is Ottilie's moral sense and sense of guilt which prevent it. At this point, one may find a close agreement with Berthollet. For there is good reason to interpret Ottilie's *character*, which prevents the divorce, in terms of an analogy with cohesion; and it was, according to Berthollet, precisely the cohesion of a substance which was one of the several factors that might determine the outcome of a reaction.

In his book on colour theory, *Zur Farbenlehre*, when discussing Newton's personality, Goethe recommends that one should adopt the terminology of 'cohesion' for grasping human 'character' (*LA*, I, 6, pp. 298f.). The novel makes no such explicit transference, yet it hints at a similar parallel. A little while after talking of oil, water and quicksilver, the Captain introduces a fourth example of cohesion, without parallel in the standard chemical literature: that of the molten lead which assumes the form of a drop as it solidifies (*Wv*, p. 36). The fact that this example occurs a little after the first three may reflect the fact that Ottilie joins the company last of all; the fact that the lead undergoes a change of state from liquid to solid may mirror Ottilie's oft-noted development during the course of the novel, from the point of least 'cohesion', when overpowered by her love, to her rigid invocation of morality; and the word which is used for the solidification, *erstarren*, is later used more than once for Ottilie (*Wv*, pp. 189ff.). All of this strongly suggests an analogy between her character and an extremely 'cohesive' body. In other words, just as Berthollet argued that, for example, cohesion could prevent a reaction from taking place, Goethe's novel implies how an analogous factor could operate in the moral sphere. Thus, as Goethe – echoing Spinoza – expressed it in his advertisement for the novel, *Elective Affinities* reveals how 'there is everywhere only *one* Nature' (*HA*, VI, p. 639). At a time when both the empiricist and the rationalist philosophies were seen to have alienated humanity from Nature through the pursuit of science, serious fiction could offer a different, no less differentiated, but holistic vision of the world.

## NOTES

1   Quotations from Goethe's works are from the following editions: *Werke* (*Works*), 4 parts, 133 vols. in 143 (Weimar, 1887–1919) (= *WA*, for *Weimarer–Ausgabe*); *Gedenkausgabe der Werke, Briefe und Gespräche* (*Memorial edition of the Works, Letters, and Conversations*), ed. Ernst Beutler, 24 vols. (Zurich, 1948–66) (= *GA*, for *Gedenkausgabe*); *Werke* (*Works*), ed. Erich Trunz *et al.*, 14 vols. (Hamburg, 1948–64; Munich, 1981) (= *HA*, for *Hamburger Ausgabe*); *Die Schriften zur Naturwissenschaft* (*The Writings on Natural Science*), ed. by the Akademie der Naturforscher (Leopoldina) zu Halle; (Weimar, 1947–   ) (= *LA*, for *Leopoldina Ausgabe*); *Die Wahlverwandtschaften* (*The Elective Affinities*), ed. Helmut Praschek (Berlin, 1963) (= *Wv*). All translations are my own. For an English version of Goethe's novel, see Goethe, *Elective Affinities*, trans. R. J. Hollingdale (Harmondsworth and Maryland, 1971).

2   Sir Isaac Newton, *Mathematical Principles of Natural Philosophy* (1687), trans. Andrew Motte and rev. Florian Cajori (Berkeley, 1934); *Opticks*, with a Foreword by Albert Einstein and an Introduction by I. B. Cohen (reprint of the 4th edn, 1730: New York, 1952).

3   Pierre-Simon Laplace, *Exposition du systême du monde* (*An Explanation of the System of the World*), 2 vols. (Paris, 1796).

4   Marsilio Ficino, *Commentaire sur le Banquet de Platon* (*Commentary on Plato's 'Symposium'*), ed. Raymond Marcel (Paris, 1956), p. 22; Giovanni Battista della Porta, *Magiae naturalis libri IIII* (*On Natural Magic*) (Naples, 1558), p. 12.

5   I shall refer to the following treatments of affinity: Etienne François Geoffroy, 'Tables des différents rapports observés en chimie entre différentes substances' (*Tables of the Different Relations Observed in Chemistry between Different Substances*), *Histoire de l'Académie Royale des Sciences, Mémoires, 1720* (1722), 32–3; Pierre Joseph Macquer, *Elémens de chymie-théorique* (*The Elements of Theoretical Chemistry*) (Paris, 1749; 2nd edn, Paris, 1753); Pierre Joseph Macquer, *Dictionnaire de chymie* (*A Dictionary of Chemistry*), 2 vols. (Paris, 1766; 2nd edn, Paris, 1778); Torbern Olof Bergman, 'Disquisitio de attractionibus electivis' ('A Dissertation on Elective Attractions'), *Nova acta regiae societatis scientiarum upsaliensis*, 2 (1775), 159–248; Claude Louis Berthollet, *Recherches sur les lois de l'affinité* (*Researches on the Laws of Affinity*) (Paris, 1801).

6   See William Cullen, end of fragment of student's fair-copy notes to Cullen's Introductory Lecture, Glasgow, 2 November 1748(?), Cullen Papers, Glasgow University Library, Box 1, no. 18, fol. 17; 'Chemical Lectures. Lecture First',

278        JEREMY ADLER

First Glasgow Course, 1748, Cullen Papers, Glasgow University Library, Box 1, no. 26, fol. 5$^r$; Joseph Black, *Experiments upon Magnesia Alba, Quick-Lime and other Alcaline Substances* (1755, printed 1756; Alembic Club Reprint no. 1, new edn, Edinburgh, 1963), pp. 12 and 24.

7   Antoine Lavoisier, *Traité élémentaire de chimie (Treatise on the Elements of Chemistry)*, 2 vols. (Paris, 1789), I, p. xiii.

8   The possible sources for Goethe's knowledge of affinity include: Pierre Joseph Macquer, *Chymisches Wörterbuch (A Dictionary of Chemistry)*, trans. J. G. Leonhardi, 6 vols. (Leipzig, 1781–3); Torbern Olof Bergman, 'Von der Attraction' ('A Dissertation on Elective Attraction'), in his *Kleine physische und chymische Werke (Shorter Works on Physics and Chemistry)*, trans. Heinrich Tabor, 6 vols. (Frankfurt-on-Main etc., 1782–90), III, pp. 360–602; Johann Samuel Traugott Gehler, *Physikalisches Wörterbuch (A Dictionary of Natural Philosophy)*, 6 vols. (1787–95); Johann Carl Fischer, *Physikalisches Wörterbuch (A Dictionary of Natural Philosophy)*, 5 vols. (Göttingen, 1798–1804); Johann Friedrich August Göttling, *Versuch einer physischen Chemie . . . (An Essay on Physical Chemistry . . .)* (Jena, 1792), *Handbuch der theoretischen und praktischen Chemie (A Handbook of the Theory and Practice of Chemistry)*, 3 vols. (Jena, 1798–1800) and *Der physisch-chemische Hausfreund (The Physico-Chemical Companion)*, 3 vols. (Jena, 1804–7).

9   E. G. Fischer, 'Berthollets neue Theorie der Verwandtschaft' ('Berthollet's New Theory of Affinity'), *Scherers Journal*, 7 (1801), 503–25; Ludwig Schnaubert, *Untersuchung der Verwandschaft der Metalloxyde zu den Säuren. Nach einer Prüfung der neuen Berthollet'schen Theorie (An Investigation of the Affinity of the Metallic Oxides to the Acids. After an Examination of the New Theory of Berthollet's)* (Erfurt, 1803); J. F. A. Göttling, *Der physisch–chemische Hausfreund*, I (1804), pp. 296ff.; Johann Bartholomäus Trommsdorff, *Systematisches Handbuch der gesammten Chemie (A Systematic Handbook of the Whole of Chemistry)*, 2nd edn, 6 vols. (Erfurt, 1805–15), I, pp. 30–55.

10  On Goethe's wish to extend affinity theory to colour, see Göttling's letter of 17 June 1797, *LA*, II, 3, p. 101; for Schelling's views, see F. W. J. Schelling, *Ideen zu einer Philosophie der Natur (Ideas on a Philosophy of Nature)*, in his *Sämmtliche Werke (Collected Works)*, 14 vols. in two parts (Stuttgart and Augsburg, 1856–60), I, 2, pp. 1–343, esp. pp. 75, 80, 94, 136, 163, 170, 259–66 and 333; and Ritter's views on the parallelism between electricity and affinity are in Johann Wilhelm Ritter, *Schreiben an A. Volta bey Übersendung des Beweises, dass ein beständiger Galvanismus den Lebensprocess im Thierreiche begleite (Letter to A. Volta on the Occasion of Sending him the Proof that a Continuous Galvanism Accompanies the Processes of Life in the Animal Kingdom)* (1798), in *Physisch–chemische Abhandlungen (Physico-Chemical Studies)* (Leipzig, 1806), I, pp. 59–90, esp. pp. 59, 78ff. and 85.

11  On the problems associated with the chemical analogy, see Robert T. Clark Jr, 'The Metamorphosis of Character in *Die Wahlverwandtschaften*', *The Germanic Review*, 29 (1954), 243–53; John Milfull, 'The "Idea" of Goethe's *Die Wahlverwandtschaften*', *The Germanic Review*, 47 (1972), 83–94; H. B. Nisbet, '*Die Wahlverwandtschaften*: Explanation and its Limits', *Deutsche Vierteljahrsschrift für Literaturwissenschaft und Geistesgeschichte*, 43 (1969), 458–86; E. L. Stahl, '*Die Wahlverwandtschaften*', *Publications of the English*

*Goethe Society*, new series, 15 (1945), 71–95; F. J. Stopp, '*Ein wahrer Narziss*: Reflections on the Eduard–Ottilie Relations in Goethe's *Wahlverwandtschaften*', *Publications of the English Goethe Society*, new series, (1959–60), 52–85; Waltraud Wiethölter, 'Legenden. Zur Mythologie von Goethes *Wahlverwandtschaften*', *Deutsche Vierteljahrsschrift für Literaturwissenschaft und Geistesgeschichte*, 56 (1973), 1–64.

## FURTHER READING

Adler, Jeremy, *"Eine fast magische Anziehungskraft". Goethes 'Wahlverwandtschaften' und die Chemie seiner Zeit* ("*An Almost Magical Attraction*". *Goethe's* Elective Affinities *and the Chemistry of its Time*) (Munich, 1987)
—'Newton, Goethe and *Die Wahlverwandtschaften*. On the Virtue of contradictory Hypotheses', *Wissenschaftskolleg zu Berlin. Jahrbuch*, 1985–6, 211–21 .
Duncan, Alistair Matheson, 'Some Theoretical Aspects of Eighteenth-Century Tables of Affinity', *Annals of Science*, 18 (1962), 177–95 and 217–32
—'Eighteenth-Century Theories of Chemical Affinity and Attraction' (unpublished University of London Ph.D thesis, 1971)
Gould, Robert David, 'Elective Affinities. An Investigation of the Influence of Goethe's Scientific Thinking on *Die Wahlverwandtschaften*' (unpublished Princeton University Ph.D thesis, 1970)
Howe, Richard Herbert, 'Max Weber's Elective Affinities: Sociology within the Bounds of Pure Reason', *American Journal of Sociology*, 84 (1978), 366–85
Levere, Trevor Harvey, *Affinity and Matter: Elements of Chemical Philosophy 1800– 1865* (Oxford, 1971)
Nisbet, H. B., *Goethe and the Scientific Tradition* (London, 1972)
Reiss, Hans, *Goethe's Novels* (London, 1969)
Thackray, Arnold, *Atoms and Powers. An Essay on Newtonian Matter Theory and the Development of Chemistry* (Cambridge, Mass., 1970)

# 19

# Kleist's Bedlam: abnormal psychology and psychiatry in the works of Heinrich von Kleist

## NIGEL REEVES

What feelings assail us as we see this horde of unreasoning creatures, some of whom may once have stood beside a Newton, Leibniz or Sterne? Where is our faith in our ethereal origin, in the immateriality and independence of our mind and in the other hyperboles of the poetic imagination that were invented in the turmoil between hope and fear?

(J. C. Reil, *Rhapsodies* . . . , 1803)

Henrich von Kleist's disturbingly violent and unpredictable short stories and plays are direct witness to their author's loss of faith in the power of the human individual to control his destiny or to understand his own personality. As a young man Kleist had adhered almost obsessively to an Enlightenment belief in the supreme part to be played by reason in directing men's lives. It was only after this belief had collapsed that he began to write works of imaginative literature. If his works are to be categorized at all, they have to be ranked among those of German Romanticism. But they belong to that rare strain that eschews any ultimate vision of a higher existence, that cannot explain the mysterious and horrifying by reference to any metaphysical and indeed quasi-Manichaean framework, and cannot even find solace in the beauty or power of Nature. His most substantial story, 'Michael Kohlhaas' (1808, 1810) tells of a horse-dealer so eaten up with a thirst for justice that he is prepared to lead a rebel band and burn a whole city to the ground to gain legal satisfaction for the theft of three horses before himself submitting to execution. One of his best-known (and certainly most notorious) tragedies, *Penthesilea* (1808) culminates in the heroine, Penthesilea, Queen of the Amazons, joining her hounds in tearing her former lover to pieces in battle. In the short story 'The Betrothal in Santo Domingo' (1811), a love story is transformed into an account of savagery perpetrated by a band of liberated slaves while the 'betrothal' ends in the murder of the beloved by her lover and his own suicide; and in the short story 'The Foundling' (1811), an orphan taken into a family out of kindness turns out to be a source of such evil and destructiveness that the adoptive father finally murders him with nothing in his heart but the desire to continue his quest for revenge down into hell.

These few examples suggest a *prima facie* case for considering Kleist's sources of inspiration in the context of abnormal psychology and even psychopathology. Certainly there has been no lack of critics who have looked for the provenance of his work in his own unstable personality or, in a curiously anachronistic manner, have attempted to interpret him with the assistance of Freudian theory. Yet the simplest and most satisfactory explanation for Kleist's repeated use of what must be described as psychopathological materials would surely be a personal acquaintance and fascination with the theories, observations and writings of contemporaneous psychiatry. Admittedly, much of Kleist's biography is shrouded in mystery and, as the account that follows will indicate, it is often difficult to establish unequivocal connections between his work and that of the medical writers of his time. Nevertheless the evidence that can be assembled strongly suggests that Kleist, like a number of Romantic thinkers and literary figures, was deeply indebted to men of medicine whom he met and whose works he read.

Perhaps the earliest decisive intellectual influence on Kleist was his reading of Wieland's *Sympathien* around 1794. This work contained three elements of seminal significance to the young Kleist – a cosmological view of the world as part of a harmonious universe, a teleological understanding of man, whose highest purpose is to fulfil God's plan by realizing the potential of his individual nature, and a presentation of friendship on earth as part of a magnetic network of relationships preceding and extending beyond earthly existence. Such eighteenth-century philosophies were the origin of that most characteristic brand of Romantic scientific thought, *Naturphilosophie*.

When Kleist encountered Wieland's optimistic work he was a young officer in the Prussian army. In one of his earliest and most revealing letters, he explained to his former teacher, Martini, in March 1799 why he had decided to abandon his military career to go and study at the University of Frankfurt-on-Oder: it was because it was not possible to serve as an officer *and* to act as a free human being. In his essay of 1784 'Beantwortung der Frage: Was ist Aufklärung?' ('The Answer to the Question: What is Enlightenment?'), Immanuel Kant had twice specifically mentioned the army officer (as a veiled reference to Frederick the Great himself). He is the man who orders his underlings to 'stop thinking, get marching' ('räsonniert nicht, sondern exerziert'), while he can himself never question orders that come from above. But the mature man, the man who is 'mündig', throws off his dependence on external alien decisions. That was precisely how Kleist explained to his sister Ulrike, only weeks after he had commenced studies, that he, as 'a free, thinking human being', intended to cast off the shackles of intellectual and moral minority and, with the aid of *Vernunft* (Reason), withstand the menace of chance or fate by creating his own life plan.

Small wonder, then, the enormous impact made on Kleist, the student, by the ideas of the physicist Christian Ernst Wünsch as expressed both in the lectures that Kleist attended and in his multi-volumed work *Kosmologische Unterhaltungen für die Jugend* (*Cosmological Diversions for Young People*) (1778–80). In this early example of *Naturphilosophie*, Wünsch saw man as the link in a physical and spiritual chain connecting the whole of earthly Nature and the cosmos. He amplified this Leibnizian view of Nature by adding to Leibniz's common denominator in all phenomena, consciousness (as contained in the monad), the idea that natural harmony rested on a balance between dynamic opposites or, better, polarities, magnetism being but one (if a fundamental) example. For Wünsch goodness and virtue rested on the exercise of rationality (*Verstand*) and happiness consisted in a process of self-perfection that continues beyond the grave as the spirit proceeds from heavenly sphere to heavenly sphere. But the man of science who lovingly traced the intimate relationships between all things in God's universe already had a foretaste (especially through astronomy) of that blessed state. Since all worldly manifestations symbolize divine realities and are evidence of Cosmic Unity, the scientific investigation of Nature brings man ever closer to God, so fulfilling both his spiritual and his intellectual purpose at the same time. Natural science is thus the most exalted of occupations and of supreme religious significance. It is, however, vital to note that Wünsch saw *absolute* truth as beyond man's *earthly* grasp. Scientific method demanded a process of doubting and careful verification through experiment, as he insisted to his young disciples in the *Unterhaltungen*. For Kleist at this time, as we see in the early essay significantly entitled 'Aufsatz, den sichern Weg des Glücks zu finden und ungestört – auch unter den grössten Drangsälen des Lebens – ihn zu geniessen' ('Essay on How to Find the Certain Path to Happiness and How to Enjoy it without Disturbance, even amid Life's Greatest Tribulations'), the danger of erroneous judgement can still be avoided by the scrupulous application of the intellect. Upon that follows joy and the possibility of determining one's own destiny.

It was at Frankfurt-on-Oder that Kleist also became acquainted with Wilhelmine von Zenge, whom he rapidly persuaded to become his fiancée. Following Wünsch's educationalist example, and perhaps prompted by Kant's view that the whole of the female sex was 'unmündig', morally immature, Kleist set about Wilhelmine's 'education'. In his essay to her, entitled by Kleist's editor, Helmut Sembdner, 'Über die Aufklärung des Weibes' ('On the Enlightenment of Woman'), he stressed precisely these Wünschian ideals of self-perfection, happiness deriving from good deeds and evil being punished by fate. He also asked his fiancée to write down what she expected of marriage. This seems to have been his undoing.

In August 1800 he embarked on a mysterious journey accompanied by Ludwig von Brockes, the purpose of which he refused to reveal but which was intended to rescue 'someone's happiness, honour and perhaps life' and which would certainly affect Wilhelmine's happiness. The two young men set out for Vienna to fulfil what Kleist called 'a *very serious* purpose', obtaining in Dresden passports from the English ambassador, Lord Elliot. Elliot advised them against going on to Vienna but recommended Würzburg or Strasbourg instead. This is not the place to go into detail about the medical reasons for Kleist's visit to Würzburg but all the evidence suggests that Kleist was suffering from an embarrassing sexual impediment. Vienna, Würzburg and Strasbourg were all medical centres and the surgical school founded by Karl Kaspar Siebold in Würzburg was particularly renowned. Indeed, in a letter, Kleist specifically mentions Siebold, who had widely travelled before settling in Würzburg in 1766, having studied in Leyden, Paris and at St Bartholomew's and St Thomas's Hospitals in London. Siebold lectured on anatomy and widely on surgical topics. Kleist also mentions Hermann Joseph Brünninghausen, formerly Siebold's surgical assistant, and by 1800 a professor in charge of all assistant surgeons working in the Julius-Spital. There is, however, no extant record of Kleist's having been treated in the hospital under his own name or the pseudonym we know he used.[1]

Whatever happened in Würzburg, it was a drastic turning point in Kleist's life. Despite his early optimism that all was well after the Würzburg visit and a dream-like account in a letter to Wilhelmine of rearing a family with her as the perfect wife and mother, Kleist hardly again spent any time with Wilhelmine and broke off the engagement a year and a half later.

The precarious teleology that Kleist derived from Christoph Martin Wieland and Wünsch has generally been seen to have shattered upon Kleist's reading of Kant in March 1801. But this experience was foreshadowed by Kleist's first encounters with the mentally sick and with psychiatry.

In 1800 the lunatic asylum at the Julius-Spital in Würzburg was under the direction of Anton Müller. In a letter Kleist describes seeing four patients: a professor who was reduced to gabbling constantly in Latin as a result of studying too much; a monk who had once produced an unintentional word in a sermon and, believing he had thus falsified God's word, now unceasingly warned all visitors 'mit einer schwachen, aber doch tönenden und das Herz zermalmenden Stimme' ('in a weak, yet resonant and heart-rending voice') against the perils of pleasure; a merchant who had gone mad on discovering that his father's nobility was not hereditary and that he would not therefore inherit the title; and an eighteen-year-old youth dying of a wasting disease as a result of 'an unnatural vice'. In Anton Müller's own account of his years at the Julius-Spital, published in 1824, it is not possible positively to identify these four patients from the sixteen admitted between February 1798 and October 1801. The Youth however can be identified directly from still existing documentation at the Julius-Spital. Nor does that exclude the possibility that the three older men, at least, were admitted prior to February 1798. Müller himself mentions two patients who suffered mental derangement (allegedly) through the vice of self-abuse. He also recites among the most common causes of insanity false or misguided religiosity, excessive studying and an obsession with justice and the processes of law.[2] There can be no doubt that whatever Kleist actually saw at the Julius-Spital it reminded him of a vivid passage in Wünsch's *Kosmologische Unterhaltungen* where he depicts the terrifying corruption of the flesh in men or women who indulge in extra-marital sex or 'clandestine crimes'. It seems likely that Kleist's vision of the young man was heightened by his own deep-rooted fear of the disastrous consequences of masturbation and it throws light on his personal sexual difficulties.

No identical vision was to recur in his works but the shocking results of excessive and uncontrollable sexuality were to feature in *Penthesilea* and, in more measured form, in 'The Marquise of O . . .' (1808), not to mention the comic variation in the earlier drama *The Broken Jug* (1802–11). The strange, heart-rending incantations of the monk are utilized in 'St Cecilia, or the Power of Music' (1810) in the description of the three insane brothers whose iconoclasm has switched to a horrifying obsessive piety. And it is also suggestive of the possibility that Kleist met Müller that aberrations through excessive religiosity also occur in his earliest short story 'The Earthquake in Chile' and an obsession with justice in 'Michael Kohlhaas'. If Kleist did meet him he would have met a man who was finding his own way forward in psychiatry at a time when there were no extensive writings on the subject available in German. Müller describes how at the time he combed all available journals for material in German on mental illness. Who better to introduce Kleist to the latest developments in psychiatric thinking – other than perhaps Christoph Wilhelm Hufeland, the editor of the well-known *Journal* that carried his own name, whom Müller quotes as one of his own sources of reference at the time and who was staying in Würzburg at the same time as Kleist.[3]

All the cases cited in Kleist's letter share a deviation from the natural balanced order, whether it be social in the case of the merchant, intellectual (the professor), religious (the monk) or sexual (the young man). Their behaviour departed in each case from their normal 'Bestimmung' or 'Zweck' (purpose) in life.[4] These patients were living examples, then, of the breakdown of the young Kleist's fervent but fragile teleology.

The conscious collapse of Kleist's teleology came in March 1801, just six months later, after an intensive reading of further works of Kant and, according to Ernst Cassirer, Fichte's *Die Bestimmung des Menschen* (*The Vocation of Man*). Ludwig Muth's account of the crisis with his step-by-step reconstruction of Kleist's progressive dismay as he read the section of the *Critique of Judgement* entitled 'Critique of teleological judgement' is certainly masterly and convincing.[5] Kant painstakingly removes the bases for any objective demonstration of a teleological explanation of Nature and rejects the argument for the divine purpose and cause of the world (which had so attracted Kleist) on the grounds that such things are unknowable and not open to objective judgement. The quasi-religious comfort that Kleist had found in teleology vanishes, together with his faith in the ultimately religious value of scientific research. The scientific search for truth could contribute nothing to the education of Man on the cosmological level that Kleist had found in the work of his beloved teacher, Wünsch. Following Kant closely, Kleist laments to Wilhelmine that we cannot decide whether our knowledge is true or delusory. The image he uses is that of coloured spectacles through which our eye believes it sees coloured objects but which it cannot know to be coloured or not. Significantly, it is the image that Wünsch himself used when discussing scientific method in the context of observing the sun through a glass. Scientific results, it is concluded, reflect the instruments of investigation as much as the object itself. In other words, without Wünsch's metaphysical underpinning natural science, as Kleist understood it, lost its claim to absolute validity. Further, a – for Kleist fatal – discontinuity occurs between the divine and the world. Once truth in Nature can no longer be discerned, both Man and Nature become a mystery, a 'Geheimnis' (secret) or a 'Rätsel' (riddle), favourite words of the mature Kleist. Nor can an individual control his own destiny; he becomes the plaything of chance just like the wretched individuals of whom he had previously written who had no life plan.

But Kleist does not succumb to an interpretation of Nature as a mere chaos. It retains some of the inner coherence of *Naturphilosophie* (which by now had found eloquent formulation in Schelling's early works) as a dynamic inter-relationship of polarities infused by an electrical magnetic life force.[6] In an essay of 1810, in which his words at times closely echo those of Schelling's *On the World-Soul*, Kleist writes of a 'curious law' affecting human behaviour. In accordance with this quasi-natural force the moment a human being comes into contact with another they both become magnetically polarized: if the one assumes a negative polarity, the other assumes the positive, and vice versa.

Two vital elements of his early philosophy have vanished: the power of the individual to determine his own direction, and the link with God. It is in the disappearance of a regular relationship between Man and the higher reality that Kleist differs from the mainstream of German Romanticism. The cosmic oneness is missing. Mars, the god of war, does not intervene to prevent the fatal collision between Penthesilea and her lover–victim Achilles. If there is an intervention, as possibly in the long-term restoration of justice in 'The Duel', when the guilty party eventually dies and the initial outcome of the duel is reversed, it remains uncertain, unexplained and mysterious. Or where a god actually appears in the shape of Jupiter, in *Amphitryon*, it is in a curiously human form. We need go no further at this stage: with the breakdown of his cosmology, Kleist was ready to turn to psychic abnormality as the very stuff of his work.

After the Kant crisis Kleist entered one of his restless phases, in which he undertook monumentally long journeys the length and breadth of Germany and France. Travelling with his sister Ulrike (who preferred to travel in the guise of a man) he went again to Dresden and on to Leipzig, Halle, Göttingen, Mainz, Strasbourg and Paris. We know that Kleist did not now abandon the world of natural science: perhaps it was in a desperate search for his lost cosmology that he visited Ernst Platner, the physiologist, Karl Friedrich Hindenburg, a professor of mathematics in Leipzig, Georg Simon Klügel, a professor of mathematics and physics in Halle, Johann Friedrich Blumenbach, the natural historian, and Heinrich and August Wrisberg, both anatomists in Göttingen. All of these he describes as 'teachers of humanity', suggesting that it was a sense of unity between natural science and philosophy that he was seeking.

From Kleist's correspondence of the time an avid interest in the Swiss, Jean-Jacques Rousseau, emerges, commencing almost from the moment of the Kant crisis and characterized by an acceptance of Rousseau's cultural pessimism. Kleist was profoundly shocked at the decadence of Paris, which he reached in early July 1801. It was this pessimism that must ultimately have led him to retreat a few months later to Switzerland where his creative career was to begin. But he did not withdraw to nature until he was finally convinced that natural science could neither make us happy nor give insight into the essential connection between all things. Despite an apparently close acquaintance with Alexander von Humboldt who, with the help of the Prussian ambassador the Marquis Lucchesini, introduced him to men of science and intellectuals writing in Paris, scientific specialization now seemed to him fragmentation. As he put it, when Newton saw a girl's bosom he noticed nothing but its line of incidence, while her heart was only of interest to him for its cubic capacity. But if Reason could not lead us safely, *nor* could feeling, the inner voice. Here we again find a radicalism in the post-Enlightenment Kleist that distinguishes him from many of the Romantics. 'The very same voice which tells the Christian to forgive his enemy, tells the Pacific Islander to roast him and then to eat him in great piety', he writes. The horrors of behaviour that might seem to belong in the lunatic asylum now appear to Kleist to be entirely human, normal and merely relative, indeed characteristic of cultural differences.

Incest, the murder of friends, suicide and death, he writes, are everyday boring events in Paris. Again Kleist touches on themes that he was to evolve in his literary work.

Kleist stayed on in Paris until the autumn of 1801 and then carried out his resolve to live as a peasant in a rustic retreat, despite Wilhelmine's refusal to join him. It was on an island in the river Aar near Thun that Kleist started his first dramatic work, the tragedy *Die Familie Ghonorez*, which was to be adapted and then published anonymously the following year under the title *Die Familie Schroffenstein*. His creative career had begun within months of the collapse of his philosophy. A work in the tradition of *Romeo and Juliet*, it centres on the deadly feud between two branches of a single family and the tragic fate of the children, Rodrigo and Ignez, who fall in love. The tragedy strikingly features a number of psychic phenomena that were under discussion at the time. Raimund, the father of Rodrigo, is obsessed with the desire to inherit the wealth of the entire family. That greed, the initial *idée fixe*, turns to vengeance when his younger son is found dead. For a moment he is as if transfixed by fury in ecstatic immobility. Assuming his relatives were responsible, he thirsts for the blood of Alonzo's daughter. Alonzo himself falls into a trance or catalepsy believing Raimund to have succeeded in his plan. Of this trance or catalepsy he can remember nothing but a sense of great well-being as if his spirit had visited the very source of life in God. Rodrigo and Ignez meet secretly in a cave where they are trapped by Raimund and his vassal Santin. Desperately attempting to deceive their pursuers, Rodrigo, in a sequence of words and gestures inducing a trance-like obedience in Ignez (a state identical to that of magnetically induced sleep), exchanges clothes with her. The gestures removing her clothes occur in a slow, deliberate order resembling the manipulations of magnetization. But this strategy (which Kleist's contemporaries claim was the scene that Kleist first imagined) fails and Rodrigo is killed by his own father in error for Ignez; thereupon Ignez, still wearing Rodrigo's clothes, is killed by Alonzo, her own father. As these horrible mistakes are uncovered by the blind grandfather, Raimund's illegitimate son, Juan, who had earlier become deranged through his unrequited love for Ignez, makes ribald comments. Looking aside from the contrivance of the last macabre scene and its obvious derivations from *King Lear*, it is as if Kleist had composed a virtual catalogue of abnormal states registered by the psychiatry of the time.[7] But perhaps the most significant is the first scene envisioned by Kleist, for it can be taken as clear evidence of a knowledge of Mesmer and his hotly debated techniques, indeed that it was the phenomenon of manipulative hypnotism that inspired Kleist's very first published lines.

Mesmer was, of course, well known, indeed notorious, and had practised in Paris which Kleist had just left. Hufeland, who had in earlier days been an opponent of Mesmer, had come round to a considerable interest in therapy through magnetization after he became director of the Berlin Charité hospital and asylum. Articles on the subject began to appear in his *Journal* after 1801.

Kleist's stay in Switzerland ended with a visit to his new doctor, Carl Wyttenbach, for the treatment of an unrevealed sickness. Kleist then commenced a fresh bout of travels during which he attempted to write the drama *Robert Guiscard*. Guiscard's ursurpation of his nephew's rightful position as Duke of the Normans is ironically frustrated by the plague which strikes him down while on a military expedition to Constantinople. Anticipating the equally ironic and wretched end of Jakob Rotbart in 'The Duel', the sole surviving scene of the drama (which Kleist never completed and may largely have destroyed in autumn 1803) depicts Guiscard emerging from his tent as if, far from dying, he were not sick at all. In fact he is only enjoying a moment of remission and the medical content of what Kleist wrote is suggestive of an interest in the psycho-somatic theories of the time and their particular attention to the progress of fevers.[8]

Kleist's return to Switzerland with Ernst von Pfuel, their travels across the Alps, their stay in Paris and Kleist's bizarre attempt to join Napoleon's abortive invasion of England (autumn 1803) have been examined in detail by Samuel and Brown, together with the period from late 1803 to mid-1804 when Kleist went to live in Mainz with Dr Georg Wedekind, one of the most famous physicians of the times.[9] Of significance in this context is that once again Kleist spent a considerable period of time sharing the house of a doctor (whom he might already have met when visiting Mainz in 1801). Wedekind published regularly, since he was clearly anxious to develop his medical reputation as his political fortunes waned. Interestingly, one of his very first lectures had concerned the intimate link between mind and body in sickness and health. Still more significantly he had published in Moritz's *Magazin der Erfahrungsseelenkunde* (*Journal of Empirical Psychology*) an article entitled 'Actions without Awareness of their Motivation – or the Power of Obscure Ideas': here Wedekind speaks of premonitions of death that he had himself experienced and of a case of somnambulism. He states that obscure ideas, the origins and relations of which we are unaware, often cause us to act and that if these ideas are powerful enough they can overwhelm conscious decision. Our actions, he concludes, are involuntary, as the suicide of Karl Wilhelm von Jerusalem (the inspiration for Goethe's Werther) showed.

It seems likely that staying with Wedekind must have strengthened Kleist's fascination with non-conscious states of mind and their concomitant actions. It also seems probable that he had access to the journals in which Wedekind had continued to publish. If so, he would have found in Moritz's periodical a three-part article by one P. Pockels on dreams and somnambulists (1788–9); in *Baldingers Magazin* of 1789, to which Wedekind also contributed, an article by D. Gmelin on somnambulism and clairvoyance; and in *Hufelands Journal* of 1800 a publication by a Dr G. Schmid classifying concepts of mental diseases. In 1802 Schmid explained in detail, in the same journal, connections between awareness in sleep with animal magnetism, galvanism and somnambulism.

From the autumn of 1804 until the summer of 1806 Kleist worked as a Prussian civil servant. It was the first time he had been in something resembling firm employment since his days as an officer. But he then asked for and obtained sick-leave, which lasted until January 1807. In this time we know that he worked on two comedies, *Amphitryon* and *The Broken Jug*. The manuscript of *Amphitryon* was finished by the end of 1806. Discussion of this play has usually centred on the confusion of feeling aroused in Alkmene, Amphitryon's wife, who finds herself betrayed in the most intimate of ways, for it is her very adulation of her husband that leads her to commit adultery with Jupiter, believing his powers as a lover to be the proof of his identity as Amphitryon. Here we have Kleist's first mature exploration of the failure of intuition to replace Reason as a guide.

But the further dimension of this confusion of feeling is the experience of Amphitryon and his servant Sosias on finding themselves confronted by their doubles and seemingly ousted from their own positions and true selves. Sosias rapidly accepts that he has a double who is 'sein zweites Ich' ('his second self'). Dispatched by Amphitryon from the battleground to report victory to his wife, Sosias comes upon Merkur (Mercury). Walking through the deepest of nights as if he were asleep, Sosias is in a state of obvious suggestibility. Merkur proceeds to beat him with a stick until he capitulates and admits that he has lost his own identity. Merkur then completes the mission. Upon his telling his master of the inexplicable episode, Amphitryon believes Sosias must be suffering from 'Gehirnverrückung' ('cerebral derangement'). But it is soon his turn to be termed a madman when, returning home, he is informed by Merkur that Alkmene is in bed with Amphitryon, whereupon he is locked out of the house.

What can have inspired this tragicomic exploration of the double personality and the treachery of the senses? Our previous discussions and reconstruction of possible sources of inspiration from the psychiatry and mesmeric theories of the day throw light on this topic. But as early as 1901 Spiro Wukadinovic[10] had noted striking parallels between Kleist's works and topics mentioned in J. C. Reil's *Rhapsodien über die Anwendung der psychischen Curmethode auf Geisteszerrüttungen* (*Rhapsodies on the Application of Psychic Therapy to Mental Disturbances*) that appeared in Halle in 1803. In 1973 Maria Tatar[11] convincingly argued that from the opening somnambulist scene in *The Prince of Homburg* of 1810 the whole range of cruel experiences to which the disobedient prince is subjected by his uncle, the Elector – the use of rigid adherence to the law to obtain clarity of mind (*Besonnenheit*) from the distracted young man, the threat of death, letting him see his own future grave, calling for a personal written statement about error and future behaviour, even the firing off of cannon beside him and the fake execution at the end – are all therapeutic methods specifically recommended by Reil for jerking the deranged back into a rational state of mind.

The play ends on a note of ambiguity: there is no evidence that this drastic treatment has succeeded. Rather it leaves the Prince still more confused than in his state of distraction following the opening somnambulist sequence. Now it seems at least possible that even in the comedy *Amphitryon*, Merkur's beating of Sosias until he is reduced to confessing that he has a double is a deliberate reversal of Reil's 'Curmethode', a method of deranging the mind rather than reordering it. It is suggestive that in Reil we also find an extensive discussion of the double personality, including express reference to confusing oneself with other persons and their characteristics.

With the exception of *The Broken Jug*, which was conceived in early 1802, the remainder of Kleist's work may have been written after he had read Reil, the ground having been prepared through his prior knowledge of Mesmer and Wedekind and his emotional and intellectual interest in psychiatry having been awoken by Müller and perhaps Hufeland in Würzburg. Other features of Reil's work also recur in Kleist. Michael Kohlhaas, whose vengeance knows no limit, deserves to be considered in this category, strengthened by his delusion at one stage that he is the Archangel Michael. Reil also deals with the 'Zerstreuung' ('absent-mindedness') of the mentally sick. This plainly is Homburg's state and the word 'zerstreut' ('absent-minded') is repeatedly used of him. The opposite is a morbid state of concentration, 'Vertiefung', in which the patient hears no one. This we find in Penthesilea and also in the state of the three converted iconoclasts in 'St Cecilia or the Power of Music'. Reil deals with the aberrations that follow religious fanaticism, a state that Kleist had already observed or claimed to have observed in the mad monk of the Julius-Spital. Reil goes further to speak of the massacres prompted by such fanaticism. 'The Earthquake in Chile', which was certainly written in a French prison, centres on this phenomenon of religious mob violence, while the previously normal cobbler who dashes out the brains of a baby against a cathedral pillar is a further frightening example of the transformations that the insane can undergo. The iconoclasts' destruction of churches in 'St Cecilia' also belongs here. A sudden shock, however, can also unhinge the mind, as we find in the ghost scene in 'The Beggarwoman of Locarno'.

But most telling of all is Reil's introduction to his *Rhapsodien*, where he eloquently speaks of the proximity of madness to the normal. When we see the inmates of an asylum who may once have stood at the side of Newton or Leibniz, where is our faith in our ethereal origin and the immateriality and independence of our mind, he asks? These were words that must have seemed to Kleist only too true in their pessimism. A fever, a stroke of bad fortune affecting our family or our country and we can find ourselves for ever in the madhouse, Reil continues. 'Fortune plays a strange game with us humans . . . The deranged who cannot counsel themselves and cannot counter deception with deception, suffer from a frailty [*Gebrechen*] that is founded in mankind itself, to which everyone of us is more vulnerable than to any other defect and which we cannot resist either through Reason or through rank and riches.'[12]

We have reached the crux of the matter. Here, Kleist could read that madness is an inherent part of the human condition, afflicting rich and poor alike. Above all it derives from man's own frailty. After she has murdered Achilles, and is recovering from her fury, Penthesilea believes she is in Elysium. Her sister tells her, 'It is I, your Prothoe, who is clasping you in her arms and what you may glimpse here is still the world, the frail [*gebrechliche*] world, upon which the gods look but from afar'. The expression that later recurs in both 'Michael Kohlhaas' and in 'The Marquise of O. . .' as the explanation for the absence of justice and the extreme unpredictability of human nature is 'die gebrechliche Einrichtung der Welt' ('the world's frail [or flawed] structure'). The impact that Reil made on Kleist is all the more explicable when we read the opening page of Reil's *Rhapsodies*:

No matter how many divine inclinations to high and noble deeds with which Nature has endowed us – the desire for fame, for personal perfection, the power to determine our actions and to keep self-control, and passions that through their very tempestuousness protect us from deadly lethargy; yet has she also, at the same time, and thanks to those very qualities, implanted in us just as many seeds of madness. Step by step we are coming closer to the madhouse as we continue down the path of physical and intellectual civilization.

For Kleist the distinction is not between the mad and the healthy, for potentially all men are mad. Hence meeting Gotthilf Heinrich von Schubert in the autumn of 1807 must have been the final rather than the first stage in Kleist's induction into the world of psychopathology. Schubert, whose cosmic *Naturphilosophie* had first been formulated in his *Ahndungen einer allgemeinen Geschichte des Lebens* (*Intimations of a Universal History of Life*) of 1806, was almost certainly encouraged by Kleist, among others, to explore the phenomena of somnambulism and animal magnetism further in his lectures held that winter on the 'Nachtseite der Naturwissenschaften' ('The Night-side of the Sciences'). For Schubert, however, states of insanity were not necessarily states of sickness but rather of heightened perception inspired by transcendental cosmic reality. Certainly, we have a hint of that optimism rubbing off on Kleist in 'Kätchen von Heilbronn' of 1808–10, which Kleist called the 'Kehrseite' ('reverse-side') of Penthesilea. Kätchen's love of Graf Friedrich Walter vom Strahl could be either of hypnotic or cosmic origin. Certainly the telepathic communication between them in a double dream is capable of both interpretations and is equally in keeping with mesmeric magnetization as with a metaphysical subconscious explanation. The subconscious link protects her against fire, poisoning and conscious misunderstandings on the part of Graf vom Strahl. Indeed it seems to be related to the phenomenon of the *idée fixe* but instead of its having baleful consequences the victim (or patient?) is spared, reflecting, it seems, a temporary benevolent influence exercised on Kleist by Schubert's intact *Naturphilosophie* with its characteristically mainstream Romantic view of life as 'the accord with the harmony of the universal interrelationship of world forces'.

But it was only an interval. Kleist could not long subscribe to a philosophy the bases of which Kant had removed with ruthless logic. Indeed Kleist could not subscribe to the metaphysical optimism of the orthodox Romantics. He was left with a human world in which the stuff of relationships lay in the dynamic of magnetic-like polarities – the stronger the signal, the stronger the reaction, the plus and the minus. And under the pressure of these magnetic forces the fragile dividing line between the normal and the lunatic could easily shatter, so that a man like Kohlhaas, with the most delicate sense of justice, could be rendered an arsonist and murderer. But unlike the psychiatrists of his time Kleist saw no effective therapy. There can be no doubt that for him the world itself was the madhouse. Homburg, at the end of his cruel and exhausting 'Kur' does not see the light of rationality, or Reil's goal of 'Besonnenheit' but rather confusion, a loss of all orientation, common also to Alkmene in *Amphitryon*. But disorientation was the very opposite of Reil's intention.

For many of Kleist's other characters death alone waited. Indeed this was Kleist's true therapy. He had often spoken of committing suicide after his life plan, his teleology, had collapsed. On 21 November 1811 Heinrich von Kleist followed the example of his own Penthesilea, or perhaps more aptly, of the still lucid if desperate lover, Gustav, in 'The Betrothal in Santo Domingo', shooting first his companion in death, Henriette Vogel, and then himself on the banks of the Wannsee outside Berlin.

## NOTES

This chapter is based substantially on the previous article by the author, 'Kleist's Indebtedness to the Science, Psychiatry and Medicine of his Time', *Oxford German Studies*, 16 (1985), 47–65, where full references are given. *Oxford German Studies* have kindly granted permission for the article to be reproduced in this revised form.

1  For details of the investigation carried out in the archives of the Julius–Spital, see Diethelm Brüggemann, *Drei Mystifikationen Heinrich von Kleists* (New York, 1985), pp. 81–2.

2  Anton Müller, *Die Irren-Anstalt in dem Julius–Hospital zu Würzburg und*

*die sechs und zwanzigjährigen ärztlichen Dienstleistungen an denselben* (Würzburg, 1824).

3  See Brüggemann, *Drei Mystifikationen*, pp. 82–5.

4  Interestingly Kleist's view, as implied in the letter to Wilhelmine of 13 September 1800, is echoed in an article by Hoffbauer which appeared about three years later in Reil and Autenrieth's *Archiv für die Physiologie*, 5:3, pp. 448–87.

5  Ludwig Muth, *Kleist und Kant. Versuch einer neuen Interpretation* (Cologne, 1954).

6  See Kleist's essay of 1810, 'Allerneuester Erziehungsplan' ('My latest educational plan'), *Sämtliche Werke*, ed. H. Sembdner (Darmstadt, 1962), II, pp. 329f. Cf the similar views in his *Über die allmähliche Verfertigung der Gedanken beim Reden (On the Gradual Formation of Thought in Speaking)*, probably of 1807, *ibid.*, pp. 321ff.

7  See Hoffbauer, quoted above, n. 4, who discussed bloodlust, catalepsy, ecstasy.

8  For a detailed account of these in relation to another major writer, see Kenneth Dewhurst and Nigel Reeves, *Friedrich Schiller, Medicine, Psychology, Literature* (Oxford, 1978), esp. pp. 117ff., 245ff., 253ff. 294f.

9  R. H. Samuel and H. M. Brown, *Kleist's Last Year and the Quest for 'Robert Guiscard'* (Leamington Spa, 1981).

10  Spiro Wukadinovic, 'Max Morris, "Henrich von Kleists Reise nach Würzburg"', *Euphorion*, 8 (1901), 771–9.

11  Maria M. Tatar, 'Psychology and Poetics. J. C. Reil and Kleist's *Prinz Friedrich von Homburg*', *The German Review*, 48 (1973), 21–34

12  J. C. Reil, *Rhapsodien über die Anwendung der psychischen Curmethode auf Geisteszerrüttungen* (Halle, 1803), pp. 8–11 (my translation, here and below).

# FURTHER READING

## TRANSLATIONS

The most easily available translation of Kleist's short stories is by the present author and F. D. Luke: Heinrich von Kleist, *The Marquise of O. and other Stories*, translated by F. D. Luke and N. Reeves, Penguin Classics series (Harmondsworth, 1978). *Penthesilea* is available in translation in *Five German Tragedies*, translated by F. J. Lamport, Penguin Classics series (Harmondsworth, 1969).

## PRIMARY LITERATURE (not included in the notes)

*Heinrich von Kleists Lebensspuren. Dokumente und Berichte der Zeitgenossen*, ed. H. Sembdner (Munich, 1969)

Schubert, Gotthilf Heinrich, *Ansichten von der Nachtseite der Naturwissenschaften* (Dresden, 1808)

Wünsch, Christian Ernst, *Kosmologische Unterhaltungen für die Jugend*, 3 vols. (Leipzig, 1778–80)

SECONDARY LITERATURE (not included in the notes)

Cassirer, Ernst, 'Heinrich von Kleist und die Kantische Philosophie', in *Idee und Gestalt* (Berlin, 1921)

Thomas, Ursula, 'Heinrich von Kleist und Gotthilf Schubert', *Monatshefte*, 51 (1951), 249–61

# 20

# Coleridge and the sciences

## TREVOR H. LEVERE

## INTRODUCTION

Samuel Taylor Coleridge was often a deeply troubled man, repeatedly accusing himself of sloth, bemoaning his lack of self-discipline and wretched at the loss of his poetic creative imagination. Subsequent critics have often been tempted to accept his self-deprecatory evaluation, and to regard him as one who, after an initial youthful burst of poetic activity, succumbed to opium and to metaphysical befuddlement. Thomas Carlyle's portrait of Coleridge is unkind but not untypical:

He began anywhere: you put some question to him, made some suggestive observation: instead of answering this, or decidedly setting out towards answer of it, he would accumulate formidable apparatus, logical swim-bladders, transcendental life-preservers and other precautionary and vehiculatory gear for setting out; perhaps did at last get under way, – but was swiftly . . . turned aside by the glance of some radiant new game . . . into new courses; and ever into new; and before long into all the Universe . . .[1]

Both the substance and the form of Coleridge's reasoning have been ridiculed;[2] they have also, in this century, received sympathetic treatment.[3]

A more sympathetic treatment has been assisted by the new editions of Coleridge's works still appearing.[4] It has become clear that Coleridge was engaged in a long-standing intellectual endeavour, of which the first stage was to clarify the rules of thought, to formulate a system of logic governing the operations of the understanding, similar in many ways to Kant's enterprise in *The Critique of Pure Reason*. Coleridge's *Logic* was not printed in his lifetime, but an edition has recently been published.[5] When Coleridge composed it is, as the editor remarks, 'largely a matter of educated guesswork. As early as 1803 Coleridge outlined a treatise that resembles it in some respects, and less than a month before he died he was still talking about it enthusiastically as work in progress . . .'[6] Coleridge was reluctant even to lecture on subjects about which he had not already thought for years, and it is little wonder that sceptical contemporaries became impatient of his claims to have written works that never appeared. But Coleridge did write them, not only the *Logic* but even his *Opus Maximum*,[7] the culmination of his philosophical struggle, wherein the facts and laws of science were to be brought into a synthetic unity with human and divine reason.

As we learn more about what Coleridge wrote but did not publish, the search for a union of science, philosophy and theology emerges more clearly as the core of Coleridge's mental life. The figure we can now see, thanks to editorial labours, is 'not a poet who dwindled into a philosopher, but a thinker who happened to be a poet'.[8] It is in this light, that we shall consider Coleridge here; as we do so, we shall find that the sciences play an essential role in his thought.[9]

This was not because he identified the sciences with philosophy. As early as 1804, he was anxious to distinguish 'keen hunters after knowledge and Science' from true philosophers, lovers of wisdom; to call philosophers those who had made 'a lucky experiment' was to prostitute and profane the name of philosopher.[10] He was consistent in this attitude. The year before he died, he visited Cambridge for the meeting of the British Association for the Advancement of Science, and rose to forbid them to call philosophers those who merely studied the material world. It was on that occasion that the word 'scientist' was first proposed.[11]

## STUDENT OF THE SCIENCES

Coleridge was a philosopher, in precisely the sense that enabled him to distinguish between natural philosophy and science; by the same token, he was in no sense a scientist. He was, however, a committed student of the sciences. This appears in his early educational plans, in his reading and note-taking, in his acquaintance among men of science, in his attendance at scientific lectures, and in his constant incorporation of scientific lore into almost every aspect of his thought and writing.

While at school, he had contemplated a career in medicine, in emulation of his brother Luke. His soon-terminated period of study at Cambridge was unfruitful and comical, but afterwards he began to find his feet in the West Country, with the congenial society of Southey and for a time of Wordsworth. Also important in these years, years of great poetic activity, were the company and library of Dr Thomas Beddoes,[12] friend and correspondent of several members of the scientifically vigorous Lunar Society of Birmingham. Coleridge and Beddoes probably met through their shared political views; both participated in local democratic protest against the government's attempts to curtail dissent during the early years of the war with France.

Beddoes had taught chemistry at Oxford, was widely read in French and

German philosophy and science, and seems the individual most likely to have urged upon Coleridge the desirability of visiting Germany. Coleridge proposed various plans, involving literature, philosophy, theology and science. In May 1796 he wrote to a friend that he thought of going to Jena, 'a cheap German University where Schiller resides', there to translate the works of Schiller for a London bookseller, while studying chemistry and anatomy. He would return 'with all the works of Semler & Michaelis, the German Theologians, and of Kant, the great German metaphysician.'[13] He received financial help from the Wedgwoods, and early in 1799 matriculated not at Jena but instead at Göttingen University. There he attended lectures by J. F. Blumenbach on anatomy and physiology; walking in the Harz mountains provided an introduction to geology. When Coleridge returned to England later that year, it was to find the then unknown Humphry Davy freshly in charge of the chemical laboratory in Beddoes's new chemical-cum-medical centre, the Pneumatic Institution.

Coleridge observed and participated in Davy's experiments on the effects of nitrous oxide, and promptly became equally enthusiastic about Davy and chemistry. Coleridge's intellectual sympathies often coincided with his personal ones. When Davy moved to the Royal Institution in London, Coleridge was eager for his friend's success on the march to glory. He even proposed to study chemistry with Wordsworth in the Lake District – a plan that failed to meet with sufficient welcome for it to be realized. But Coleridge none the less asked Davy's advice about chemical apparatus and books, while admitting that his passion for science might be 'but *Davyism*!'[14] Coleridge, on this as on many occasions, was expressing a half-truth ungenerous to himself. He attended Davy's lectures in 1802, making extensive notes; then and subsequently, he pursued chemistry more systematically than any other science, and as poet and philosopher, he transmuted his experience and knowledge of chemistry into metaphors and imagery that provide a key to his intellectual and imaginative constructions. In poetry, philosophy, science and theology, he repeatedly made use of his experience of nature and science; that experience, apprehended symbolically and metaphorically, conferred a unity on his entire intellectual enterprise. The search for interconnectedness, of reticulation[15] not only within a given realm of discourse, but between different realms, is surely one of the distinguishing characteristics of Romantic striving, and it informs Coleridge's life's work.

## SCIENCE IN PHILOSOPHY

That work, more than adumbrated before the interruption caused by opium and signalled in 1804 by his voyage to Malta, was to gain momentum after his return. By the time he published *The Friend* in 1818 he had developed a method and a coherent philosophical apparatus for tackling a host of interrelated problems. How could we apprehend the developmental aspects of nature? How were history and nature connected? To what extent were our perceptions of nature creative, and dependent upon ideas? How were the laws of nature related to ideas, and were they merely regulative or also constitutive – how, in other words, was mind related to nature? What were symbols, and what did they tell us about nature? How were the languages of science and poetry respectively related to nature, and what was the importance of metaphor in each? How did God's creativity and governance operate?

Coleridge addressed all these questions, repeatedly and in different ways between his return from Malta and his death in 1834. For much of that time, and especially in the years around 1820, he read systematically in the natural sciences, seeking knowledge that would illuminate a philosophy of nature.

I shall not seek here to trace the development of his thought, nor the details of his growing knowledge of the natural sciences. What I shall do is to explore the importance of the sciences for Coleridge by examining their role in his exploration and application of three interconnected sets of concepts: powers and the role of productivity in nature; development, and the distinction between historical sequence and genetic form; and ascent, whether through a hierarchy of levels, or through a spiral.


## POWERS AND PRODUCTIVITY

In *The Friend*, Coleridge wrote of the chemical elements of his day, 'the assumed indecomponible substances of the LABORATORY', that they were 'the symbols of elementary powers, and the exponents of a law'.[16] Laws accounted for relations in nature that, properly understood, were necessary. For Coleridge, to understand a law meant to perceive that the consequences of the operation of that law were necessary. Laws were causal and constitutive. Coleridge was here closer to Aristotle than to the majority of his scientific contemporaries, who were quite willing to regard empirical generalizations as laws, as in the case of John Dalton's laws of chemical combining proportions by weight, or of Gay-Lussac's law of the ratio by volume in which gases combined. But although in a minority in Britain, Coleridge was by no means alone. John Herschel was probably the most popular philosopher of science in mid-nineteenth-century Britain,[17] and in his *Preliminary Discourse on the Study of Natural Philosophy* (London, 1831) he related laws to causes through the idea of inherent necessity: it was 'the predisposal of what shall happen, that impresses us with the notion of a *law* and a *cause*' (p. 36). But if Coleridge's discussion of law would at least have been intelligible to his readers, there remains the problem of what he meant by presenting chemical elements as 'symbols of elementary powers'.

He once wrote that all our questions about nature could be generalized into just two questions:

First Question: What are the POWERS that must be assumed in order for the thing to be that which it is: or what are the primary Constituent POWERS of Nature . . .?
Second Question: What are the Forms, in which these Powers appear or manifest themselves to our Senses?[18]

That tells us how important the concept of powers was for him: powers *constituted* nature. They did it by producing nature, while being self-limiting. Coleridge here has travelled much the same road as F. W. J. von Schelling and other post-Kantian philosophers. The opposition between the productive and self-limiting aspects of powers appears in what Coleridge called the 'universal Law of Polarity', that 'EVERY POWER IN NATURE AND IN SPIRIT must evolve an opposite' in order to manifest itself. And he illustrated this not immediately intelligible doctrine by discussing the composition of water, which arose not from the mere addition of hydrogen to oxygen, but by their polar opposition and synthesis as water.[19]

Powers are constitutive. What appears in nature is produced by the synthesis of polar powers. But what does Coleridge mean by describing chemical elements as the symbols of powers? For Coleridge, a symbol was a sign included in the idea that it represented; it was a part of the unity of nature through which one could contemplate underlying essential truths. Symbols enabled one with the help of imagination or reason to pierce through the sensible appearances of phenomena to their insensible but more real essences.[20]

> For all that meets the bodily sense I deem
> Symbolical, one mighty alphabet
> For infant minds . . .[21]

Nature presented to us 'the lovely shapes and sounds intelligible' of God's language in his creation, teaching 'Himself in all, and all things in himself':[22]

Chemistry revealed relations between substances apparently distinct:

so water and flame, the diamond, the charcoal, and the mantling champagne, with its ebullient sparkles, are convoked and fraternized by the theory of the chemist . . . Hence the strong hold which in all ages chemistry has had on the imagination. If in SHAKSPEARE we find nature idealized into poetry, through the creative power of a profound yet observant meditation, so through the meditative observation of a DAVY . . ., we find poetry . . . realized in nature: yea, nature itself disclosed to us . . . as at once the poet and the poem.[23]

Chemistry, like poetry, revealed underlying unities not apparent to the senses. Both involved the exercise of imagination, although in different degree. Coleridge in the *Biographia Literaria* had offered a definition of imagination as either primary or secondary. The former he regarded as 'the living Power and prime Agent of all human Perception, and as a representation in the finite mind of the eternal act of creation in the infinite I AM'. The secondary imagination added conscious will to the primary imagination: 'It dissolves, diffuses, dissipates, in order to recreate it; or where this process is rendered impossible, yet still at all events it struggles to idealize and to unify.'[24] Poets clearly were concerned with re-creation; but they shared with natural philosophers the struggle to 'idealize and to unify'.

## DEVELOPMENT: NATURE AND HISTORY

Unification meant uncovering relations. These relations were dynamic, not static, deriving from the polarities of underlying powers. Nature had been created according to a process of development – the realization of God's ideas.[25] Natural philosophy would help to elucidate that creation and development. But this was to ask a lot of natural philosophy, certainly more than was contained in the sciences as they had been practised since the late–seventeenth-century mechanization of the world picture.

Coleridge clearly understood what was needed: 'In order to contemplate the laws even of nature, in order to refer the phaenomena of the perishable world to a permanent law, we are constrained to consider each . . . minute element as a living germ in which the present involves the future and in the finite the infinite abides potentially.'[26] The teleology of this position is clear. The present involves the future, to which it is in a sense directed. As for the relation between infinite and finite, it is predicated in God's relation to his creation, and also in the reticulated complexity of nature. Coleridge, like Blake, knew what it was

> To see a World in a Grain of Sand,
> And a Heaven in a Wild Flower . . .[27]

But unlike Blake, he presented that insight in philosophical context, and tied it to a view of natural philosophy:

Are we struck at beholding the cope of heaven imaged in a dew-drop? – The least of the animalcula to which that dew drop is an ocean presents an infinite problem, of which the omnipresent is the only solution . . . [T]he philosophy of nature can remain philosophy only by rising above nature . . .[28]

There was an enormous gap between the philosophy of nature on the one hand, and empirical science and natural history on the other. A true conception of the natural world required the exercise of reason within a framework of philosophy. What is more, by leading one from the finite to the infinite, natural philosophy verged on the threshold of theology. This is not the place to seek to cross that threshold. But we are left with the immediate problem of how, within Coleridge's philosophy of nature, the present could involve the future.

Much of science was historical by the early nineteenth century. A. G. Werner's geology and then Georges Cuvier's palaeontology based their account of the earth and its creatures on a succession of processes and events. Evolutionary or developmental theories were well known in the life sciences, in the work of the Comte de Buffon, Jean-Baptiste Lamarck, Erasmus Darwin and others. Kant and Laplace had offered evolutionary cosmologies to account for the structure of the heavens. Historical thought, through the work of Herder and others, was becoming firmly part of contemporary culture, especially in Germany. Coleridge was imbued with German as well as British and classical sources; he had been powerfully attracted by Schelling's philosophy of nature, with its emphasis upon organic metaphors, notably including growth and productivity. Philosophical cosmologies and geogonies proliferated, like those of Henrik Steffens, Schelling's scientifically literate follower. Steffens made ingenious use of Werner's historical model, intertwining it with Schelling's philosophy of nature. History in science was, in short, well established by Coleridge's day, and was particularly clearly so for those conversant with the latest in German thought.[29]

Elucidating the nature of time was a particular problem for Coleridge. As he wrote to a friend in 1801, he had been working hard at metaphysics, had 'completely extricated the notions of Time, and Space' and had also overthrown 'the irreligious metaphysics of modern Infidels . . .'[30] It is important to note that logic served to clarify metaphysics and theology. In the *Logic*, Coleridge gave a treatment of space and time similar to Kant's, presenting them not as properties of the material world, but rather as forms of the understanding, whereby our perceptions of the world were organized. Could one then say nothing about the role of history in nature, except with reference to our understanding? Coleridge believed that one could go further. In a difficult but revealing note of 1819, he observed that

the proper objects of knowledge . . . are Nature and History – or Necessity and Freedom. And these attain their highest . . . perfection, when each reveals the essential character of the other in itself . . . Thus Nature attains its highest significancy when she appears to us as an inner power . . . when . . . she reveals herself as a plastic Will, acting in time and . . . finitely. {This maintains the necessary distinct[ion] of [Nature] from Deity, and consequently the reality of both . . .} Here is Process and Succession . . . – and the Knowledge of Nature becomes Natural History. History . . . had . . . her consummation, when she reveals herself to us in the form of a necessity of Nature . . . – But here is *Law*, and the *Ever-present* in the moving Past, the eternal as the Power of the Temporal – and the Historic Science becomes . . . a transcendent Nature . . .[31]

Nature is the product of will, of mind acting with a purpose and producing change with time. That is how we see it, and how we record our perceptions of it. History, a record of past time, embodies laws, which are eternal, the ever-present in the moving past. Because nature is subject to laws, it is analogous to science; and in so far as these laws are constitutive, they make of history 'a transcendent Nature'.

Coleridge sought to follow the prescription implicit in his view of the nature of history and of nature in constructing a host of metaphysical accounts, some tied primarily to theology, others to the sciences. There is development in Coleridge's thought, and there is also a remarkable consistency in the principles informing that thought. Coleridge had his admirers and even disciples, as well as those who ridiculed his intellectual schemes. Carlyle's satire, quoted above, must have met with much approbation. The *British Review*'s response to Coleridge's *Aids to Reflection* was more concise: 'We can recollect no instance, in modern times, of literary talent so entirely wasted, and great mental power so absolutely unproductive, as in the case of this eminent author.'[32] One aspect of his thought that has proved especially forbidding is his repeated and varying construction of schemes of powers, of sciences and of living organisms. Coleridge was as fond of constructing such schemes as Enlightenment philosophers were of constructing taxonomies of living beings and classifications of human knowledge. But Coleridge's schemes were central and crucial to his thought. Eighteenth-century naturalists might debate the respective virtues of natural and artificial classifications. For Coleridge, only natural classifications had any meaning. They revealed relations between aspects of nature that were part of the essence of nature. Nothing existed in isolation; everything, because of the laws governing and the ideas constituting it, led to the contemplation of eternity in the present, and of multeity in unity.

## ASCENT, AND THE WIDENING GYRE

Many of Coleridge's schemes represented an ascent through a hierarchy. He presented hierarchies of powers symbolizing the steps in the creation and construction of the physical world; hierarchies of sciences arranged according to the powers that were both representative and active in them – for example, Humphry Davy had argued from the chemical effects of an electric current to the suggestion that chemical affinity might be identical to galvanic electricity, and Coleridge particularly identified the galvanic power with chemistry; and arranged hierarchies of living beings according to their degree of individuation, or to the ascent of life.[33] Coleridge's most extensive account of the ascent of life is in a posthumous publication, *Hints towards the Formation of a more Comprehensive Theory of Life*,[34] which contains an account of his doctrine of powers, and a sequential classification of living beings. It is tempting to treat this sequence as evolutionary; after all, it progresses from simple to complex organisms, and Coleridge makes it clear that he is talking about an ascent. But he is not talking about an ascent in time. We have to remind ourselves of Coleridge's emphasis upon laws of nature and the ever-

present in the moving past. We order organisms sequentially, and our understanding contemplates them under the form of time. But reason, which has for
Coleridge the same role in mind as have laws in nature,[35] presents these same
relations under the aspect of the ever-present. They are simply not governed
by time.

We must, however, remember the relation between nature and history.
Relations arising from laws of nature are perceived under the form of the
history of nature. The metaphors with which Coleridge described the ascent
of powers, and their culmination in the ascent of life, become increasingly
historical. In his critical reading of Henrik Steffens, who applied Schelling's
philosophy of nature to a range of geological, chemical and biological
knowledge, Coleridge borrowed and modified Steffens's concept of the
compass of nature.[36] Steffens took the four cardinal points of the compass,
and associated them with a quartet appropriate to different sciences – the four
races of man in anthropology, the four principal elements occurring in organic
compounds (carbon, nitrogen, hydrogen and oxygen), and so forth. Coleridge
took this compass to represent the cardinal powers in nature at its different
levels – physical, chemical and biological, with appropriate subdivisions.
Here was a graphic image for the unity of nature, the same metaphor
expressing essential relations at every stage. Coleridge emphasized the interaction of poles to produce intermediates, such as the halogens, or their
synthesis, as when hydrogen and oxygen, symbolizing powers of the west–
east axis of the compass, yielded their synthesis in water. The compass of
nature, in the hands of the nature philosophers of Germany, was a useful but
limited device. Coleridge made it less limited, but even in his hands it was too
rigid. He wanted a more flexible metaphor to interpret the ascent of powers
and life, for this ascent did not always occur smoothly and uniformly.

Coleridge was troubled by the inadequacy of a simple hierarchy. How, for
example, were crystals related to the simplest organisms? His answer, around
1820, was that the power of life manifested itself in its lowest form in crystals,
then sank back, re-emerging in altered kind in the lowest living organism.[37] In
the *Theory of Life*, Coleridge allowed for this form of ascent by adding to the
compass of nature another model, an ascending and expanding spiral.[38] He
remarked,

for this is one proof of the essential vitality of nature, that she does not ascend as links
in a suspended chain, but as the steps in a ladder; or rather she at one and the same time
*ascends* as by a climax, and expands as the concentric circles on the lake from the point
to which the stone in its fall had given the first impulse.[39]

## CONCLUSION

Nature had furnished Coleridge with the image of an ascending spiral, which
he used to organize and extend his knowledge, and then applied to the

construction of his philosophical and religious works.[40] And as he demonstrated in the *Biographia Literaria*, his philosophy was intimately connected with his literary criticism, central to an understanding of his aims as a poet.

To say this is to say no more than has often been said before, that Coleridge, in the organic unity of his thought, sought to achieve the Romantic goal of bringing all knowledge into harmony. Facts of nature were a touchstone for his metaphysics, just as scientists' discoveries of laws of nature helped him with the construction of critical, philosophical and religious systems of thought. Knowledge of nature validated other forms of knowledge. Since nature was for him symbolic of eternal truths, it is scarcely surprising to find, over and again, traces of imaginative insight in his philosophy, of philosophy in his theology and in his poetry, and of scientific lore in all. The symmetry that he found between Shakespeare's transmutation of nature into poetry, and Davy's realization of poetry in nature, was a symmetry informing his life's chief endeavour, the search for unity in multeity, the making whole of knowledge. And as facts of science were the common ground between him and those materialist thinkers with whom he most disagreed, his almost lifelong concern with the nature and progress of the sciences was fundamental and essential throughout his intellectual life.

## NOTES

1 T. Carlyle, *Life of John Sterling* (London, 1851), p. 73.
2 E.g. by T. L. Peacock, *Nightmare Abbey*, ed. H. F. B. Brett and C. E. Jones, in Peacock's *Works*, 10 vols. (London, 1924–34), III, p. 49.
3 O. Barfield, *What Coleridge Thought* (London, 1972).
4 *The Collected Works of Samuel Taylor Coleridge*, general ed. K. Coburn, Bollingen Series, 75 (Princeton and London, 1969–  ); *The Notebooks of Samuel Taylor Coleridge*, ed. K. Coburn, Bollingen Series 50 (Princeton and London, 1957–  ); *The Collected Letters of Samuel Taylor Coleridge*, ed. E. L. Griggs, 6 vols. (1956–71).
5 *Logic*, ed. J. R. de J. Jackson (London and Princeton, 1981), vol. XIII in *The Collected Works*.
6 *Ibid.*, p. xxxix.
7 To be published as vol. XV in *The Collected Works*.
8 H. J. Jackson, '"Turning and Turning": Coleridge on Our Knowledge of the External World', *Publications of the Modern Language Association* (Oct. 1986), p. 848.
9 T. H. Levere, *Poetry Realized in Nature. Samuel Taylor Coleridge and Early Nineteenth-Century Science* (Cambridge and New York, 1981).
10 *Collected Letters*, II, p. 1032.
11 W. Whewell, *Quarterly Review*, 51 (1834), 59–60.
12 D. A. Stansfield, *Thomas Beddoes M.D. 1760–1808* (Dordrecht, Boston, Lancaster, 1984). For Coleridge's biography, see W. J. Bate, *Coleridge* (New York, 1968).

13 *Collected Letters*, I, p. 209.
14 *Collected Letters*, II, p. 735. Coleridge's acquaintance with Davy is discussed in Levere, *Poetry Realized in Nature*, ch. 1.
15 This concept is the starting point for T. McFarland, *Coleridge and the Pantheist Tradition* (Oxford, 1969).
16 *The Collected Works*, vol. IV, *The Friend*, ed. B. Rooke, 2 vols. (1969), I, p. 470.
17 A useful discussion of Herschel in context is given in ch. 4 of D. Oldroyd, *The Arch of Knowledge* (New York and London, 1986).
18 British Library Egerton MS 2801 ff. 143–4.
19 *The Friend*, I, p. 94.
20 Levere, *Poetry Realized in Nature*, p. 96.
21 Coleridge, *Poetical Works* (London, 1967), p. 132.
22 *Ibid.*, p. 242.
23 *The Friend*, I, p. 471.
24 *Biographia Literaria* (first published London, 1817), in *The Collected Works*, VII, 2 vols. ed. J. Engell and W. Jackson Bate (1983), I, p. 34.
25 Coleridge, *Commonplace Book. On the Divine Ideas*, Huntington Library (San Marino, California) MS HM 8195.
26 *Ibid.*, f. 7. I have inserted Coleridge's revisions without special indication.
27 W. Blake, 'Auguries of Innocence'.
28 *Ibid.*, f. 9.
29 For the incorporation of historical thinking in science, the best introduction is still S. Toulmin and J. Goodfield, *The Discovery of Time* (London, 1965). Useful in the wider context of the history of ideas is M. Mandelbaum, *History, Man and Reason. A Study in Nineteenth-Century Thought* (Baltimore and London, 1971). For Schelling, Steffens and science, see Levere, *Poetry Realized in Nature*.
30 *Collected Letters*, II, p. 706.
31 Coleridge MS notebook 28 ff.12–13, in the British Library, quoted in Levere, *Poetry Realized in Nature*, p. 103.
32 *British Review*, 33 (1825), p. 486, reprinted in *Coleridge, the Critical Heritage*, ed. J. R. de J. Jackson (New York, 1970), p. 485.
33 Levere, *Poetry Realized in Nature*.
34 Coleridge, *Hints towards the Formation of a more Comprehensive Theory of Life*, ed. S. B. Watson (London, 1848).
35 D. Emmet, 'Coleridge on Powers in Mind and Nature' in J. Beer, *Coleridge's Variety* (London, 1974), pp. 166—82.
36 Levere, *Poetry Realized in Nature* pp. 114–21.
37 *Ibid.*, p. 209.
38 H. J. Jackson, '"Turning and Turning"'. I am indebted to this paper for the recognition of the spiral and of its importance in organizing Coleridge's picture of nature.
39 Coleridge, *Theory of Life* (London, 1848), p. 41.
40 H. J. Jackson, '"Turning and Turning"', p. 855.

## FURTHER READING

Coleridge deserves to be approached through his own writings. An initial approach could well be through *The Oxford Authors. Samuel Taylor Coleridge*, ed. H. J.

Jackson (Oxford, 1985). Those wishing to explore further in Coleridge's unpublished manuscripts should go to K. Coburn (ed.), *Inquiring Spirit* (London, 1951,
rep. Toronto, 1979). Indispensable for serious study is *The Collected Works of
Samuel Taylor Coleridge*, general editor K. Coburn, Bollingen Series, 75 (Princeton
and London, 1969–    ). Much of that monumental edition has now appeared, but
until it is complete readers will need to go to earlier editions, of which the *Complete
Works*, ed. W. G. T. Shedd (New York, 1853, rep. 1871, 1875, 1884), may be
recommended, and includes Coleridge's longest scientific or rather meta-scientific
work, *Hints towards the Formation of a more Comprehensive Theory of Life* (first
published London, 1848, ed. S. B. Watson). The best introduction to Coleridge's
thought is Owen Barfield, *What Coleridge Thought* (London, 1972). I have
discussed Coleridge in relation to science in my *Poetry Realized in Nature. Samuel
Taylor Coleridge and Early Nineteenth-Century Science* (Cambridge and New
York, 1981). W. J. Bate, *Coleridge* (New York, 1968) is the best modern biography.
A fine set of essays on many aspects of Coleridge is *Coleridge's Variety. Bicentenary
Studies*, ed. J. Beer (London, 1974).

# Nature's book: the language of science in the American Renaissance

## DAVID VAN LEER

Although American Romantics never toasted 'confusion to mathematics' as did their English counterparts, they generally seemed hostile to the ways in which Newtonian science undermined the spirituality of nature. Ralph Waldo Emerson rejected scientists as crypto-materialists whose 'spirit is matter reduced to an extreme thinness'. Edgar Allan Poe characterized science as a 'vulture' that 'preyest . . . upon the poet's heart'. And Nathaniel Hawthorne regularly portrayed the scientist as a villain whose relentless pursuit of some spurious ideal violated the inner sanctity of the human heart.

Yet these summary dismissals disguise the seriousness with which American authors confronted 'natural philosophy'. It is difficult, from our post-Freudian, post-Darwinian perspective, to imagine an intellectual environment in which neither of these figures is central. In general the American Romantics did not find in contemporary geological or evolutionary theories a congenial expression of their organicism, as did the English Victorians. They regularly, however, enlisted earlier theories of the physical sciences to reinforce their belief in the unity of nature: if the names of Lamarck, Lyell and the two Darwins figure but slightly in their accounts, Bacon, Newton, Laplace, Swedenborg and Paracelsus loom larger. And even when they did not welcome the new emphases science introduced, they were deeply concerned with the degree to which scientific thinking altered contemporary ways of talking.

In his pursuit of the spiritual at the expense of the material, Emerson may seem an unlikely spokesman for any form of science. The tone of his early masterpiece *Nature* (1836) is characteristic. From Mme de Staël Emerson learns that 'the axioms of physics translate the laws of ethics'.[1] Dissatisfied with mechanistic interpretations that reduce nature to 'carpentry and chemistry', Emerson turns to Kant's demonstration of the creative dimension of human perception (CW, I, 37). Nature's subservience to man suggests the 'noble doubt' that the world may not exist apart from his perception of it (p. 29). In such an anthropic context, 'empirical science is apt to cloud the sight, and, by the very knowledge of function and processes, to bereave the student of the manly contemplation of the whole'. Quite simply, 'the savant becomes unpoetic' (p. 39).

Emerson's relative disdain for science in *Nature*, however, does not fully represent the subtleties of his position. Natural philosophy is not the primary subject of *Nature*, or indeed of any of Emerson's most famous essays. In attempting to describe man's access to universal power, Emerson more characteristically turned to philosophy than to science, preferring Kantian epistemology to Newtonian physics.[2] Yet science plays a role in Emerson's larger metaphysical project, implicitly in *Nature*, and explicitly in the works that preceded this first published statement. The initial result of his decision to write a 'little book about nature' was a lecture series on science, delivered in late 1833 and early 1834, his first lectures after leaving the ministry in 1832. The topic was dictated in part by the interests of his audience. But Emerson shared those interests, and early in his career saw in scientific law a promise of power, only gradually supplanted by the greater promise of the Kantian Reason. In his journal a few months before the lectures, Emerson recorded his reaction to a visit to the *Jardin des Plantes* in Paris:

13 July. I carried my ticket from Mr Warden to the Cabinet of Natural History in the Garden of Plants. How much finer things are in composition than alone . . . Ah said I this is philanthropy, wisdom, taste – to form a Cabinet of natural history . . . Here we are impressed with the inexhaustible riches of nature. The Universe is a more amazing puzzle than ever as you glance along this bewildering series of animated forms, – the hazy butterflies, the carved shells, the birds, beasts, fishes, insects, snakes, – & the upheaving principle of life everywhere incipient in the very rock aping organized forms. Not a form so grotesque, so savage, nor so beautiful but is an expression of some property inherent in man the observer, – an occult relation between the very scorpions and man. I feel the centipede in me – cayman, carp, eagle, & fox. I am moved by strange sympathies, I say continually 'I will be a naturalist.'[3]

This ecstatic passage, one of Emerson's first attempts to discover a vocation to replace his discarded divinity, summarizes his early position on science. Science is 'philanthropic', more concerned with man than with a truly objective Other. The natural world is interesting only to the extent that it reveals unsuspected substrata in man, especially (in the phrase that echoes throughout the lectures and *Nature*) man's 'occult relation' to nature. But most important, in his notion of 'man the observer', Emerson reduces science to what he later called 'dedication to observation'. As he explains more fully in the lectures,

The state of mind which nature makes indispensable to all such as inquire of her secrets is the best discipline. For she yields no answer to petulance, or dogmatism, or affectation; only to patient, docile observation. Whosoever would gain anything of her, must submit to the essential condition of all learning, must go in the spirit of a little child. The naturalist commands nature by obeying her. (*EL*, I, 20)

By emphasizing the innocence of the scientist, Emerson does not simply adapt a familiar Romantic theme to empirical pursuits. He elevates the psychological effects of observation over the actual knowledge gained. And the process of experimentation itself becomes the result sought.

This focus on the process of science informs the whole lecture series. In the first and most important of the lectures, 'The Uses of Natural History', Emerson employs an argument for science's usefulness which will reappear as the first half (or 'lower argument') of *Nature*. Summarizing the 'advantages which may flow from the culture of natural science', he adds to his initial list of health, useful knowledge, delight and improvement of mind and character a fifth use: 'to explain man to himself' (*EL*, I, 23). He concludes:

Is there not a secret sympathy which connects man to all the animate and to all the inanimate beings around him? Where is it these fair creatures (in whom an order and series is so distinctly discernible,) find their link, their cement, their keystone, but in the Mind of Man? It is he who marries the visible to the Invisible by uniting thought to Animal Organization.

The strongest distinction of which we have an idea is that between thought and matter. The very existence of thought and speech supposes and is a new nature totally distinct from the material world; yet we find it impossible to speak of it and its laws in any other language than that borrowed from our experience in the material world . . . And this, because the whole of Nature is a metaphor or image of the human Mind. (*EL*, I, 24)

The 'secret sympathy' (a variant of the 'occult relation'), the marriage of the visible and the invisible, even the nascent idealism implied in the metaphoricity of the natural world: all are motifs explored more fully in *Nature*. What is interesting in this passage is the order, which differs significantly from *Nature*'s lower argument. In the essay, the uses of nature progress from commodity through beauty and language to discipline; this universal commitment to teaching man then suggests in the essay's second half the 'noble doubt' of idealism. In the lecture, however, language, not idealism, is the final goal; the 'science' of observation only flirts with idealism to establish thought and speech as 'a new nature totally distinct from the material world'. For all his images of relation and marriage, Emerson in fact depicts two unrelated worlds, of which the natural world, not a source of knowledge in itself, merely provides a language for talking about a spiritual world it does not understand.

In his early career, then, Emerson is primarily interested in ransacking science for terms and analogies that establish the insufficiency of the material world. In later years, however, Emerson's approach to science recognizes that the 'discipline' of science is not simply passive and preliminary. The mature essays regularly acknowledge science's role in overthrowing traditional authority, the admission that 'experiment is credible, antiquity is grown ridiculous'.[4] One of the earliest signs of this shift is Emerson's implicit preference for Swedenborg's science over his spiritualism in *Representative Men*. Yet the belief that science provides more than terminology is most evident in the late lecture series, published posthumously as 'Natural History of Intellect'.[5] His goal is the very traditional post-Kantian one of approaching philosophy with the mathematical precision of the sciences. In his enthusiasm for lectures on physiology by Robert Owen and on magnetism by Michael Faraday, Emerson wonders:

Could not a similar enumeration be made of the laws and powers of the Intellect, and possess the same claims on the student? Could we have, that is, the exhaustive accuracy of distribution which chemists use of their nomenclature and anatomists in their descriptions applied to a higher class of facts, to those laws, namely, which are common to chemistry, anatomy, astronomy, geometry, intellect, morals and social life; – laws of the world? (W, XII, 3–4)

The familiar assumptions remain. Morals and the intellect are still the laws of a 'higher' world. And the naturalist's language continues to teach the more significant language of mind. Yet now the 'usefulness' of science strikes Emerson less than its 'sufficiency'. Earlier physics merely translated morals; here it anticipates and even originates them: 'In short, the whole moral of modern science is the transference of that trust which is felt in Nature's admired arrangements, to the sphere of freedom and of rational life' (p. 87). Or, in a more direct statement cut by his editor, 'if natural philosophy is faithfully written, moral philosophy need not be, for it will find itself expressed in these theses to a perceptive soul' (W, XII, 426n).

Along with a new respect for the power of science comes a new humility about philosophy. Science is not simply a repository for facts but a model for philosophical inquiry. The very title of the series is suggestive. Before, natural history provided the intellect with terminology; now, intellect has itself a natural history. 'These powers and laws [of the world] are also facts in a Natural History. They also are objects of science and may be numbered and recorded, like stamens and vertebrae' (p. 4). If observation is still the aspect of science most relevant to philosophy, the limitation rests in philosophy's methodological naivety, not science's prosaic character. His new philosophic method is only 'natural history', comparable to Francis Bacon's catalogues at the beginning of modern science. Satisfied like Bacon merely to list phenomena, he offers not a unified theory of mind, but merely 'anecdotes of the intellect', a 'catalogue of natural facts' (pp. 11, 41). Evidently, however much it may admire modern achievements, the science of intellect is not yet ready to imitate the more advanced physics of Newton or Faraday.

Emerson's growing conception of science as a language rather than a dictionary is both incomplete and unselfconscious. Even as Emerson eschews

organization, he secretly searches for structuring principles. Nor is he willing
to push his best insights far enough. He accepts that 'every creation, in parts or
in particles, is on the method and by the means which our mind approves as
soon as it is thoroughly acquainted with the facts' (p. 4). Yet he does not
realize, as his reading in Coleridge should have taught him, that the mere
collection of facts presupposes a principle of selection structuring as it selects.
Nor does he see that these implicit principles of 'approval' can tell as much
about the psychological and cultural situation of man as Kant's categories.
Nevertheless his willingness to accept science as a form of knowledge represents an advance over his early subordination of science to epistemology. For,
as Emerson is beginning to suspect, science cannot provide spirit a vocabulary
without providing it a syntax as well.

Poe might seem no better a philosopher of science than Emerson.[6] Never like
Emerson a student of traditional metaphysics, Poe appears in his self-conscious Byronism and proleptic decadence the chief American spokesman for
poetic irrationalism, even 'obscurantism'. His 'Sonnet–To Science' is custom-
arily read, along with Keats's *Lamia*, as the quintessential expression of
Romantic anti-empiricism. And his stories regularly subordinate logic to a
Dionysian (even drug-induced) frenzy. Yet in fact Poe begins much where
Emerson ends, with an explicit recognition of the role of empiricism in any
future metaphysics. And in his writings, familiar and unfamiliar, he extracts
from contemporary science, especially cosmology, both a language and a plot
for his own fiction.

The sonnet 'To Science', far from a critique of empiricism, is in fact the
earliest sign of Poe's interest in science as a model of thought:

> Science! true daughter of Old Time thou art!
>     Who alterest all things with thy peering eyes.
> Why preyest thou thus upon the poet's heart,
>     Vulture, whose wings are dull realities?
> How should he love thee? or deem thee wise,
>     Who wouldst not leave him in his wandering
> To seek for treasure in the jewelled skies,
>     Albeit he soared with undaunted wings?
> Hast thou not dragged Diana from her car?
>     And driven the Hamadryad from the wood
> To seek a shelter in some happier star?
>     Hast thou not torn the Naiad from her flood,
> The Elfin from the green grass, and from me
>     The summer dream beneath the tamarind tree?[7]

Despite the disenchantment of the poem's persona, Poe himself is less interested in blaming science than in understanding its effect on future poetic
projects. True, science has somewhat brutally dragged Diana from her car.

But Poe is not especially bothered by this deposition. Diana, the Hamadryad, the Naiad and the Elfin were never the subjects of his poetry but of the neo-classical tradition against which he rebels. Rather than continue in self-indulgent 'wandering' through a 'jewelled' landscape, Poe wants to assimilate the lesson of change taught by Time's 'true daughter'. And even as the narrative voice whines, 'How should [I] love thee', Poe asks the question straightforwardly, serious in his choice of both verb and mood.

Unlike Emerson, then, Poe understands at the start of his career the centrality of science in any search for a new poetic language. He is less successful in writing a verse informed by that understanding. His early attempt in *Al Aaraaf*, the epic poem introduced by the sonnet in that early collection, presents not a fiction of science but simply science fiction. The setting, a remote star awaiting annihilation, incorporates some of the findings of contemporary astronomy and cosmology. But these facts serve merely as the backdrop for a conventional Neoplatonic account of primal innocence lost and regained, no more seriously scientific than costume drama is histori-cal. Only when, in the later prose, Poe confronts the logic of science as well as its content, does he begin to find a suitable replacement for the dethroned Diana. And only when he explicitly addresses the relation between this logic and his own poetic Neoplatonism, does he begin to write the kind of scientific literature to which the sonnet can stand as unironic preface.

The invention of the tale of ratiocination is the most famous result of Poe's interest in science. Although subsequent writers of detective stories take the genre in other directions, Poe emphasizes the logical, even mechanical, nature of detection. The Dupin stories, especially 'The Mystery of Marie Roget', unravel mysteries through a 'Calculus of Probabilities', and the application 'of the most rigidly exact in science . . . to the shadow and spirituality of the most intangible in speculation' (*CWP*, III, 724). As the detective Dupin himself explains, only the man who is both 'poet *and* mathematician' can reason well (*CWP*, II, 986). The hyper-rational language of these tales is, as Poe well knows, a pose. The success of the stories depends less on their content than on the form in which they present it, not on what they deduce but on their faith in logical analysis itself. As Poe admits to a friend, 'their method and *air* of method' make them seem more ingenious than they really are: 'where is the ingenuity of unravelling a web which you yourself (the author) have woven for the express purpose of unravelling?'[8]

The circularity of Poe's analytic method, however, does not distinguish him from the scientist. For Poe realizes, as Emerson did not quite, the hypothetical quality of all constructs, mathematical or literary. His language takes on a scientific tone not in empty mockery but because he sees science itself as primarily a way of talking: his fictional creations merely imitate the fictiveness of mathematics itself. As Dupin explains:

Mathematical axioms are *not* axioms of general truth. What is true of *relation* – of form and quantity – is often grossly false in regard to morals, for example. In this latter science it is very usually *un*true that the aggregated parts are equal to the whole . . . There are numerous other mathematical truths which are only truths within the limits of *relation*. But the mathematician argues, from his *finite truths*, through habit, as if they were of an absolutely general applicability . . . In short, I never yet encountered the mere mathematician who could be trusted out of equal roots, or one who did not clandestinely hold it as a point of his faith $x^2 + px$ was absolutely and unconditionally equal to $q$. (CWP, III, 985–6)

Poe's interest in science as a language and mathematics as a relation receives its fullest expression in his late work *Eureka* (1848). A cosmology heavily indebted to Newton's notion of universal gravitation and Laplace's nebular hypothesis of planetary formation, *Eureka* attempts 'to speak of the *Physical, Metaphysical and Mathematical – of the Material and Spiritual Universe: of its Essence, its Origin, its Creation, its Present Condition and its Destiny*' (WP, XVI, 185). The essay argues that '*In the Original Unity of the First Thing lies the Secondary Cause of All Things, with the Germ of their Inevitable Annihilation*' (pp. 185–6). In this account the history of matter involves an initial dispersion from and ultimate reintegration back to an indivisible atomic point of absolute unity. Although early scholars found this long work an anomaly, later critics have accepted Poe's evaluation of the essay as a masterpiece. And reversing early attempts to read its science allegorically, recent criticism has treated the work as a recognizable form of cosmological writing, roughly comparable to the post-Newtonian accounts of Colin Maclaurin, Alexander von Humboldt, John Nichol and the Bridgewater treatises. Opinion is still divided on Poe's success as cosmologist. Most agree, however, that the essay's attempt to find a place for God within a largely mechanistic model of the universe is well within the tradition of late-eighteenth-century physico-theology, best represented by the Rev. Thomas Dick.[9]

*Eureka*'s history of the unparticled particle obviously represents a more successful (and more convincingly scientific) version of the project to reconcile empiricism and Neoplatonism that has been with Poe since *Al Aaraaf* and 'To Science'. Yet the implications of this reconciliation are more evident in the essay's digressions than in its main account of the universe. Poe prefaces his analysis of the universe with a ten-page 'letter' from the future on the flaws of current scientific methodology. Written in very broad humour, the letter has troubled those who take seriously the cosmology of the later sections. Yet despite the bad puns and mean jokes of 'Pundita', the letter's author, this introductory section raises astute objections to contemporary philosophy of science. At issue is the epistemic status of scientific knowledge. Pundita attacks both the deductions of Aristotle ('Ram') and the inductions of Bacon ('Hog') to insist that true science is closer to the intuitions of Kepler. Her point is not the Romantic one that the heart intuits what the mind cannot know.

Instead she emphasizes the extent to which all science is a combination of the three methods: deduction requires some minimal knowledge of experience; induction presupposes categories of classification; and intuition is really an unconscious (even instantaneous) combination of the other two.[10]

Poe's primary concern is not science's method but its underlying notion of truth. Attacking Mill's definition of axioms, Pundita insists that '*no such things as axioms ever existed or can possibly exist at all*' (p. 192). His point is not metaphysical but logical. Mill rightly asserts that '"ability or inability to conceive . . . is *in no case* to be received as a criterion of axiomatic truth"' (p. 193).[11] But, in fact, even the most fundamental of 'axioms' presupposes the ability of conception as a criterion; nothing is necessarily true in an absolute sense, outside a particular conceptual system. The principle, for example, of the excluded middle – that a statement must either be true or not true – really means only that ' "we find it *impossible to conceive* that a tree can be anything else than a tree or not a tree." . . . That a tree can be both a tree and not a tree, is an idea which the angels, or the devils, *may* entertain, and which no doubt many an earthly Bedlamite, or Transcendentalist, *does*' (pp. 194–5).

In affirming the dependence of axioms on man's 'conceptions', Poe explicitly acknowledges that science takes place within a prior mental construct from which it can never escape. But the denial of axioms undermines more than the notion of scientific objectivity. It implicitly challenges the correspondence theory of truth. Angrily rejecting the dogmatism of both deductive and inductive approaches, Pundita concludes:

> that in spite of the eternal prating of their savans about *roads* to Truth, none of them fell, even by accident, into what we now so distinctly perceive to be the broadest, the straightest and most available of all roads – the great thoroughfare – the majestic highway of the *Consistent*[.] Is it not wonderful that they should have failed to deduce from the works of God the vitally momentous consideration that *a perfect consistency can be nothing but an absolute truth*? (pp. 195–6)

The phrase, repeated later in the essay, defines truth as simply that which is fully self-consistent (we now say 'coherent'). Rather than imitate or 'correspond to' some external, unverifiable 'reality', truth is that which answers with the greatest consistency most of the questions we now think to ask about a topic. Subsequent experimentation may undermine currently universal laws by introducing new facts and questions. A new more 'self-consistent' truth then will evolve, capable of addressing the new questions as well as the old.

Poe's belief in truth as coherence of course only partially anticipates the full-fledged pragmatism of William James and later philosophers. And it probably acknowledges too little the very quirks of personal preference that so influenced the early theories of his heroes Copernicus and Kepler. But whatever its metaphysical sophistication, Poe's sense of truth as consistency does explain his troublesome position about the relation of literature to truth.

In general, Poe favours form over content, labelling a preoccupation with theme as the 'heresy of *The Didactic*'. This emphasis does not make him an early proponent of art for art's sake, as the French sometimes claim. Sceptical about the creative power of the imagination, he ridicules the Romantic idea of the poet as prophet. And his hostility to the moralizing of some fiction is not a statement about truth in literature but one about truth in general.[12]

*Eureka* begins with an apparent denial of the truth-value of the work:

> I offer this Book of Truths, not in its character as Truth-Teller, but for the Beauty that abounds in its Truth; constituting it true . . . *What I here propound is true* . . . Nevertheless it is as a Poem only that I wish this work to be judged after I am dead. (p. 183)

This apparent preference for beauty over truth is misleading. The work regularly insists on its value as truth. Moreover, Poe increasingly denies the possibility of distinguishing between the two evaluative principles. Of Laplace's nebular hypothesis he says, 'From whatever point we regard it, we shall find it *beautifully true*. It is by far too beautiful, indeed, *not* to possess Truth as its essentiality' (p. 252). His point is one less of congruency than of identity. Like *Eureka* itself, Laplace's theory is not a truth accompanied by beauty, but known by its beauty as true. At least within the limits of human knowledge, its truth and its beauty are one. And so, at the end of the essay, he combines the ambiguities of his preface with the rhetoric of Pundita's letter to claim:

> It is the poetical essence of the universe – *of the Universe* which, in the supremeness of its symmetry, is but the most sublime of poems. Now symmetry and consistency are convertible terms: – thus Poetry and Truth are one. A thing is consistent in the ratio of its truth – true in the ratio of its consistency. *A perfect consistency, I repeat, can be nothing but an absolute truth.* (p. 302)

It is within the context of this theory of truth that we must understand Poe's emphasis on the craft of fiction. Beauty, like truth, is a matter of symmetry and inter-relation. And just as all post-Newtonian cosmologies subordinate universal law to divine power, so Poe's aesthetics of science imagine God as the only true poet:

> Not only is this Divine adaptation, however, mathematically accurate, but there is that about it which stamps it *as divine*, in distinction from that which is merely the work of human constructiveness. I allude to the complete *mutuality* of adaptation . . . The pleasure which we derive from any display of human ingenuity is in the ratio of *the approach* to this species of reciprocity. In the construction of *plot*, for example, in fictitious literature, we should aim at so arranging the incidents that we shall not be able to determine, of any one of them, whether it depends from any one other or upholds it. In this sense, of course, *perfection* of *plot* is really, or practically, unattainable – but only because it is a finite intelligence that constructs. The plots of God are perfect. The Universe is the plot of God. (pp. 291–2)

In the last ringing sentences, Poe does more than revive the familiar Puritan commonplace of America as 'God's plot'. For in making literature an imperfect form of natural history, he makes science a perfect form of storytelling. The aesthetic dimension of science is clearest in the work's treatment of the ether. Scientists have frequently posited the existence of an invisible universal ether to explain the mechanics of forces, first gravitational and later electromagnetic, not inherent in material bodies themselves. Poe rejects the ether less for empirical than for structural reasons:

Had an end been demonstrated, however, from so purely collateral a cause as an ether, Man's instinct of the Divine *capacity to adapt*, would have rebelled against the demonstration. We should have been forced to regard the Universe with some such sense of dissatisfaction as we experience in contemplating an unnecessarily complex work of human art. Creation would have affected us as an imperfect *plot* in a romance, where the *denoument* is awkwardly brought about by interposed incidents external and foreign to the main subject; instead of springing out of the bosom of the thesis – out of the heart of the ruling idea – instead of arising as a result of the primary proposition – as inseparable and inevitable part and parcel of the fundamental concept of the book. (p. 306)

His conclusion – that, even if necessary, the ether must be rejected as a *deus ex machina* – does not reveal a preference for poetry over science. It suggests that extent to which scientific explanation has always, in its preference for the 'simple' or 'elegant' demonstration over the convoluted, valued the aesthetic as one of its unexamined 'conceptions'. And if all cosmologists knew Lucretius to be a poet, Poe is the cosmologist who most fully recognizes that Occam is one as well.

And what of Hawthorne? Emerson comes to accept science as the language of experience; Poe to celebrate plot and law as two versions of the same universal relation. Hawthorne never rises above his suspicions of a discipline that subordinates the individual to the universal. Nor do all readers find thematic significance in his settings. Even in his most scientific work, the short story 'The Birth-mark', the protagonist Aylmer's attempt to eradicate the facial blemish of his otherwise perfect wife Georgiana has been read apart from Hawthorne's intellectual presuppositions: as a nobly Romantic pursuit for perfection; an unpardonably sinful misuse of one individual by another; and (most recently) an unconsciously chauvinist exploitation of the female by both husband and author.[13] Yet the specificity of the setting suggests at least that these moral issues may be read in terms of Aylmer's science. And his willingness to consider the linguistic and epistemological implications of Aylmer's method as well as its morality shows Hawthorne no less insightful a student of science than Emerson or Poe.

Hawthorne's emphasis is clear from the opening sentence: 'In the latter part of the last century, there lived a man of science – an eminent proficient in every branch of natural philosophy – who, not long before our story opens, had made experience of a spiritual affinity, more attractive than any chemical one.'[14] The moral thrust of the story is implied in Aylmer's insensitive view of his love and subsequent marriage as an 'experience' (that is, 'experiment'). More interestingly, however, virtually every word in the sentence imitates the language of the scientific world it depicts. So, thoughout the story, Hawthorne describes Aylmer's moral problems in the terms of his empiricist philosophy.

Both Aylmer and Georgiana attribute his character to his scientific development. Aylmer offers as self-explanation 'a history of the long dynasty of the Alchemists' (p. 46). To understand this tradition further, and more critically, Georgiana examines the books in her husband's library.[15] She rejects the alchemists for imagining themselves 'to have acquired from the investigation of nature a power over nature, and from physics a sway over the spiritual world' (p.48). Scarcely more impressive are the *Philosophical Transactions*, for the early virtuosi, 'knowing little of the limits of natural possibility, were continually recording wonders, or proposing methods whereby wonders might be wrought' (p. 48). Georgiana's evaluation of her husband's own laboratory notebook appears positive: 'he handled physical details, as if there were nothing beyond them; yet spiritualized them all, and redeemed himself from materialism, by his strong and eager aspiration towards the infinite. In his grasp, the veriest clod of earth assumed a soul' (p. 49). This last sentence, however, epitomizes the case against Aylmer. His willingness to overinterpret Georgiana's mole, to invest all clods with soul, strikes most readers as foolhardy. And the story seems a critique of science's pseudo-religious tendency to spiritualize the material.

Yet 'The Birth-mark' is not a simplistic attack on Aylmer's inadequate spirituality, a dramatized version of Emerson's claim that science's spirit is only 'thin matter'. Instead it examines more generally the relation of Nature's processes to those of the scientist. Early in the story the narrator, probably echoing Aylmer's thoughts, explains why he gave up looking for the secret of life:

The latter pursuit, however, Aylmer had long laid aside, in unwilling recognition of the truth, against which all seekers sooner or later stumble, that our great creative Mother, while she amuses us with apparently working in the broadest sunshine, is yet severely careful to keep her own secrets, and, in spite of her pretended openness, shows us nothing but results. She permits us indeed, to mar, but seldom to mend, and, like a jealous patentee, on no account to make. (p. 42)

Evidently Nature's laboratory is shut to man. In trying to understand Nature, Aylmer becomes a Romantic over-reacher, whose 'spirit was ever on the march – ever ascending – and each instant requir[ing] something that was beyond the scope of the minute before' (p. 52). But this negative reading ignores one of the basic structural elements of the story: Georgiana stands to Aylmer as Aylmer does to Nature herself, forbidden to 'pry' into the secrets of his laboratory. And our sympathy for Georgiana's feelings of exclusion – our exhilaration when she finally does gain access to the lab – suggests that Aylmer's error is not simply, as the narrator suggests, his desire to know secrets.

The amorality of Aylmer's science only suggests its more fundamental logical incoherence. Science, says the narrator in the opening paragraph, 'seemed to open paths into the region of miracle', a project taken up later in the credulous virtuosi's 'proposing methods whereby wonders might be wrought'. The futility of such hopes does not delineate a Romantic chiasm between expectation and reality: the idioms of miracle and law are simply mutually exclusive. Scientific law attempts to explain the conditions under which a particular cause will have a particular effect. Miracle suggests that a given event has no natural cause. Supernatural miracles are by definition inaccessible to natural standards of evaluation. Thus the virtuosi's idea of a 'method' to achieve wonders is semantically impossible: any wonder so methodically 'wrought' would by definition not be 'wonderful'.

The claim, then, that early scientists misunderstood the 'limits of natural possibility' may imply the extent to which the meaning of 'natural' is itself open to question. Poe suggests that 'natural' really means 'humanly conceivable' and is thus in some senses more ideal than real. Hawthorne reapplies this general notion to suggest paradoxically that the concept of natural law is itself unnatural, lying not in nature but outside it. To the extent that laws are taken as ontologically real, science is forced into questions about the location of its properties, the very sort of 'where is gravity' debate that Newton so studiously avoided. To the extent, however, that laws are meant only as logical short-hands and structural conveniences, science is confronted with Emersonian 'noble doubts' about what difference is made by belief in an objective reality.

The uncertain epistemic status of law is compounded by an equal uncertainty about experimentation. Georgiana and the narrator agree that scientists are wrong to think the investigation of nature provides a power over nature. At the same time, it is hard to determine from Aylmer's example exactly what experimentation can offer. Most of his experiments are mere displays of power, attempts to convince Georgiana of his ability to cure her by demonstrating other fantastic tricks. To this extent his experiments do recall Christ's miracles; according to the 'higher criticism' of nineteenth-century theologians, both are minimally relevant magical feats offered as proof of powers themselves not illustrable. The more sanctioned use of experimentation – to check after the fact the validity of a hypothesis – is less common in the story. Georgiana explains this absence when she rejects one of the few such experiments – Aylmer's curing of the diseased geranium to illustrate the power of the liquid that will purify Georgiana's cheek: '"There needed no proof . . . I joyfully stake all upon your word"' (p. 53). Her point is not simply that her trust knows no limits. It is also that any such proof is only approximate; until she actually takes the potion, nothing can be known about its effect on her.

Georgiana's scepticism, and Hawthorne's behind it, identifies a troublesome gap between science and nature: she stakes all on Aylmer's 'word' because there is no other reality on which to rely. Even before science generalizes its findings into law, it translates events into experiments. The framing process that singles out an event as significant, or even identifies certain characteristics as an event, is itself an act of interpretation one step removed from reality. And in this remove lies the real failure of Aylmer's laboratory notebook. The problem is less that his achievements fall short of his ideals than that his record is written. Every account of nature is already a story; the more interpretable the story, the more interpretation stands between text and reality. This gap may even explain the narrative's odd reticence, its unwillingness to define or at times even describe the situation. The narrator concludes that Aylmer should have 'woven his mortal life of the self-same texture with the celestial' (p. 56). But the narrator's unwitting use of the loaded word 'texture' reveals the limitations of his moral. Choosing a moral for a story, like positing a law for an experimental result, is the very act of interpretation that Hawthorne suspects. There is no absolute truth apart from interpretation, no 'texture' in which a 'text' is not embedded. Nature, we are told early, shows 'nothing but results'. Any redefinition – even so elementary a one as that of 'effects' for 'results', on which science depends – involves an unverifiable interpretation. The relation between Georgiana's mark and Aylmer's symbolic interpretation only illustrates the more problematic relation between every occurrence and its redefinition as event. And the real interpretive gap rests not between Aylmer's results and his ideals but between Nature and her 'book'.

Hawthorne's scepticism about science is merely a corollary of his more general problems with interpretation. But in his cursory identification of certain epistemological paradoxes in nineteenth-century science, he in some ways understands science better than either Emerson or Poe. Emerson's fairly traditional enthusiasm takes him no further than considerations of science as a language for talking about intellect. Poe understands that apparently comparable languages are in fact identical. However much experience is subdivided into matter and spirit, truth and poetry, cause and effect, there is really only a single language of relation or 'plot'. Taking Poe's perception one step farther, Hawthorne refuses to stand outside controlling the unity of cause and effect. His 'twice-told' tales offer inter-relation as texture; in his world plots are imperfect not because man is not so skilful as God, but because unlike God he is necessarily inside discourse. Whatever his opinion of empiricism, by writing in the logic of science as he writes in its language, Hawthorne evinces a better understanding of science's role in modern epistemology than many more sympathetic with its immediate goals.

## NOTES

1   *The Collected Works of Ralph Waldo Emerson*, ed. Robert Spiller, Alfred Ferguson *et al.*, 4 vols. to date (Cambridge, Mass., 1971–    ), I, p. 21. Hereafter this edition will be cited as 'CW' parenthetically in the text. In the lecture passage that stands behind this sentence the subordination of science to ethics is even clearer: 'It is a most curious fact that the axioms of geometry and of mechanics *only* translate the laws of ethics' (emphasis added). See *The Early Lectures of Ralph Waldo Emerson*, ed. Stephen Whicher, Robert Spiller and Wallace E. Williams, 3 vols. (Cambridge, Mass., 1959–72), I, p. 25. Hereafter this edition will be cited as 'EL' parenthetically in the text. For the source in de Staël, see *Germany*, 3 vols. (London, 1813), III, p. 151 (cited in CW, I, 249).

2   For a fuller account of the scientific sources of *Nature*, see B. L. Packer, *Emerson's Fall: A New Interpretation of the Major Essays* (New York, 1982), esp. pp. 73–81. In some senses, even the essay's most notorious metaphor, that of the 'transparent eye-ball', alludes to this Newtonian influence. For more general treatments of Emerson's relation to science, see Harry Hayden Clark, 'Emerson and Science', *Philosophical Quarterly*, 10 (1931), 225–60, and Leon Chai, *The Romantic Foundations of the American Renaissance* (Ithaca, N.Y., 1987), pp. 141–55. I have discussed the centrality of philosophy, and especially post-Kantian epistemology, to Emerson's major essays in *Emerson's Epistemology: The Argument of the Essays* (Cambridge, 1986).

3   *The Journals and Miscellaneous Notebooks of Ralph Waldo Emerson*, ed. William Gilman *et al.*, 16 vols. (Cambridge, Mass., 1960–82), IV, pp. 198–200. The passage is incorporated almost verbatim into the first of the science lectures (EL, I, 1). Brief phrases from the journal description also surface in *Nature* (for example, CW, I, 39).

4   *The Complete Works of Ralph Waldo Emerson*, ed. Edward Waldo Emerson, 12 vols. (Boston, 1903–4), X, pp. 329; see also pp. 130–1, 335–6. Hereafter this edition will be cited as 'W' parenthetically in the text.

5   As with most of the late works, it is difficult to determine the role of Emerson's editor James Elliot Cabot in constructing the 'Natural History'. Emerson delivered at least three series of lectures on this or related topics between 1848 and 1870.

6   In general, treatments of Poe's relation to science focus on one specific aspect, such as his relation to mesmerism or evolution. For a good overview, see Chai, *Romantic Foundations*, pp. 103–41. For general intellectual histories that touch on Poe's relation to science, see Margaret Alterton, *The Origins of Poe's Critical Theory* (Iowa City, 1925); Killis Campbell, *The Mind of Poe and Other Studies* (Cambridge, Mass., 1933); and Edward H. Davidson, *Poe: A Critical Study* (Cambridge, Mass., 1957).

7   *Collected Works of Edgar Allan Poe*, ed. Thomas Olive Mabbott, 3 vols. to date (Cambridge, Mass., 1969– ), I, p. 91. Henceforth references will cite this edition as 'CWP' parenthetically in the text.

8   *The Letters of Edgar Allan Poe*, ed. John Ward Ostrom (Cambridge, Mass., 1948), II, p. 328. For related discussions of the insufficiency of mathematics, see CWP, II, 527–31; and *The Complete Works of Edgar Allan Poe*, ed. James A. Harrison (New York, 1902), XIV, pp. 6–37. Hereafter this edition will be cited as 'WP' parenthetically in the text.

9   For a good overview of current cosmological theory, see Stephen W. Hawking, *A Brief History of Time: From the Big Bang to Black Holes* (New York, 1988). For Poe's place in this tradition, see Richard P. Benton (ed.), *Poe as Literary Cosmologer: Studies on Eureka, a Symposium* (Hartford, 1975); and Harold Beaver (ed.), *The Science Fiction of Edgar Allan Poe* (Harmondsworth, 1976), pp. 395–403. The fullest treatment of Poe's relation to Thomas Dick is in Alterton, *Origins*, pp. 112–31, 138–41.

10   Poe's position is perfectly consistent wtih the scientific theory of his time. See, for example, John F. W. Herschel, *A Preliminary Discourse on the Study of Natural Philosophy* (1830; repr. Chicago, 1987), pp. 164–75. Poe knew this work, at least indirectly in his reading of John Nichol.

11   For the probable reference, see John Stuart Mill, *A System of Logic Ratiocinative and Inductive* (1843), Bk II, ch. 5, sect. 6, ed. J. M. Robson and R. F. McRae (Toronto, 1973), VII, pp. 238–51. Poe's quotations do not seem verbatim, and his general point does not so much disprove Mill's own argument as take the same argument in a different direction.

12   For the famous passage on the 'heresy of *The Didactic*', see 'The Poetic Principle' (WP, XIV, 272). For Poe's oft-repeated attack on the Romantic notion of the poet as creator, and specifically on Coleridge's distinction between the fancy and the imagination, see WP, XII, 37; XV, 13n.

13   For a general study of Hawthorne's knowledge of science, see Taylor Stoehr, *Hawthorne's Mad Scientists: Pseudoscience and Social Science in Nineteenth-Century Life and Letters* (Hamden, Conn., 1978). This work, however, treats 'The Birth-mark' only briefly. For an influential early account of Aylmer's nobility, see Cleanth Brooks, Jr, and Robert Penn Warren, *Understanding Fiction* (New York, 1943), pp. 103–6. Most readings that view Aylmer as artist see him as noble to the degree that he realizes his aesthetic ideal. The best general attack on his character is R. B. Heilman, 'Hawthorne's "The Birthmark": Science as Religion', *South Atlantic Quarterly*, 48 (1949), 575–83. For a feminist reading, see Judith Fetterley, *The Resisting Reader: A Feminist Approach to American Fiction* (Bloomington, 1978), pp. 22–34.

14   *The Centenary Edition of the Works of Nathaniel Hawthorne*, ed. William Charvat *et al.*, 18 vols. to date (Columbus, 1962– ), X, 36. All references to Hawthorne's 'The Birth-mark' will cite parenthetically the appropriate page in the tenth volume of this edition.

15   This famous passage has received much attention as a clue to Aylmer's intellectual sources. See, for example, Alfred S. Reid, 'Hawthorne's Humanism: "The Birthmark" and Sir Kenelm Digby', *American Literature*, 38 (1966), 337–51. I have profited as well from Charles Swann's commentary on similar passages in *Septimus Felton*, 'Alchemy and Hawthorne's *Elixir of Life Manuscripts*' (forthcoming).

# The shattered whole: Georg Büchner and Naturphilosophie

## JOHN REDDICK

It is curiously apt that Georg Büchner should bring up the rear in this varied parade, and as the tailpiece in a section inaugurated by Goethe. Goethe was the towering giant of his age, a figure with whom Büchner might seem to bear no comparison. He lived to be eighty-two, wrote vigorously for over sixty years, published Complete Edition after Complete Edition, each fatter than the last, and long before his death had become the awesome high priest of a massive temple of German classicism that he himself had very largely created. In this perspective, Büchner is a figure both puny and profane. He died at twenty-three (an age at which Goethe had not even produced *Werther*); he left the barest handful of texts; and he impinged scarcely at all on the consciousness of the century in which he so briefly lived. Not for him the succession of definitive editions, the Eckermanns eager to immortalize each crumb of wisdom from his mouth. His *œuvre*, already slender enough, was further decimated by the disappearance, probably even the physical destruction, of an entire play – the putative *Pietro Aretino* – along with all his diaries, and many if not most of his letters. Just one of his writings was published in his lifetime in authentic and definitive form: his doctoral dissertation, in French, on the cranial anatomy of an obscure fish. For the rest, his work survives only in a more or less conjectural, fragmentary or reconstituted form; and even today, a century and a half after his death, we wait in exasperation for a proper historical–critical edition.

In sheer stature, then, Büchner is utterly dwarfed by the monumental figure of Goethe. But monumental stature can exact a heavy price. Goethe looms on his plinth like Nelson on his column, but he is equally frozen and remote. Whatever his true merits, whatever quick vitality lies trapped within the cold stone, he has acquired the status of a curiosity, a monument greatly respected but largely ignored. With Büchner it is quite the reverse. This slender, provocative, sharp-edged figure lives more vitally amongst us than ever before. No other German writer before Kafka and Brecht so vividly catches the modern imagination. It is an extraordinary phenomenon that, whereas the nineteenth century was almost wholly deaf to this man's voice, he seems to speak to us now with 'incendiary' force (in the words of Günter Grass)[1] and 'remarkable relevance' (according to Heinrich Böll)[2] as if he were alive and well in Munich or Berlin. Political protesters in the Federal Republic have daubed the war-cry of *The Hessian Messenger* across a thousand banners and squatted houses: 'Peace to the people! War on the oppressors!' ('Friede den Hütten! Krieg den Palästen!') His plays are regularly staged in the two Germanies and abroad. Werner Herzog scored a particular success with his film of *Woyzeck* (and chose a quotation from *Lenz* as the epigraph for *The Enigma of Kaspar Hauser*). In the East German eyes of Christa Wolf, 'German prose begins with Büchner's *Lenz*'; it is her 'absolute ideal', her 'primal experience' ('Ur-Erlebnis') in German literature.[3] And we should not forget the revolutionary and revelationary impact that the newly discovered Büchner had on earlier writers: on Gerhart Hauptmann and his fellow Naturalists; on the German Expressionists; above all, perhaps, on Wedekind and Brecht.

Undisputed though the strength and immediacy of Büchner's voice may be, however, there are wild and bitter disputes about what that voice is actually saying. This is not surprising, for several factors conspire to make him a natural focus of controversy. Most obviously, there is the sheer smallness of scale and uncertain state of his output. Imagine the jousting ground that would have been afforded to critics if Goethe had left only *Götz, Urfaust* and *Werther*, let us say, and if these had survived only in scrawled, incomplete, often illegible manuscripts, or in printed versions that were variously mutilated, bowdlerized or garbled, as well as being largely posthumous and wholly unauthorized. Then there is the richly provocative nature of his concerns. Sex, religion, politics, those taboo topics amongst all decent folk, are amongst his most urgent preoccupations. From the very first lines of *Danton's Death* with their image of the pretty lady who gives her heart to her husband and her cunt to her lovers, Büchner's 'obscenities' have ensured him the status of *enfant terrible*, and in the process have served to betray the blinkered perspective of countless critics. Gods, God and spirits are insistently invoked by his characters, to be denied, condemned, entreated – and thus to serve as a constant challenge to believer, agnostic and atheist alike. Büchner's politics are of course an especially fulminant issue. Here is a man who was one of the most radical left-wing thinkers of his age within the German lands, a proto-Marxian revolutionary who, although he entered the fray as a political publicist and activist for only the briefest of periods, remained wholly committed throughout his life to the overthrow of what he saw as an illegitimate, parasitical and effete ruling class, and the resurgence and emancipation of the popular masses. Given the paucity both of direct evidence, such as Büchner's correspondence, and of indirect evidence, such as reminiscences

of friends and acquaintances, police files, court records, there is much scope
for argument even about his precise activities and stance within the political
micro- and macro-realities of the time. The fiercest controversy, however, is
inevitably provoked by his writing. At one extreme is the view exemplified by
Georg Lukács: Büchner as an unswerving Jacobin essentially unaffected by
the grim fiasco of *The Hessian Messenger*: 'Büchner was at all times a rigorous
revolutionary.'[4] At the opposite extreme, there is the view exemplified by
Robert Mühlher: Büchner as a man whose abrupt and bitter insights propelled
him into 'extreme or absolute nihilism', and in the process depoliticized him
and 'thrust him for ever from the liberal and democratic camp'.[5]

The scant and uncertain status of the texts, the inflammatory nature of the
issues they contain: these features of Büchner's work are themselves conducive to controversy. But their effect is greatly compounded by a third decisive
element: the *manner* of Büchner's art, the language, modes and structures that
he uses to express his concerns. For the flickering image of the world that he
evokes is profoundly un- and anti-classical, and consciously remote from the
prevailing conventions and expectations of his age. Whether in language,
mood, plot or personae, he offers no steady development, no sense of anything
resolved or unified. Instead of unfolding in measured rhythm, his works
progress through a succession of kaleidoscopic convulsions, enacting what
Walter Jens has called a 'law of discontinuity'.[6] It is *particles* that loom large:
discrete elements that he highlights in startling isolation, or in disparate
clusters and combinations that create a constant sense of paradox,
multivalence and mystery. This is a chief mark of his spectacular modernity:
already in the 1830s he is doing the kind of thing that will seem outrageously
new when practised by the most avant-garde painters, composers and writers
of the early twentieth century. But it also makes him especially difficult to
interpret. In particular, it entails the problem of perspective: being so disparate and discrete, the elements in his work change their aspect and apparent
importance quite radically when viewed from different vantage points.

How are we to deal with this systematic discontinuity? As a first step,
perhaps, we need to take it seriously. This might seem an easy and obvious
measure, but it has eluded a great many critics.

The most unsubtle way of not taking it seriously is that favoured by certain
critics in the English tradition, who have patronizingly applied the yardstick
of good old common sense, and have typically declared his 'disunity' to be a
'symptom' of 'philosophical and . . . poetic–dramatic immaturity'.[7] This plain
man's approach is severely reductive: the more challenging a complexity, the
more likely it is to be branded a defect or mistake – a tendency hair-raisingly
exemplified when A. H. J. Knight touches on the Marion episode, one of the
most powerful and extraordinary moments in the mosaic of Büchner's work,
and baldly dismisses it as 'contribut[ing] nothing to the theme of the play'.[8]

A more subtle and far more common way of failing to do justice to Büchner's disjunctive and paradoxical mode is to behave as though it did not exist. It is all too easy to don spectacles of this hue or that, and to believe that the particular pattern that they reveal is the only one, or the only one that matters. The basic trouble perhaps is that critics have traditionally been the products of academe who were schooled chiefly or wholly in the traditions of classicism. We are accustomed to seeing works of literature as programmatic and exemplary, as vehicles purpose-built to embody and demonstrate an already fully developed view or ethos. We recognize in the late plays of Schiller, for example, a magnificent complexity, but a complexity like that of a baroque fugue with its rich and balanced elaboration of lucidly stated themes. Such an approach can only be reductive to the point of distortion when applied to Georg Büchner. He never writes to communicate conclusions. Instead, his writing is a kind of happening, a constant search, a dynamic enactment of the very process of argument, of the collision and interaction of contrary possibilities. His works begin, but never at a beginning, and they come to an end, but not to a conclusion. This means that we should never be tempted to seize on a particular discrete element and single it out as a summation of the whole, or as the definitive fixing of a position – though many critics have done so, hence the persistent misrepresentation of Büchner as being variously a programmatic pessimist and nihilist, a programmatic fatalist, a programmatic Christian, a programmatic Jacobin revolutionary. There is indeed an underlying consistency and unity in Büchner; but it will be found only *within* and *through* the paradoxes and multiplicities of his work – not despite them.

It helps for us to recognize what is perhaps the paradox of paradoxes in Georg Büchner: his disjunctive mode with its relentless insistence on fragments and particles is always the expression of a radiant vision of *wholeness*. Again and again, in every area of his existence – his politics, his science, his aesthetics, his art – we find an ardent sense of wholeness, but a wholeness that is almost always poignantly elusive: it *was* but is no longer; or *will* be but isn't yet; or – most poignant of all – it *is* in the present, but can be perceived or possessed only partially or transiently. Büchner is thus forced to be a maker of mosaics. But the more jagged the fragments in these mosaics, the more strident they are in their invocation of the whole – a pattern that is perfectly thematized and epitomized in the earliest pages of his work when he has Lacroix define the quest of Danton amongst the whores in just such terms: 'He's trying to get the Medici Venus together again piece by piece, he's making a mosaic, as he puts it', 'It's a crying shame that nature has broken beauty into pieces and stuck it in fragments like that into different bodies' (I. 20f.).[9] Minutes later the theme is echoed and intensified in Danton's yearning response to Marion with its double stress on 'totality': 'Why can't I take your beauty wholly into myself, wholly enfold it?' (I. 22). In *Leonce and Lena* it is the wholeness of love that is fragmented: split asunder into the separate colours of a rainbow. But as always the emphasis on fragments implies the conviction of wholeness – which is made explicit here in the image of love *beyond* the differentiated spectrum of the rainbow: as a single shaft of white-hot radiance ('de[r] weiss[e] Gluthstrahl der Liebe'; I. 112). And it is precisely Leonce's experience of a love-inspired totality of being that is so intensely celebrated in the fleeting climax of the play: 'All my being is in this single moment' (I. 125).

The force and central importance of Büchner's vision of wholeness becomes clearer still when we realize that it also lies at the heart of his work as a scientist–philosopher. Büchner spells this out in his Trial Lecture 'On Cranial Nerves', which he delivered on 5 November 1836 (less than four months before his absurdly early death from typhus), and which – astonishingly – he must have written at the very same time that he was working on *Woyzeck*. Trial Lectures were a ritualistic affair, not unlike the modern Inaugural except that they constituted a final hurdle *before* the victim's confirmation in a teaching post, in Büchner's case as a *Privatdozent* in comparative anatomy at the brand new University of Zurich. They encouraged a contender to demonstrate his stance as well as his standing; and with an audience of dignitaries that included Lorenz Oken, the university's founding *Rektor*, and one of the most influential and most controversial scientist–philosophers of the age within the German lands, Büchner goes to considerable lengths to define his general standpoint and frame of reference, before launching into his particular argument concerning the skull. And Büchner puts a quite remarkable emphasis in these prefatory pages on his sense of the natural world as an *organic whole* characterized by *order, proportion, unity* and essential *simplicity*. The study of the natural world, he says (II. 293), has taken on a new shape. Previously, botanists and zoologists, physiologists and comparative anatomists had been confronted by a monstrous chaos of data – 'a huge mass of material, laboriously heaped up over the centuries, that had scarcely even been systematically catalogued', 'a confusion of weird forms under the wildest names', 'a mass of things that previously weighed heavily on one's memory as so many separate, unconnected facts'. But 'major progress' has at last been made, and the chaos and confusion have resolved in consequence into 'simple, natural groups' having the 'most beautifully even proportion'. The essential thrust of this new understanding, in comparative anatomy as in the various kindred subjects, was towards a kind of unity, with all the forms being traced back to the supremely simple primordial type, or archetype, from which they were developed ('In der vergleichenden Anatomie strebte Alles nach einer gewissen Einheit, nach dem Zurückführen aller Formen auf den einfachsten primitiven Typus').

As we might expect, Büchner tells us that while the whole picture in all its richness is not yet fully understood, coherent parts of it have nevertheless taken shape ('Hat man auch nichts Ganzes erreicht, so kamen doch zusammenhängende Strecken zum Vorschein'). In the similar but more vivid image conveyed earlier in the same paragraph, 'Even though one had not found the well-spring, there were nevertheless many places where one could hear the river roaring down below.' This has a familiar ring, for it echoes the kind of pattern generated in the poetic writing: there, too, the river is never reached, but its roaring can be heard. We have only the fragments of a mosaic, the separate notes of the scale, the scattered colours of the spectrum; but they imply and betoken a vibrant if elusive whole.

The supreme importance to Büchner of this sense of wholeness is evinced even more remarkably a little earlier in the Trial Lecture (II. 292). What is it that paved the way for this new understanding of the physical world? It is the fundamental postulate that all things in nature are part of a *single organic complex*, a 'gesammte Organisation', and that this rich complex is governed and patterned according to a *single natural law*, a 'Grundgesetz'. Büchner had already begun to define his scientific–philosophical credo in the brief last paragraph of his doctoral dissertation earlier in 1836: his conviction that the grand richness of nature is not due to any kind of arbitrary functionalism, but is the elaboration of a design or 'blueprint' of supreme simplicity:

La nature est grande et riche, non parce qu'à chaque instant elle crée arbitrairement des organes nouveaux pour de nouvelles fonctions, mais parce qu'elle produit, d'après le plan le plus simple, les formes les plus élevées et les plus pures. (II. 125)

This is directly echoed, and intensified, in the Trial Lecture: the 'Grundgesetz', the all-informing law of nature, is a 'primal law [*Urgesetz*] that produces the highest and purest forms according to the simplest patterns and designs'. On this view, everything – form, matter, function – is governed by the one law ('Alles, Form und Stoff, ist . . . an dies Gesetz gebunden'; 'Alle Funktionen sind Wirkungen desselben [Gesetzes]'). The astonishingly positive nature of Büchner's stance is radiantly clear in these lines. The 'primal law' that he sees as the matrix of all things is to him nothing less than a 'law of *beauty*' ('ei[n] Geset[z] der Schönheit'). And since this law is so benign, and since all things in nature – functions as well as form and matter – are generated by it, so the myriad workings of nature not only never conflict with each other, they interact with one another to yield a *'necessary harmony'* ('nothwendige Harmonie').

This paean of faith in a universal order of rich simplicity, engendered by beauty and resonant with harmony, can seem bewildering indeed, coming as it does from the pen of a man whom critics of different eras and very different persuasions have been wont to describe as an exponent of extreme nihilism.

Even when we disregard this inveterate but now largely discredited critical tradition, we are still faced with the strident paradox within the texts themselves: the beauteous harmony and order postulated with such faith and confidence in the Trial Lecture – and the bleak visions so frequently and so eloquently evoked in the poetic work: the terrifying cold isolation of Lenz at the end of the story, and of the child in the un-fairytale in *Woyzeck* (I. 100f.; I. 151); the famous, or infamous, cry of Danton that the world is chaos, and nothingness its appropriate deity (I. 72); the fear of Leonce that all we see may be mere imaginings masking a reality of blank, bare vacuity (I. 118). Such examples could be multiplied.

Again we face the problem of how to cope with the unremittingly paradoxical nature of Büchner's writing; and once again I would insist that we can only begin by taking it seriously. It evades the issue to suppose, as Knight did, that Büchner was simply changing his ground in the last phase of his life, and moderating from his alleged 'total pessimism'.[10] Hans Mayer does not get us much further when he claims that there is a regrettable discrepancy, a 'dissonance', between Büchner's (radical) view of society and his (conservative) view of nature.[11] And it is quite mistaken, I believe, to allege that the radiant faith expressed in the Trial Lecture is some kind of bogus remedy, a 'nostrum', as J. P. Stern has asserted, hastily contrived 'to repair the shattered fabric of existence'.[12] Büchner's faith in abundant, vibrant wholeness was not a sudden new stance, nor a strange aberration, nor a convenient refuge in adversity: it was fundamental to his existence and to all his doings; and even the most raucous anguish in his writings – *especially* the most raucous anguish – is always born of it.

One of the most extraordinary paradoxes in Büchner is that, whereas the *manner* of his poetic writing is inexorably un- and anti-classical, the faith and vision that underlie it are classical almost to the point of anachronism. We get an inkling of this once we register the gross discrepancy between Büchner's reception as a writer, and his reception as a speculative scientist. As a writer, he notoriously remained unrecognized throughout most of the nineteenth century: there was no framework of reference or of expectations that could begin to accommodate the radical unconventionality of his work. *Danton's Death*, the only one of his poetic works to appear during his own lifetime, could do so only after it had been morally and politically sanitized, and it met with almost no response at all, except for instance to be savagely castigated by a pseudonymous reviewer as 'filth', 'degeneracy', 'pestilential impudence', 'excrescences of immorality', 'blasphemy against all that is most sacred'.[13] As a comparative anatomist, on the other hand, Büchner was instantly admitted into the fold of international orthodoxy. His doctoral dissertaion, *Mémoire sur le système nerveux du barbeau*, published in Strasbourg in 1836, was immediately welcomed in authoritative circles in France, Germany and Switzerland: its 'Partie descriptive' was hailed as 'very thorough' and 'entirely

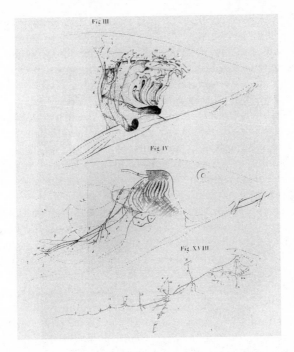

25   Anatomical drawings from Georg Büchner's doctoral dissertation on the anatomy of the barbel. From *Mémoires de la Société du Muséum d'Histoire Naturelle de Strasbourg*, vol. 2, Paris, 1835.

correct', and as 'extending knowledge with all desirable precision',[14] while its 'Partie philosophique', containing Büchner's specific argument, was endorsed by Johannes Müller, a like-minded comparative anatomist, but also a physiologist who was to make advances fundamental to nineteenth-century science.[15] Büchner's Trial Lecture, in its turn, met with 'the widest approval' among its local but distinguished audience from established academia, and Oken himself not only made a point of recommending his new young colleague's classes, but even sent along his own son.[16]

The problem with the particular scientific–philosophical position that Büchner embraced as wholeheartedly as he rejected its literary counterpart, is that it was rapidly losing its predominance and credibility even as he entered upon it. There is an eloquent irony in the fact that, during the period in which Büchner was immersing himself in his dissections and concretizing his (received) wholist vision of the natural world, Charles Darwin was busily collecting his data on the *Beagle*; by the time his ship returned to England in October 1836, just a few weeks before Büchner's lecture, Darwin had all the ingredients of a theory that would help to make the world-view of which Büchner was such an ardent exponent seem antiquated and irrelevant, an apparent by-water remote from the mainstream of scientific progress – and, moreover, a deeply suspect one. This suspect status is both demonstrated and sharply reinforced in Thomas Henry Huxley's famous Croonian Lecture of 1858, when he set out to discredit precisely the central tenet that lies at the heart of both Büchner's doctoral dissertation and his Trial Lecture: Goethe's and Oken's Vertebral Theory of the skull. The triumphant, sabre-rattling tone is unmistakable when Huxley ridicules 'the speculator' for his conjuror's ability to 'devise half a dozen very pretty vertebral theories, all equally true, in the course of a summer's day', and calls for support from 'Those who, like myself, are unable to see the propriety and advantage of introducing into science any ideal conception, which is other than the simplest possible generalized expression of observed facts'.[17]

We need to appreciate the real enormity of the problems faced by life-scientists in the half-century or so before Darwin did for biology what Newton had done for physics almost two centuries earlier; even the very word 'scientist' – not coined until 1834 (by William Whewell) – is an anachronism that tends to beg essential questions. Büchner is not exaggerating when he speaks in his Trial Lecture of a monstrous chaos of disorderly, undifferenti-ated data. Referring to the great systematizer John Stuart Mill and his attendance at zoology lectures at Montpellier University in 1820, Sir Peter Medawar has remarked that 'there seems no doubt that his thought on methodology was strongly influenced by the study of a subject overwhelmed by a multitude of "facts" that had not yet been disciplined by a unifying theory'; and he continues:

Coleridge described it [in 1818] as 'notorious' that zoology had been 'fully abroad, weighed down and crushed as it were by the inordinate number and multiplicity of facts and phenomena apparently separate, without evincing the least promise of systematizing itself by any inward combination of its parts'.[18]

Biologists of this period found themselves battling through a teeming jungle of new knowledge. Behind them lay Aristotle-Land, with its clear but no longer adequate model of a fixed and static 'Ladder of Nature'; ahead of them somewhere was that magnificent vantage point that Darwin was ultimately to construct, with its momentous spectacle of an evolutionary pattern in nature both dynamic and explicable. Many crucial stations were established along the way by great speculative and/or systematizing minds like Linnaeus, Bonnet, Buffon, Lamarck, Cuvier; but being on the whole too closely mod-elled on the Aristotelian 'fixed-and-final' scheme, none of them could suffi-ciently order or accommodate the riotous growth of new facts and discoveries.

That there was a real fear in this period of being overwhelmed by the riot of 'facts and phenomena' is clear from the comments of Coleridge, and implicit in Büchner's use of language in the Trial Lecture ('a huge mass of material . . . scarcely even systematically catalogued', 'a confusion of weird forms under the wildest names', 'a mass of things . . . so many separate, unconnected facts'). The fear was a complex one. At its most banal there was no doubt the professional fear of all scholars in all eras that their minds may not be equal to their material. At a much deeper level, the gathering confusion was bound to generate anxious perplexity in an age conditioned by the Enlightenment to believe that all things in existence are systematically patterned, and that man's mind can discern that pattern. At its deepest, however, the fear was existential: what was threatened was man's whole sense of the world in which he lived, and of his place within that world.

Both Büchner and Coleridge point to the gravest particular cause of fear: the real threat lay not so much in the sheer extent or bulk of the new 'facts and phenomena', but rather in their disorderliness and, above all, their discreteness: they were 'separate, unconnected facts'; they were 'apparently separate' and showing no promise of an 'inward combination of [their] parts'. This 'inordinate . . . multiplicity' not only resisted Enlightened assumptions about an orderly progression towards the discernment of order in the *natural* world; it also very readily seemed to echo, to symbolize and even to compound the atomistic forces that were tending to disrupt progress towards a better order in the *human* world (as seen from a progressive point of view), or alternatively to disrupt the human order already prevailing (as seen from a conservative point of view). For approximately a century, cataclysm after cataclysm ensured that no thinking person could easily sustain a clear and stable picture of the world: the devastating Lisbon earthquake of 1755 that so profoundly affected Voltaire; the incessant revolution in scientific data; the Industrial Revolution with its colossal social and economic repercussions; the French Revolution with its magnificent aspirations and horrific reality – precisely the setting of Büchner's first play; the international turmoil of the Napoleonic wars; the bloody spasms of social revolution in 1830 and 1848. This was by far the greatest and the most obsessive age of taxonomy and system-building in human history: like Büchner's Danton, they were 'making mosaics', feverishly trying to assemble the exploded pieces into sensible, significant order. And still today we live in the long shadow of the twin texts that were the towering culmination of these endeavours, texts that were both dedicated to the ordering of classes, and to the modes of change to which those classes are subject: *The Origin of Species* – and *The Communist Manifesto*.

Nowhere was the atomistic threat felt more acutely than in Germany – not least because there *was* no Germany: in sharp contrast to France or England there was no kind of political, economic, social or cultural entity characterized by an 'inward combination of its parts' (to borrow Coleridge's words once again), but instead an 'inordinate number and multiplicity' of states and statelets – the atomization and attendant backwardness of particularism. The hundred years from about 1750 to 1850 saw an astonishing outpouring of genius in German thought and literature (not to mention music) that is without parallel in Europe at least since the Renaissance. But this great torrent welled up in a fragmented landscape that had lain barren in many respects since time immemorial, and which enjoyed nothing of the clear, well-established, centralized system of channels and reservoirs that afforded an instant sense of direction, context and common endeavour to the gifted Frenchman or Englishman. Referring particularly to literature, W. H. Bruford has observed that

It is remarkable, as has often been pointed out, that Germany succeeded, in the absence of . . . a national tradition and of political institutions to support it, in producing a literature that came to be looked upon as classical, though it was, in Freytag's phrase, 'the almost miraculous creation of a soul without a body'.[19]

It *was* remarkable. But it would have been even more remarkable if the mighty talents of the age had *not* energetically built themselves elaborate constructs to compensate for what was missing.

But what *kind* of constructs? A fundamental disparity at once begins to appear between the German pattern and the pattern elsewhere. At its simplest, it is the contrast between empiricism and idealism; between the inductive and deductive modes; between progression from matter to mind and progression from mind to matter. Given the increasingly evident backwardness and atomization of the German reality, and the almost non-existent role of the emergent intelligentsia, and its ideas, in the prevailing political structures and processes, it is scarcely surprising that the constructs of thinkers and writers came more and more to be built as it were on stilts, at a deliberate remove from narrow, intractable reality.

The decisive figure here was Immanuel Kant (a man who himself moved from 'matter' to 'mind' in the sense that he taught physics and mathematics before he changed to philosophy). What could the mind know, and how could it securely know it? Kant posited on the one hand a realm of essential reality, of 'things in themselves', of which we can know nothing whatever. But the position is profoundly different with regard to the world of phenomena, of things as they *appear*. Kant's 'Copernican revolution', as he himself described it, was truly revolutionary. Just as Copernicus had shown that the apparent motion of the heavenly bodies was due to the motion of the beholder on his mobile planet, so Kant argued that the world as we know it, the world of appearances, is a function of our vantage point, and is constituted solely by the interaction of our senses and our intelligence: every feature of it, even its thereness in space and time, is ascribed to it by the mind.

The radical epistemology of this 'Transcendental Idealism' took the giant first step towards establishing the mind as the giver of meaning, as the creator, in a certain sense, of the knowable world – and it thereby prepared the ground for a whole plethora of systems, philosophies and ideologies that set the mind or spirit ever more intensely at the centre of the universe. In terms of philosophy itself, there are the 'Absolute Idealist' systems of Fichte, Schelling and, above all, Hegel. In the domain of art, there is the High Romanticism of Friedrich Schlegel and Novalis (Friedrich von Hardenberg). And in the realm of the natural world, there is *Naturphilosophie* – which is what particularly concerns us in the context of Georg Büchner.

We can enter the fray at a conspicuously benign moment. It is 1794. Goethe and Schiller, these twin giants of Weimar Classicism, are gravely estranged, thanks largely to Goethe's conviction that they are at 'diametrically opposite poles', and separated by such an 'enormous gulf' that there can be 'no question' whatsoever of their reconciliation (X.540).[20] But they happen to find themselves emerging together from a meeting of J. G. K. Batsch's 'Naturforschende Gesellschaft' in Jena – and it is Schiller's criticism of the analytical, atomistic tone of this meeting that suddenly sparks their famous friendship. What Schiller objects to is the treatment of nature as so many separate fragments ('eine so zerstückelte Art die Natur zu betrachten', X. 540). Goethe agrees in deeply characteristic terms: instead of nature being regarded as an assemblage of separate bits and pieces, it can readily be shown to be vibrant and alive, carrying its wholeness through into all its parts ('nicht gesondert und vereinzelt . . . sondern . . . wirkend und lebendig, aus dem Ganzen in die Teile strebend', X. 540). What serves to unite these two diametrically different men, therefore, is a shared hostility to what they regard as an excessive and barren empiricism: a concentration on the part, on the particularity of discrete data, that forfeits all sense of the whole. Again there is the Coleridgean spectre of man's ordering mind being swamped by 'facts and phenomena'. This is why Goethe utterly rejects Baconian science: it purports to collate the particular only in order to discern the universal ('Partikularien'/'Universalien'); but it loses itself so completely in individual data that 'life goes by and all energies are exhausted before any simple essence or any conclusion can be arrived at' (XIV. 91). Goethe's crucial point, indeed one of his central articles of faith, is that the whole is always present within the part, and a single fact can therefore serve for thousands in that it contains their essence within itself, all being equally manifestations of the primordial type, the particular 'Urphänomen', that wholly informs them (XIV. 91–2).

In rejecting the Baconian absorption in empirical data, Goethe is by no means turning his back on reality. On the contrary, it is in his view Baconian (and Newtonian) empiricism that is remote from true reality, for it forfeits any chance of seeing the wood through its exclusive concentration on the trees. It is only through a wholist approach, he believes, that one can apprehend reality. He insists on his 'obstinate realism' ('hartnäckigen Realismus', X. 541); and when he conjures up for Schiller his vision of nature as 'vibrant and alive, carrying its wholeness through into all its parts', this is to him no abstraction, but something he sees and experiences with his own eyes, as he might a table or chair. Having meanwhile been carried by their conversation into Schiller's house, he not only *expounds* his notion of the metamorphosis of plants, but makes it palpable by actually drawing a 'symbolic plant' (X. 540) for the other to physically see with his eyes. But this very nearly ends their friendship before it has begun, for it touches on that 'enormous gulf' that had always yawned between them: Schiller as a 'fully fledged Kantian' (X. 541) cannot accept this assertion of experiential reality: '"That is not an experience, it is an idea."' (X. 540) This was Goethe's self-confessed and blessed 'naivety': that his thoughts and ideas were literally, palpably visible to his eyes (XIII. 26–7). We do not have to be learned Kantians to recognize that Schiller is of course quite right: what Goethe envisions is to his own eyes a lived reality, but it nevertheless remains a product of his mind, an idea, an ideal – not an inherent and necessary quality of reality itself.

This begins to define the central thrust of that contentious but widespread and persistent mode of scientific inquiry that was *Naturphilosophie*. In essence, and at its best, it was an attempt to syncretize the epoch's two antithetical attitudes of empiricism and idealism, to establish a middle ground in which mind and matter, instead of dominating and diminishing each other, came fully and equally into their own in a kind of rich interplay. In its fundamental wholist tenets it is unquestionably idealist, even metaphysical. But the *Naturphilosophen* used their idealist–metaphysical–mystical postulates as a vantage point from which to comprehend the real workings of the real natural world, to descry the order within an otherwise unmanageable chaos of data. Throughout all his scientific–philosophical speculations, Goethe never strayed from detailed and painstaking observation and analysis of specimens; Oken published no fewer than thirteen volumes of 'straight' descriptive natural history. But their approach in their laborious observation was always deductive and integrative, never inductive and atomistic: whereas the Baconian starts with the part (and in the view of the *Naturphilosophen* can never get beyond it), they begin with the whole – which indeed they believe to be always immanent in every least particle.

By the time it came to Georg Büchner's brief spell in the realm of the sciences in the mid-1830s, *Naturphilosophie*, although still predominant, was under severe threat, for the parallel and alternative mode of empiricism was rapidly moving towards that position of supremacy that it still holds to this day. In the process of the gradual discrediting of *Naturphilosophie* it attracted much disparagement and even ridicule. This is partly because of the hyperbolization in both expression and conviction that the battle of attitudes forced upon its

participants. This was engagingly symbolized on the occasion of Oken's
Inaugural Lecture at Jena in 1807, a sensational and polemical affair in which
Oken felt driven to the climactic and preposterous assertion that 'The entire
human-being is but a vertebra' ('Der ganze Mensch ist nur ein Wirbelbein').[21]
The polemical intensity of the battle can be readily gauged from the scathing –
and no less hyperbolic – tone of Justus von Liebig, the great chemist, who in
1840 attacked the *Naturphilosophen* and their doings as 'the pestilence, the
Black Death, of the nineteenth century'.[22] Much later, Liebig pronounced on
*Naturphilosophie* in more moderate, but more devastating, terms: 'We look
back on German *Naturphilosophie* as though on a dead tree that bore the
most beautiful foliage and the most magnificent flowers – but no fruit.'[23] It is
no doubt true that some of the extravagances of *Naturphilosophie* were
'fantastic to the verge of insanity'.[24] And one of its most radical exponents was
J. B. Wilbrand, professor of comparative anatomy, physiology and natural
history at Giessen from 1809 – and the young Liebig's particular *bête noire*
from the moment in 1824 that he took up his own chair in Giessen, where both
men were still very active in their antagonistic camps when Büchner arrived to
continue his studies in 1833.

It is easy to mock at the excesses of *Naturphilosophie*: at Wilbrand's
dogged and absolute denial of the circulation of the blood, and likewise of the
interchange of oxygen and carbon dioxide in respiration; at his treatise
succinctly entitled 'On the Connection between Nature and the Supernatural,
and How a Thorough Study of Nature and its Phenomena Points Ineluctably
to the Continuance of Spiritual Life after Death'; at the belief of Oken and
others that light is the consciousness of God, ether his self-positing activity
and objects his concretized thoughts; at Goethe's repudiation of Newton's
theory of light. It is just as easy for us, too, to follow the polemicists and
propagandists of the time, and distinguish neatly and categorically between
sheep and goats, to see deductivists and speculative idealists on the one hand,
and inductivists and rigorous empiricists on the other. In reality, as Popper
and Medawar have persuasively argued, there are no two distinct camps in
science, but a continuous spectrum linking the opposite and equally unfruitful
extremes of 'an inventory of factual information' and 'a totalitarian world
picture of natural laws'[25] – with all significant advances in science being made
in that mid-range of the spectrum that involves the most fruitful interaction
between speculative, imaginative intuition and careful empirical testing. At its
zaniest, *Naturphilosophie* did veer towards a ludicrously absolute and
unproductive extreme. Even in its normal, median condition as a classical
mode of scientific inquiry enshrined in all the universities of the land, its
subordination of experiment to *a priori* ideology ensured that it could not
survive against the professional scepticism, the questioning subjection of
hypotheses to experimental testing, that increasingly became the hallmark of

nineteenth-century science. Nevertheless, *Naturphilosophie* did make a very
real contribution to the development of the sciences: it considerably extended
the realm of the thinkable; and it either made, or provided the stimulus for, a
whole range of specific discoveries. In particular, its dare-to-speculate mental-
ity greatly furthered the sciences in their quantum leap from the physics-
derived fixed-mechanism model of the world in the eighteenth century, to the transformational, evolutionary model so characteristic of the
nineteenth.

This, then, is the kind of context within which Büchner and his science
belong, and within which we need to try to locate and understand him. Its
most crucial characteristic is its 'continuous spectrum' quality: Büchner
criticism has been – and continues to be – bedevilled and distorted by that
inveterate tendency to polarization that sees only separate, mutually exclusive
camps of empiricists-materialists-realists, and speculative idealists. Such a
tidy dichotomy may conceivably be relevant to the realm of abstract philosophy and literary fantasizing, both of which offered an alluring refuge from
reality during this period, but it is mischievously irrelevant to the realm of
scientific inquiry – and hence also to the writing of Georg Büchner. For the
beliefs and practices that inform his science equally inform his art, not only in
the commonplace sense that the laboratory dissector is also the literary
dissector, but in the much more important sense that there is in both a
profound and essential interaction of the Real and the Ideal. For the devotees
of polarization, it has to be either/or, and what they inevitably do is to hustle
Büchner into the empiricist–materialist camp, which is of course by definition
anti-idealist. This is a tendentious travesty, whether in the crass form exemplified by Walter Müller-Seidel with his depiction of Büchner as wholly anti-
idealist and wholly 'scientific',[26] or in the subtle form proffered by Raimar
Zons with his claim that Büchner held to the notions of *Naturphilosophie* –
but only as an expedient heuristic category, a kind of handy toolkit to help him
deal with the world.[27]

This surely is the crux, the teasing but critical question that every serious
student of Büchner has to confront. Do we credit the radiantly positive vision
– Goethean and *naturphilosophisch* in its essence – that he voices in the
preamble to the Trial Lecture? Do we believe he believed in a primal law of
beauty, in simplicity as the fount of rich complexity, in pure harmony as the
necessary outcome of the primal law throughout the natural world? Or do we
decide that the vision is uncharacteristic and illusory – a stratagem perhaps, a
piece of ritual rhetoric, a contrived nostrum, an act of calculated ingratiation,
an inexplicable and dismissible aberration? Both alternatives are problematic.
To take the latter is generally to find oneself in a familiar logical bind: Büchner
is a fearless speaker of unidealized truth, therefore it cannot be the truth when

26  Büchner, drawn from life by Alexis Muston: sketch discovered by Heinz
Fischer in 1970. From Heinz Fischer, *Georg Büchner, Untersuchungen und
Marginalien* (Bonn, 1972).

he speaks of ideals. To prefer the other alternative is to collide at once with the
fact that harmonious beauty and rich simplicity are scarcely the most resonant
message of his poetic work. Nevertheless I believe this alternative to be
incontrovertibly the right one: Büchner did mean every warm and positive
word of his preamble; and so completely was he blessed and cursed by an
inherited sense of natural harmony that even the slightest discord, the slightest
departure from received pitch, was for him an agony. His misfortune – and in
consequence our delight – is that he happened, like E. T. A. Hoffmann's Ritter
Gluck, to be a man of richest harmonies in a world increasingly dominated by
orchestrated scratchers and scrapers and mechanical contraptions grinding
out the same old cracked, broken, excruciating tunes.

Fanciful language, perhaps? But it echoes Büchner's own, in a memorable
letter to his beloved Minna in March 1834 (II. 424). Even the reference to
Hoffmann is there (though not specifically to *Gluck*): 'I could have sat as a
model for Herr Callot-Hoffmann.' So, too, is the strident duality of natural
harmony and desperate mechanical grinding. He has just been outside in the
open, he writes: 'A single resonant tone from the throats of a thousand larks

bursts through the brooding summer air, a heavy bank of clouds wanders over
the earth, the booming wind rings out like its melodious tread.' This is his
vibrant, melodious present. But until the outside air served to free him and
give him life again, he had long been transfixed by a kind of rigor
('Starrkrampf'), by a sense of being already dead ('Gefühl des Gestor-
benseins'), so that he and all around him seemed like deathly puppets with
glassy eyes and waxen cheeks. And at this point Büchner suddenly launches
into a characteristically thrilling cadenza of despair (one wonders what poor
Minna made of it all):

and then, when the whole machinery began to grind away, with jerking limbs and
grating voice, and I heard the same old barrel-organ tune go tralala and saw the tiny
prongs and cylinders bob and whirr in the organ box – I cursed the concert, the box, the
melody – oh, poor, screaming musicians that we are – could it be that our cries of agony
on the rack only exist to ring out through the gaps in the clouds and, echoing on and on,
die like a melodious breath in heavenly ears? Could it be that we are the victims roasted
in the belly of Perillus' bull, whose screams as they die ring out like the jubilant roars of
the bull-god as it is swallowed in the flames?

An unnerving antiphon: in nature – the wind and the larks and their liberating
melody; among men – a deathly mechanical rasping, and tortured screams
extracted perhaps by some distant deity for his melodious titillation. And it is
precisely this drastic antiphon that Büchner uses nine months later to ring in
the crescendo that marks the grand-opera climax of *Danton's Death*:

PHILIPPEAU. My friends, one doesn't have to stand very far above the earth to see no
    trace any more of all this shifting, shimmering chaos, and to behold instead a
    simple, great and godly outline. There is an ear for which the cacophony and
    clamour, so deafening to us, are a stream of harmonies.
DANTON. But we are the poor musicians and our bodies the instruments. Are the ugly,
    vamping sounds bashed out on them just there to rise up higher and higher and
    gently fade and die like some voluptuous puff of breath in heavenly ears? (I. 71)

It would be easy to conclude – as innumerable critics have done – that
Philippeau is simply a stooge, a foil of fatuous optimism serving to silhouette
the 'true' negativity in the ensuing chorus of grandiloquent despair. But this
would be quite wrong. Philippeau's 'stream of harmonies' is profoundly real
to Büchner. When he heard its melody among the wind and the larks in
Giessen, it brought him back from figurative death (thus inaugurating a topos
of death and resurrection that is central to his work). And it is, above all, the
measure that makes the sounds given out by the human 'instrument' seem by
contrast so ugly, so raspingly mechanical.

A crucial question arises here as to *why*, for Büchner, men have become so
agonizingly out of tune, so remote from that 'necessary harmony' that the
primal law of beauty bestows so readily upon the whole of the rest of nature.
But that is a question for another occasion.

## NOTES

This article is a revised version of a chapter to appear in the author's monograph on Büchner, forthcoming from Oxford University Press in 1991. Oxford University Press have kindly granted permission for its inclusion here.

1 *Büchner-Preis-Reden 1951–1971* [no editor] (Stuttgart, 1972), p. 162. Here, as everywhere in this chapter, the translation is by the author.
2 *Ibid.*, p. 183.
3 Christa Wolf, *Fortgesetzter Versuch. Aufsätze. Gespräche. Essays* (Leipzig, 1979), p. 64.
4 Georg Lukács, 'Der faschistisch verfälschte und der wirkliche Georg Büchner', in Wolfgang Martens (ed.), *Georg Büchner* (Darmstadt, 1965), p. 201. The essay was originally written in Moscow in 1937.
5 Robert Mühlher, 'Georg Büchner und die Mythologie des Nihilismus', in Martens (ed.), *Georg Büchner*, p. 260.
6 Walter Jens, *Euripides. Büchner* (Pfullingen, 1964), p. 46.
7 Ronald Peacock, 'A Note on Georg Büchner's Plays', *German Life and Letters*, new series, 10 (1956–57), p. 191. Peacock's approach is still wreaking its effects: his article was the explicit departure point for Dorothy James's 1982 monograph on *Danton's Death*, which is systematically vitiated by her assumption of Büchner's 'immaturity as a dramatist' (Dorothy James, *Georg Büchner's 'Danton's Tod': A Reappraisal* (London, 1982), p. 25).
8 A. H. J. Knight, *Georg Büchner* (Oxford, 1961), p. 74.
9 All bracketed page references to Büchner's work relate to Georg Büchner, *Sämtliche Werke und Briefe*, ed. Werner R. Lehmann, 2 vols. (Munich, 1972 (vol. II) and 1974 (vol. I)).
10 Knight, *Georg Büchner*, pp. 69, 174f.
11 Hans Mayer, *Georg Büchner und seine Zeit* (2nd edn, Wiesbaden, 1959), p. 372.
12 J. P. Stern, *Idylls and Realities. Studies in Nineteenth-Century German Literature* (London, 1971), p. 35.
13 See Thomas Michael Mayer, 'Georg Büchner. Eine kurze Chronik zu Leben und Werk', in *Georg Büchner I/II*, ed. H. L. Arnold (Munich, 1979; Special Number in 'text+kritik' series), pp. 403f.
14 *Ibid.*, p. 408.
15 See Jean Strohl, *Lorenz Oken und Georg Büchner. Zwei Gestalten aus der Übergangszeit von Naturphilosophie zu Naturwissenschaft* (Zurich, 1936), p. 59.
16 T. M. Mayer, 'Eine kurze Chronik', p. 419.
17 T. H. Huxley, 'On the Theory of the Vertebrate Skull', in *The Scientific Memoirs of Thomas Henry Huxley*, ed. M. Foster and E. R. Lankester, vol. I (London, 1898), pp. 584f.
18 P. B. Medawar, *Induction and Intuition in Scientific Thought* (London, 1969), p. 9.
19 W. H. Bruford, *Germany in the Eighteenth Century. The Social Background to the Literary Revival* (Cambridge, 1965), p. 292.
20 Volume and page numbers concerning Goethe refer to the 'Hamburger Ausgabe', 14 vols., ed. Erich Trunz *et al.* (Hamburg, 1948–66).

21  Strohl, *Lorenz Oken und Georg Büchner*, p. 11.
22  Cited by W. Müller-Seidel, 'Natur und Naturwissenschaft im Werk Georg Büchners', in *Festschrift für Klaus Ziegler*, ed. E. Catholy and W. Hellmann (Tübingen, 1968), p. 207.
23  Cited by R. S. Zons, *Georg Büchner, Dialektik der Grenze* (Bonn, 1976), p. 61.
24  Charles Singer, *A Short History of Scientific Ideas to 1900* (Oxford, 1960), p. 385.
25  Medawar, *Induction and Intuition*, p. 59.
26  Müller-Seidel, 'Natur und Naturwissenschaft', p. 210.
27  Zons, *Georg Büchner*, pp. 69ff.

## FURTHER READING

Benn, M.B., *The Drama of Revolt. A Critical Study of Georg Büchner* (Cambridge, 1976)

Bruford, W. H., *Germany in the Eighteenth Century. The Social Background to the Literary Revival* (Cambridge, 1965)

Golz, Jochen, 'Die naturphilosophischen Anschauungen Georg Büchners', *Wissenschaftliche Zeitschrift der Friedrich-Schiller-Universität Jena*, 13 (1/1964), 65–72

Höllerer, Walter, 'Georg Büchner', in Walter Höllerer, *Zwischen Klassik und Moderne. Lachen und Weinen in der Dichtung einer Übergangszeit* (Stuttgart, 1958)

James, Dorothy, *Georg Büchner's 'Dantons Tod': A Reappraisal* (London, 1982)

Reddick, John, '"Ihr könntet einen noch in die Lüge verliebt machen"', Georg Büchner and the Agony of Authenticity', *Forum for Modern Language Studies*, 13 (1987), 290–324

Strohl, Jean, *Lorenz Oken und Georg Büchner. Zwei Gestalten aus der Übergangszeit von Naturphilosophie zu Naturwissenschaft* (Zürich, 1936)

# Index of names

Abernethy, John 223
Adelmann, Georg 108
Agassiz, Louis 137–9
Allen, William 19
Aner, Karl 70
Appel, Toby 149, 152
Apreece, Jane 223
Arber, Agnes 158
Ariès, Philippe 20
Aristotle 298, 313
Audubon, J. J. 22

Baader, Franz von 7, 55, 91, 92, 209, 233
Bacon, Francis 1, 16, 17, 62, 64, 86, 191, 192, 265, 307, 310, 313
Baer, Karl Ernst von 128, 132, 133, 137, 139, 140
Bailly, J. S. 57
Banks, Sir Joseph 208
Barclay, John 156
Barruel, the abbé 87
Barth, Karl 70
Batsch, A. J. G. C. 146, 333
Béclard, P. A. 165
Beddoes, Thomas 15, 20, 86, 215, 217, 218, 296
Bergman, Torbern 57, 243, 266–8, 271, 275, 276
Berman, Morris 224
Berthollet, Claude-Louis 268, 269, 275–7
Berzelius, Jöns Jacob 21, 236
Bewick, Thomas 22
Bichat, François Xavier 161–5
Bisson, Mathurin Jacques 231
Black, Joseph 266
Blainville, Henri de 145, 148, 150, 152, 153, 155, 157
Blake, William 300

Blumenbach, Johann Friedrich 32, 57, 107, 120, 122, 124, 146, 183, 247–50, 252, 257, 286, 297
Boeckh, August 43
Bois-Reymond, Emil du xix, 7, 197
Böll, Heinrich 323
Bolton, Matthew 215
Bonnet, Charles 248, 330
Bonpland, Aimé 171, 179
Bopp, Franz 43
Borlase, Bingham 214
Boyle, Robert 17
Brandt, Richard 72
Brecht, Berthold 322, 323
Brentano, Clemens 27, 200
Brewster, David 235
Brockes, Ludwig von 283
Brodie, Benjamin 19, 20
Brontë, Anne 21
Brown, H. M. 288
Brown, John 61, 107, 215, 217
Brücke, Ernst von 196
Bruford, W. H. 332
Brünninghausen, Hermann Joseph 283
Buch, Leopold von 251, 252
Büchner, Georg 322–31, 333, 334, 336–8
Buckland, William 243, 246
Buffon, Comte Georges de 57, 64, 120, 146, 148, 158, 242, 246, 247, 301, 330
Burdach, Karl Friedrich 126, 127
Burke, Edmund 83–7, 89, 91
Burnet, Thomas 57
Byron, George Gordon, Lord 226

Camerarius, Rudolf Jakob 250
Camper, Petrus 242, 245–8
Carlisle, Anthony 20, 208
Carlyle, Thomas 45, 295, 302

Carpenter, W. B. 158
Carus, Carl Gustav 58, 60, 63, 155
Cassirer, Ernst 285
Chambers, Robert 135, 137
Chladni, Ernst 91, 234
Coleridge, Luke 296
Coleridge, Samuel Taylor 3, 4, 14, 15, 20,
    22, 40, 46, 90, 93, 94, 154, 213–20, 222,
    223, 295–304, 311, 330, 331
Cook, Captain James 171
Copernicus, Nicolaus 62, 195, 199, 314, 332
Correggio, Antonio Allegri da 33
Coulomb, Charles 231
Cullen, William 266
Cuvier, Georges 124, 125, 133, 134, 136,
    148–50, 152, 153, 155, 157, 242, 243, 247,
    257, 301, 330

Dähling, H. A. 273
Dalton, John 8, 298
Darwin, Charles 15, 16, 22, 59, 140, 155,
    307, 329, 330
Darwin, Erasmus 57, 215, 216, 301, 307
Daubenton, L. J.-M. 125, 146, 148, 152, 242
d'Aubuisson, J. F. 251
Davy, Humphry 8, 14, 15, 18–22, 86, 92, 93,
    154, 213–27, 297, 299, 302, 304
Davy, John 213, 224
Deluc, Jean-André 248, 252
Descartes, René 1, 57, 62, 64, 265
Desmond, Adrian 156
Dick, Thomas 313
Diderot, Denis 57, 91
Dillenberger, John 70
Dilthey, Wilhelm 48, 72
Döllinger, Ignaz 113, 164
Draper, John 69
Duchesne, A. N. 57
Dulong, Pierre Louis 236

Eble, Burkhardt 113
Eckermann, Johann Peter 189, 322
Edgeworth, Anna Maria 217
Einstein, Albert 17
Eliot, George 130
Eliot, T. S. 101, 104, 112
Elliot, Lord 283
Emerson, Ralph Waldo 40, 45, 307–12, 316,
    317, 319
Engelhardt, Dietrich von 180
Eschenmayer, Karl Adam August von 63
Esmarch, Lauritz 228, 236
Esper, Johann Friedrich 242

Faraday, Michael 21, 310
Faujas de Saint-Fond, Barthélemy 243

Fichte, Johann Gottlieb 5, 7, 27, 28, 30,
    38–40, 45–7, 49, 84, 85, 233, 285, 333
Ficino, Marsilio 265
Fischer, Heinz 337
Forbes, Edward 145, 158
Forster, Georg 85, 89, 93, 171
Forster, Johann Reinhold 173
Franklin, Benjamin 82
Frederick II, the Great, King of Prussia 281
Frederick William I, King of Prussia 105
Frederick William II, King of Prussia 39
Frederick William III, King of Prussia 38
Freytag, Gustav 332
Friedrich, Caspar David 254, 255
Fries, Jacob Friedrich 70, 72–8
Füchsel, Johann Christian 253

Galileo 62, 195, 199
Galvani, Luigi 200–2, 204
Gay-Lussac, Joseph-Louis 298
Geber (Jābir ibn Ḥayyān) 265
Gehler, J. S. T. 270
Geoffroy, E. F. 265–6
Geoffroy Saint-Hilaire, Etienne 134, 136,
    144, 145, 147–50, 152, 153, 155, 157
Gerard, Alexander 85
Gilbert, L. W. 206, 236
Glauber, Johann 265
Gmelin, D. 288
Gode-von Aesch, A. xix
Goethe, Johann Wolfgang von xix, xx, 3, 4,
    7, 16–18, 20, 22, 50, 62, 73, 82, 88, 122,
    123, 126, 127, 140, 145–7, 153, 158, 164,
    169, 176–8, 183, 189–97, 201, 208, 217,
    246, 247, 252, 263–77, 288, 322, 323, 330,
    333–5
Goldfuss, Georg August 58
Goodsir, John 145, 158, 166
Görres, Jacob Joseph von 26, 62
Göttling, J. F. A. 270
Grant, Robert Edmond 134, 135, 145, 154,156–8
Grass, Günther 323
Green, Joseph Henry 145, 154–6
Grosse, Karl 88

Haeckel, Ernst 140, 148
Hales, Stephen 119
Haller, Albrecht von 119, 120, 146, 248, 249
Hamann, Johann Georg 55, 86, 89
Hardenberg, Friedrich von see Novalis
Harris, Moses 196
Hartley, David 216
Harvey, William 119
Hasler, U. 72
Hasselblatt, Meinhard 75
Hauch, Adam 236

Hauptmann, Gerhart 323
Hawthorne, Nathaniel 307, 316, 318–20
Hegel, G. W. F. xx, 19, 28, 38, 40, 46, 55,
　56, 58–60, 70–3, 146, 189, 197, 333
Helmholtz, Hermann von 7, 196, 197
Herbert, George 18
Herder, Johann Gottfried 2, 50, 55, 83, 241,
　246, 247, 253, 254, 301
Herschel, John 15, 207, 208, 298
Herzog, Werner 323
Heusinger, Carl Friedrich 164
Hindenburg, Karl Friedrich 30, 286
Hinrichs, H. F. W. 72
Hirschel, Bernhard 112
Hoffmann, E. T. A. 244, 337
Hölderlin, Friedrich 2, 28
Horn, Ernst 110
Hoven, Friedrich Wilhelm von 110, 112
Hufeland, Christoph Wilhelm 101, 110,
　113–17, 284, 287, 290
Humboldt, Alexander von 15, 22, 50–2, 92,
　124, 169–83, 200–2, 206, 251, 286, 313
Humboldt, Wilhelm von 38, 39, 40, 41, 43,
　49, 50, 55, 169, 171
Hume, David 2
Hunter, John 19, 242
Hutton, James 248, 251
Huxley, Thomas Henry 7, 130, 131, 138,
　140, 330

Jacobi, F. H. 72
Jacobson, Ludwig 236
Jacyna, L. S. 156
James, William 314
Jameson, Robert 157, 251
'Jean Paul' *see* Richter, Friedrich
Jens, Walter 324
Jerusalem, Karl Wilhelm von 288
Joule, James 22

Kafka, Franz 322
Kämpf, Johann 103
Kant, Immanuel 2, 5, 15, 26, 27, 29, 31, 32,
　38–41, 44–6, 50, 56, 60, 69, 71, 74, 75, 78,
　83, 84, 86, 88–90, 92, 120–4, 145, 146,
　158, 169–71, 196, 204, 217, 231, 232, 238,
　281–3, 285, 286, 292, 295, 297, 301, 307,
　311, 332
Kästner, A. G. 57
Keats, John 3, 17, 20, 226, 311
Kepler, Johannes 62, 90, 195, 199, 313, 314
Kielmeyer, Carl Friedrich 32, 57, 123, 124,
　148
Kierkegaard, Sören 45
Kieser, Dietrich Georg 60
King, John 221

Kleist, Heinrich von 2, 233, 280–92
Kleist, Ulrike 286
Klügel, Georg Simon 286
Knigge, Adolf von 88
Knight, A. H. J. 324, 328
Knox, Robert 134, 135, 145, 154–7
Koreff, Johann Ferdinand 64
Kraemer, Hans 229

Lamarck, Jean-Baptiste 57, 64, 149, 155,
　301, 307, 330
Lambert, Johann Heinrich 2, 5
Laplace, Pierre 75, 264, 301, 307, 313, 315
Larkin, Philip 223
Lavater, Johann Kaspar 88, 89
Lavoisier, A.-L. 18, 20, 21, 30, 61, 222, 268
Lawrence, William 223
Lear, Edward 22
Leibniz, G. W. 57, 245, 290
Lenoir, T. 146
Lessing, G. E. 50
Lichtenberg, Georg 91, 248, 249
Liebig, Justus von xix, 7, 22, 335
Link, D. H. F. 35
Linnaeus, Carl 169, 330
Locke, John 216
Loder, Just Christian 146
Lonsdale, Henry 158
Lucchesini, the Marquis 286
Lucretius 316
Ludwig, Franz, Bishop of Ethal 108
Lukács, Georg 324
Lyell, Charles 21, 138, 307

Maclaurin, Colin 313
Macquer, P. J. 265–7, 269–71, 275
Maillet, Benoît de 57
Malpighi, Marcello 119
Malthus, Thomas 22
Marcus, Adalbert Friedrich 106, 108, 110,
　112, 114
Martin, William 22
Marx, Karl 27
Maupertius, P. L. M. de 57
Mayer, Hans 328
Meckel, Johann Friedrich 124, 125, 148,
　165, 257
Medawar, Sir Peter 330, 335
Mendelsohn, Moses 89
Merck, Johann Heinrich 246, 247
Mersenne, Marin 1
Mesmer, Franz Anton 287, 290
Mill, J. S. 314, 330
Miller, Hugh 136–8
Milne-Edwards, Henri 165
Monboddo, Lord (James Stewart) 252

Moritz, Carl Philipp 288
Möser, Justus 55, 82, 84
Mühlher, Robert 324
Müller, Adam 55, 56
Müller, Anton 283, 284, 290
Müller, Johannes 196, 329
Müller-Seidel, Walter 336
Muston, Alexis 337
Muth, Ludwig 285

Napoleon I, Emperor of the French 112,
    113, 149, 206, 288
Neve, Michael 214, 224
Newman, John Henry, Cardinal 40
Newport, George 158
Newton, Isaac 3, 4, 16, 17, 21, 50, 62, 89,
    90, 120, 189–91, 193–7, 199, 201, 222,
    224, 231, 264, 276, 286, 290, 307, 310,
    313, 330, 335
Nichol, John 313
Nicholson, William 20, 208
Niebuhr, Barthold Georg 39
Niethammer, Friedrich Immanuel 26, 28
Novalis (Hardenberg, Friedrich von) 4–7,
    33, 34, 46, 55, 64, 82, 91, 104, 200, 201,
    210, 243–5, 251, 253, 254, 333

Occam (Ockham), William of 316
Oersted, Hans Christian 19, 58, 61, 94, 206,
    228–38
Oken, Lorenz 3, 4, 8, 58, 63, 131–3, 135,
    137–9, 145–8, 153, 155, 183, 189, 254,
    257, 326, 329, 330, 334, 335
Owen, Richard 130, 133–40, 145, 147,
    153–6, 158, 310

Paley, William 14, 22, 135
Pallas, Peter (Pyotr) Simon 245, 246
Paracelsus 265, 307
Peacock, Thomas Love 88
Pestalozzi, Johann Heinrich 39
Pfaff, Christian Heinrich 236
Pfuel, Ernst von 288
Piper, H. W. 219, 221
Platner, Ernst 286
Plato 16, 30
Plotinus 274
Pockels, P. 288
Poe, Edgar Allan 307, 311–16, 318,
    319
Pohl, G. F. 56
Popper, K. 335
Porta, Giambattista della 265
Priestley, Joseph 18, 20, 57, 89, 214–16
Purkinje (Purkyně), J. E. 56, 196

Raphael 33
Rehberg, August 84
Reil, Johann Christian 111, 124, 165, 280,
    289–92
Reinhardt, Johannes 236
Reinhold, Karl Leonhard 26, 27
Reynolds, Joshua 85
Richter, A. G. 113
Richter, Friedrich ('Jean Paul') 3, 7
Riisbrigh, Børne 231
Ritter, Johann Wilhelm 19, 58, 62, 63, 91–3,
    199–210, 233, 234
Robinet, J. B. 57
Robinson, Henry Crabb 26, 27
Robinson, John 87
Roget, Peter Mark 137, 158
Röschlaub, Andreas 106, 107, 112–17
Rosenmüller, Johann Christian 243, 245,
    246, 248
Rousseau, Jean-Jacques 39, 242, 286
Runge, Phillipp Otto 196
Russell, E. S. 145

Savigny, Friedrich Karl von 39
Savigny, Jules-César 150
Scheele, C. W. 18
Schelling, Friedrich W. J. xix, 3–5, 7, 18, 19,
    27–35, 38–46, 49, 50, 55, 56, 58–60, 69,
    70, 72–4, 78, 84, 90–2, 104, 107, 132, 146,
    158, 180, 189, 200, 202–4, 207, 232, 233,
    238, 254, 257, 269, 285, 299, 301, 303, 333
Schelver, Friedrich Joseph 62
Schiller, Friedrich 3, 26, 28, 39, 45, 50, 92,
    169, 171, 178, 180, 183, 196, 297, 325,
    333, 334
Schlegel, August Wilhelm 7, 33, 34, 88, 233
Schlegel, Caroline 33, 34, 200
Schlegel, Friedrich 91, 146, 199, 200, 201,
    233, 333
Schlegel, Wilhelm 7, 33, 34, 35, 200, 333
Schleiden, M. J. 44, 147
Schleiermacher, Friedrich 39, 43, 47–50,
    70–3
Schlotheim, E. F. von 251
Schmid, G. 289
Schofield, Robert 216
Schopenhauer, Arthur 189
Schubert, Gotthilf Heinrich von 7, 63, 254,
    256, 257, 291, 292
Schwann, Theodor 147, 161, 162, 165, 166
Schweigger, J. Christoph Salomo 61
Sedgwick, Adam 137, 138
Sembdner, Helmut 282
Sepper, Dennis 17
Serres, Etienne Reynaud Augustin 144

Shakespeare, William 304
Sharrock, Roger 218–21
Shelley, Mary 20, 82, 93
Shelley, Percy Bysshe 88, 226
Siebold, Christian von 108
Siebold, Elias von 110, 113
Siebold, Karl Kasper 283
Snow, C. P. 14
Sömmerring, Samuel Thomas 247
Soulavie, Jean-Louis Giraud 57, 242
Southey, Robert 15, 20, 92, 213, 215, 217,
   219, 220, 296
Spencer, Herbert 140
Spinoza, Baruch 28, 199, 277
Steffens, Henrik (Heinrich) 3, 4, 7, 39, 50,
   57, 58, 59, 60, 73, 91, 301, 303
Stein, Karl vom 38
Steno, Nicholaus 57
Stern, J. P. 328
Struve, Christian August 110
Süssmilch, Johann Peter 252
Swainson, William 22
Swedenborg, Emanuel 89, 307, 310

Tennyson, Alfred, Lord 137
Thomann, Joseph Nickolas 108–10, 113
Thompson, D'Arcy 158
Tieck, Ludwig 7, 33, 88, 91, 200, 244
Tilesius, Wilhelm Gottfried 243
Treviranus, Gottfried Reinhold 119, 163,
   166, 183
Turner, J. M. W. 22

Viet, Dorothea 34, 200
Virchow, Rudolf 113, 166
Vogel, Henriette 292
Voigt, F. S. 57
Voigt, J. C. W. 251
Volta, Alessandro 20, 200–2, 204–6, 208,
   233, 235
Voltaire, François Marie Arouet de 331

Walther, Philip Franz von 106, 113, 165
Walton, Isaak 225

Watt, Gregory 20
Watt, James 20, 215
Wedekind, Georg 288, 290, 323
Weigard, Melchior Adam 112
Weiss, Christian Samuel 236
Welch, Claude 70
Werner, Abraham Gottlob 4, 93, 170, 244,
   250, 251, 253, 254, 301
Whewell, William 330
Whiston, William 57
White, Andrew Dickson 69
Wieland, Christoph Martin 246, 281,
   283
Wilberforce, Samuel, Bishop 133
Wilbrand, J. B. 335
Wilcke, J. C. 233
Will, Peter 88
Willdenow, Karl Ludwig 173
Windischmann, Karl Hieronymus 7
Winterl, Jakob Joseph 234
Wleugel, Peter 236
Wolf, Christa 323
Wolf, F. A . 39, 50
Wolff, Caspar Friedrich 120, 146
Woltman, Karl Ludwig 26
Woodward, John 57
Wordsworth, Dorothy 217
Wordsworth, William 4, 20, 217–21, 296,
   297
Wrisberg, August 286
Wrisberg, Heinrich 286
Wukadinovic, Spiro 289
Wünsch, Christian Ernst 282–5
Wyman, Jeffries 145
Wyttenbach, Carl 288

Young, Edward 85, 86
Young, Thomas 13, 18, 197

Zeise, William 236
Zenge, Wilhelmine von 282, 285, 287
Zimmermann, Eberhardt 170
Zons, Raimar 336